THE BUSINESS AND INVESTMENT ENVIRONMENT IN TAIWAN AND MAINLAND CHINA

A Focus on the IT and High-Tech Electronic Industries

THE BUSINESS AND INVESTMENT ENVIRONMENT IN TAIWAN AND MAINLAND CHINA

A Focus on the IT and High-Tech Electronic Industries

Chen-Min Hsu
National Taiwan University, Taiwan

Wei-Guo Zhang
Chongqing University, China

Leslie Lok
Kampar Venture, Malaysia

World Scientific

NEW JERSEY • LONDON • SINGAPORE • BEIJING • SHANGHAI • HONG KONG • TAIPEI • CHENNAI

Published by

World Scientific Publishing Co. Pte. Ltd.
5 Toh Tuck Link, Singapore 596224
USA office: 27 Warren Street, Suite 401-402, Hackensack, NJ 07601
UK office: 57 Shelton Street, Covent Garden, London WC2H 9HE

Library of Congress Cataloging-in-Publication Data
Hsu, Chen-Min.
The business and investment environment in Taiwan and Mainland China : a focus on the IT and high-tech electronic industries / by Chen-Min Hsu, Wei-Guo Zhang, & Leslie Lok.
p. cm.
ISBN-13 978-981-270-366-8 -- ISBN-10 981-270-366-7
1. Investments, Foreign--China. 2. Investments, Foreign--Taiwan. 3. Business enterprises, Foreign--China. 4. Business enterprises, Foreign--Taiwan. 5. Electronic industries--China. 6. Electronic industries--Taiwan. I. Zhang, Wei Guo. II. Lok, Leslie. III. Title.

HG5782.B87 2007
338.4'7621380951--dc22

2006042109

British Library Cataloguing-in-Publication Data
A catalogue record for this book is available from the British Library.

Typeset by Stallion Press
Email: enquiries@stallionpress.com

Printed in Singapore by World Scientific Printers (S) Pte Ltd

Contents

Chapter 1 Theoretical Analysis of Foreign Direct Investment (FDI) 1

1. Theoretical Analysis of FDI by Large Multinational Enterprises 1
2. Theoretical Analysis of FDI by Small- and Medium-Sized Enterprises in Taiwan 7

Chapter 2 FDI Development in Mainland China and Taiwan 11

1. Status Quo in Mainland China 11
2. Status Quo in Taiwan 32

Chapter 3 Investment Environment Index 53

1. R&D and Innovative Mechanisms 55
2. Human Resources and Human Literacy 57
3. Information Technology 59
4. Social Infrastructure 60
5. Relevant Indices for the Introduction of a Foreign Knowledge-Based Economy 61

Chapter 4 Analysis of Information Technology and Electronics Industry Market in Mainland China 73

1. Current Status of Electronics Information Industries in Mainland China 73
2. Economics of Electronics Information Industry's Operations 84
3. Analysis of Mainland China's IC Industry Development 88
4. Analysis of Mainland China's Automotive Electronics Market 93

5. Analysis of Mainland China's Consumer Electronics Market 108
6. Analysis of Mainland China's Mobile Storage Industry 112
7. Analysis of Mainland China's IC Design Industry 116
8. Analysis of Mainland China's IT Industry 123
9. Analysis of Mainland China's Electronic Components Distributors 131

Chapter 5 Analysis of the Information Technology and Electronics Industry Market in Taiwan 143

1. Current Status of Information Technology and Electronics Industries in Taiwan 143
2. Development Background of Industries 143
3. Industrial Structure and Characteristics 160
4. Business Opportunities 163

Chapter 6 The Business and Investment Environment of the Information Technology and Electronics Industries in Mainland China 177

1. Index of the Economic Environment 180
2. Analysis of Systematic Factors in the Investment Environment 196
3. Economic Policies and Laws 202
4. Infrastructure and Natural Factors in the Investment Environment 206
5. Comparison of Regions within Mainland China 215
6. Changes in the Investment Environment in Mainland China After its Entry into WTO 226

Chapter 7 The Business and Investment Environment of the Information Technology and Electronics Industries in Taiwan 247

1. Evaluation of Taiwan's Business Environment 247
2. Competitive Strengths and Weaknesses in Taiwan's Information Technology and Electronics Industries 249

3. Development Strategy of Taiwanese High-Tech Industries 250
4. Taiwan's Development Strategy to Become a Knowledge-Based Economy 253

Chapter 8 Analysis of Each Index 255

1. Knowledge Capital 255
2. Innovative Capability 264
3. Application of Information Technology 277
4. Social Infrastructure 280

Appendix A 319

Appendix B 329

References 433

Index 437

Chapter 1

Theoretical Analysis of Foreign Direct Investment (FDI)

1. Theoretical Analysis of FDI by Large Multinational Enterprises

1.1. The theory of imperfect competitive markets and multinational foreign direct investment

The hypothesis of the traditional theory of international economics is established under the assumptions of perfect market competition, adequate information and immobile production factors among countries. In addition, international trade is assumed to be the major or even the only method for the international economic interaction. However, the framework of this theory does not account for the rapid growth of international economic integration or the activities of foreign direct investment (FDI) as well as the international capital movements after World War II. Many economists, therefore, started to criticize the traditional theory and conduct systematic research on the behavior of direct investment in multinational enterprises.

In 1960, Stephen Hymer, an American scholar and pioneer in the theory of FDI, proposed the imperfect market theory (or industrial organization theory) in his Ph.D. dissertation at MIT. He proved in his dissertation that the advantage of monopolistic enterprises and the imperfection of domestic and international markets are crucial factors in an enterprise's FDI. With additions and improvements by Charles P. Kindleberger (1969), the theory of monopolistic advantage was then proposed and was called Hymer–Kindleberger approach of multinational enterprise. The basic concept of this theory is that the realistic market is imperfect. In the goods market, the difference in products, trademarks, and marketing capabilities, and the varied complicit behaviors among manufacturers limit market power. The monopolization of technology formed by the patent system, the internal and external scale economy, and the natural monopolization of specific resources further contribute to imperfect competition in the market. In the international market, tariffs and exchange rate policies by governments also cause market imperfection. Non-market powers such as obstructed products and factors for unrestrained flow can also cause difficulty when applying the process of the free market described by traditional economic theory. Consequently, the efficiency of resource allocation will be destroyed.

Nevertheless, even confronted with the condition of market imperfection, some enterprises may still obtain an advantage due to monopolization, and some countries may benefit from the unbalanced resource allocation or their unique locations. This is because multinational direct investment, when combined with the two advantages mentioned above, will create immense profits. From this viewpoint, the behavior of proposed multinational direct investment will improve and reallocate resources across global boundaries, as well as lead to a Pareto-improvement. Kindleberger (1969) states that the prospering of direct investment must come from the imperfection in products or factor markets.

In short, the theory of imperfect markets successfully combines the theory of neoclassical economics with the theory of multinational direct investment. It also explains the original motivation of the behavior of multinational direct investment, and establishes the foundation of theory in the research of the theory of foreign direct investment of multinational enterprises. However, this theory does not explain why direct investment should be chosen among other varieties of multinational investment methods. Thus, based on this foundation of theory, many scholars have conducted relevant research and empirical studies of the crucial factors behind multinational direct investment.

1.2. Theoretical analysis of decisive factors for multinational direct investment

1.2.1. Theory of enterprise advantage

Compared with local companies, foreign investors are likely to incur additional costs in order to bridge the differences in culture, law, system and language, and to gain sufficient knowledge of local markets. To ensure a profit from these additional investments, the multinational enterprises should have certain advantages that local competitors do not have. Based on this concept, researchers focused on one or several core advantages that enterprises held, then analyzed each one systematically, such as patent right, special permission to obtain capital and market entry, economies of scale, product differentiation, etc. In the early stages, the existence of these factors allows multinational direct investment to obtain higher profitability than other investing methods. This method, therefore, has become one of the major investing methods for investors.

In general, the advantages of multinational enterprises are as follows: firstly, the technological advantage. Almost all theories of multinational direct investment emphasize the core effect of this factor. In practice, new products and new technology are the essential elements of technological advantage for multinational corporations. The product difference is a vital factor that can be advantageous if the quality and functions of the product surpass those of competitors. The market technique and superior organizational management of enterprises can also be seen as technological advantages. Through FDI, multinational enterprises are able to internalize the technology transfer, which not only guarantees the technology nondisclosure, thereby making it hard for competitors to imitate, but also increases profit.

Secondly, the advantage of monopoly. The advantage of monopoly will not only create a great margin, but also obtain better economies of scale for enterprises. Multinational direct investment encourages enterprises to extend their monopolistic advantage through the consolidation and development of a worldwide market share; this direct investment will be more direct and effective than other investing methods. In response to this, the oligopsonistic competition model developed by American scholar Nick Brooks holds that a strategy of prevention will be carried out by other multinational enterprises to ensure an advantage of oligopoly worldwide during the process of direct investment. Thus, while one enterprise may establish a branch abroad, other enterprises will also follow suit in order to reduce any advantage that the first enterprise gained. Obtaining the advantage of monopoly is therefore a priority as enterprises practice multinational direct investment. Enterprises must also continually maintain and expand their monopolistic advantage.

Thirdly, the advantage of capital and currency. If the currency of the capital in export countries is relatively strong, there will be a currency surplus from FDI, and its investing benefit may not only be higher than that of the domestic investment, but also higher than the host countries' within the same conditions. In addition, the surplus of inflation will lead to the surplus of interest rate. According to the Abberley Model, this may enable multinational investors to reduce operation costs and become more competitive by collecting the capital with lower interest rates than the host countries'.

Finally, FDI will allow the enterprise to avoid risks from the diversification of investment and achieve the purposes of profit and diversified risks, which have been explained and proven by the theory of portfolio selection or assets choice. This factor has become extremely important after the system of fixed exchange rate collapsed and inflation and economic differences grew worldwide since 1973.

1.2.2. The internalization theory of direct investment by multinational companies

The advantages discussed above not only benefit multinational investors, but also achieve enormous profit that cannot be gained while enterprises invest only in the domestic market. However, based on this idea, enterprises are also able to achieve the advantage of economic benefits through domestic production, followed by export to other countries. The existing advantages of enterprises do not explain why enterprises are unable to achieve a method of export to gain further advantages.

In order to analyze this issue, economists refer to the equilibrium model proposed by Ronald H. Coase (Coase Theorem) in his discussion of the origin of the multinational corporation. The model brings forth the internalization theory related to activities of multinational direct investment. This theory states that transaction costs are always present in the trading market. These costs include those of price determination, contract signing, supervising the contract execution, and risk. For certain types of trading, the cost of market dealing is rather high and inefficient, which sometimes renders the deal difficult and unprofitable. At this time, enterprises should make use of internal management instead of dealing through the market, provided the cost of internal management trading is less than that of market trading. This concept could also be applied to the multinational production of enterprises. Generally speaking, export and technology transfer are the methods of marketization through which enterprises are able to apply their advantages of capital and technology. Direct investment is the method of internalization, however, if the cost of the former is higher than the latter; the difference in cost between the two methods would then create a surplus.

In comparison with general activities of market trading, exporting activities make for a more complicated trading process and create higher trading costs due to the influence across borders of countries, tariffs and long-distance transportation. However, there will be higher profit if the activities of multinational direct investment can be internalized. When the place has less cost of international resource allocated, the multinational enterprises will be established. Internalization will be useful for multinational enterprises to internalize their specific advantages. American scholars Buckley and Casson (1976) later integrated the theory of ownership-specific advantage and the theory of internalization (i.e., intra-firm trade) successfully. Magee (1977) even proposed the adequate reward theory based on the foundation of this theory; he emphasized that the advantages of multinational enterprises come from the guarantee of adequate capabilities from their investment in technologies and products. In addition, the internalized technological capability in multinational enterprises is one of many factors leading to high profit.

1.2.3. The location theory and the eclectic theory of international production

When explaining the decisive factors in the circulation of investment for multinational enterprises, researchers have also included the location factor of host countries into the theory framework. In practice, the circulation of multinational investment is influenced by the following factors: firstly, the labor cost. The obstruction of international circulation of labor due to population control for each country renders the international labor market imperfect. This results in the cost differences in worldwide labor. When industrial technologies enter the mature stage and the market advantages become dependent on low cost, arranging for production activities to take place in a location with low labor cost becomes more attractive. As a result, the labor cost is a major factor when deciding the circulation of investment. Other similar factors include the cost of raw materials and specific resources. After World War II, many enterprises in developed countries were eager to create investments in developing countries for these reasons.

Secondly, the market situation. The market size, market growth rate, the stage of development and the degree of competition in a host country can strongly influence an enterprise's decision regarding direct investment. Many surveys on investment by multinational companies express that most multinational companies invest to win large market share in host countries; several investigations conducted by FDI enterprises in Mainland China have also proven this viewpoint. When the economic effect of industrial scale is significant and the market of investors' home countries is small and unable to allow investors to develop scale advantages, then the market situation of the host countries will likely become the main issue when deciding the circulation of investment. Other factors similar to the market situation factor are free trade zone, tariff union, trade barriers and local stability. All these factors play a crucial role in deciding the circulation of multinational direct investment.

Thirdly, the policy of host countries. Generally speaking, the policies of method of entry, stock share by shareholders, and the arrangement of the market of foreign enterprises in host countries will influence the amount of revenue from foreign investors directly and indirectly. These policies have much influence when enterprises make investment decisions. The above factors and the standard of infrastructure in host countries are normally categorized as part of the investing environment that will be considered in the early stages of investment. Most investments take place in countries or areas with ideal conditions and significant regional advantages as stated above.

2. Theoretical Analysis of FDI by Small- and Medium-Sized Enterprises in Taiwan

According to data from the International Monetary Fund (IMF), inward FDI in developing countries has increased more than 12 times since the 1980s. Long-term inward FDI has provided many benefits to developing countries, one being the significant increase of infrastructure for public sectors. Long-term inward FDI also expands the capital stock for the host countries. Inward FDI became active in East Asia after the 1990s. Taiwan, South Korea, Hong Kong, Singapore and Japan are the major countries supplying inward FDI. Mainland China and Southeast Asian countries, on the other hand, are the major countries seeking inward FDI.

The Asian financial crisis, which took place from 1997 to 1999, has changed the circulation of inward FDI. This can be seen in the allocation of investments between Mainland China and the countries of Southeast Asia that were affected by the financial crisis. For example, the investment flow from Taiwan to Mainland China has been increasing sharply in the past decade, compared to the stable flow of inward FDI from the Southeast Asian countries affected by the financial crisis to China after 1997. Although there is much motivation behind the integration of the regional economies in East Asia, such as technologies, preferences, and public policies, the private enterprises seem to play the principal role in this issue.

According to the inward FDI theory proposed by the Japanese economist Kiyoshi Kojima in 1973, 1978 and 1982, the models of inward FDI show differences that are due to the difference in source of investment. He pointed out that the comparative advantages between the FDI supply in Japan and the investments of host countries are consistent. It would, therefore, result in trade promotion effects. On the other hand, the supply of inward FDI in the United States is entirely different. The comparative advantages of FDI in America normally represented are inconsistent. The weakness of trading is this: FDI supply in Japan is provided by the competitive small- and medium-sized enterprises (SMEs), whereas inward FDI supply in the US is provided by the monopolistic major enterprises. Thus, the market composition differences between Japan and the United States lead to diverse types of inward FDI.

Kojima's theory challenges the traditional inward FDI theory proposed by Hymer in 1960. The traditional theory holds that foreign investment would only be established on the intangible rather than tangible property in the competitive industries. In contrast, Kojima states that the reason competitive Japanese enterprises are investing overseas is the changing macroeconomic situation, which makes local production near impossible. The competitive enterprises, however, also hold intangible property in several particular industries; otherwise, they would have already moved into new industries in order to adapt to the new economic environment rather than transferring their capital overseas to maintain their existing businesses. Kojima also states that the changing macroeconomic environment in domestic markets may lead to less competition among those companies. The reason is that their intangible property originally comes from the non-existing economic rent in those competitive industries. When the current market is not able to obtain the rent by providing authorization or direct export, organizations will try to derive the economic rent from multinational management of FDI.

The argument Kojima holds in this theory of competitive industry originally came from his observation of the size of the Japanese organizations that had FDI; they were all SMEs. In addition, all of them were smaller than the companies in America. The enterprises in Taiwan started to have direct investment in Southeast Asia since the 1980s and in Mainland China since the 1990s. Kojima argues that the majority of this outward FDI was from SMEs, which is proven by this model. The intangible property they owned was the ability to manage the small amount and flexible production. This ability came from the support of an efficient production network, which was provided by the manufacturers with a highly specialized facility of integration.

The manufacturers in the network are also SMEs. They all possess the same characteristics of independence and the relationship of mutual competition. The sharing of production and market information also enables a prompt reaction to changes in technology and market situation. Since the mid-1990s, their percentage of outward FDI has increased in order to exploit their property, including patent rights, other technological property, goodwill, production skills and marketing. The role of inward FDI has played an important role in the international markets where local production has been gradually adopted. The SMEs in Europe and America can refer to the above-mentioned experiences before making investments in Mainland China.

The research regarding investment in developing countries like Mainland China or Southeast Asian countries by enterprises in Taiwan show that the investment in Mainland China improves when its purpose is market extension. In addition, production becomes more efficient if the investment by Taiwanese companies aims at reducing cost. In other words, either the market extension or the reduced production cost, or both, is improving the production of Taiwanese SMEs. In conclusion, the investment of Taiwan enterprises in Mainland China improves the production in cross-Straits trade. That is, the investment that Taiwanese enterprises put into Mainland China will be promotional and complementary to the production in cross-Straits business. The horizontal and vertical integration will take place in the countries with FDI.

Chapter 2

FDI Development in Mainland China and Taiwan

1. Status Quo in Mainland China

Because of globalization, the capital flowing between countries has been increasing faster than before. However, the unequal distribution of capital restricts the flow to and among only developed countries. In order to attract foreign capital, developing countries have started a fierce competition with each other. Worldwide, the quantity of foreign capital invested in Mainland China comes second just after the United States. The entry of foreign capital, especially FDI, has brought the benefits of capital formation, improved technology, more foreign exchange, higher export, scientific management, systemized reform, developing resources, industrial optimization, and an overall healthy competitiveness into Mainland China. Meantime, the foreign-invested enterprises also have much to gain. With the policy of open foreign investment and the progressive systematization and reforms of the market economy, Mainland China has become one of the most ideal locations in which to have investments. The position that FDI has in the economic development of Mainland China, therefore, is becoming increasingly important.

The definition of FDI given by the IMF is: "A continuing benefit acquired by investors whose investment in running the business aims to win the right of speak of the enterprise management."[1] In other words, FDI is "the dominion of the production information like factory and land", or "the dominion of stock that makes shareholders able to control the investment of operating activities". The investor of the FDI is to acquire the effective right to control and participate in management.

[1] IMF (1977), *Balance of Payment Manual*, Washington D.C. p. 408.

The meaning of indirect investment is "to obtain the profit from the investment or capital rather than participating in the management of enterprises".

To clarify, the FDI mentioned in this book is interchangeable with multinational direct investment. From the viewpoint of the host country, it generally means the direct investment from foreign enterprises; but from the viewpoint of the investing country, it refers to outward foreign direct investment.

1.1. Analysis of overall national investment trends

By reviewing the number of projects after implementing the "open market" policy for more than two decades, one can conclude that Mainland China has been developing rapidly through well-applied FDI. Up until the end of December in 2004, Mainland China had approved 508,941 FDI enterprises with a total amount of 1096.608 billion USD in joint venture foreign capital, of which the total amount of realized foreign capital was 562.101 billion USD (see Table 2.1).[2] China began encouraging FDI enterprises in the late 1970s. According to reports from *Fortune* magazine, 400 out of the 500 biggest companies worldwide have invested more than 2000 projects in Mainland China. They include major manufacturers of computers, electronic products, telecommunication facilities, pharmaceutical factories, petrochemical and power generation facilities, which have now extended their production network to Mainland China.

It can be seen from Table 2.1 that Mainland China started to apply FDI enterprises since 1979, and that the total amount of realized FDI has increased 45 times from 916 million USD in 1983 to 40.715 billion USD in 2000. The amount of FDI was growing every year. Three periods of rapid growth are represented significantly in the table, namely 1985–1986, 1992–1993, and 2002–2004. Especially in 1993, the amount of realized FDI nearly doubled from 1992. The total amount of realized FDI decreased slightly after 1998; however, after two years of adjustment, FDI revived in 2000. The encouragement

[2] The Department of Foreign Investment Administration of the Ministry of Commerce.

Table 2.1. Amount of FDI in Mainland China as of 31 December 2004 (100 million USD)

Year	Number of Projects	Total Amount of Foreign Capital from Joint Ventures	Total Amount of Realized Foreign Investment
1979–1982	920	49.58	17.69
1983	638	19.17	9.16
1984	2,166	28.75	14.19
1985	3,073	63.33	19.56
1986	1,498	33.30	22.44
1987	2,233	37.09	23.14
1988	5,945	52.97	31.94
1989	5,779	56.00	33.93
1990	7,273	65.96	34.87
1991	12,978	119.77	43.66
1992	48,764	581.24	110.08
1993	83,437	1,114.36	275.15
1994	47,549	826.80	337.67
1995	37,011	912.82	375.21
1996	24,556	732.76	417.26
1997	21,001	510.03	452.57
1998	19,799	521.02	454.63
1999	16,918	412.23	403.19
2000	22,347	623.80	407.15
2001	26,140	691.95	468.78
2002	34,171	827.68	527.43
2003	41,081	1,150.70	535.05
2004	43,664	1,534.79	606.30
Total	508,941	10,966.08	5,621.01

Source: Analysis of Foreign Capital from Ministry of Commerce of PRC.

of foreign capital increased greatly in 2002 and reached 39% in 2003, and in 2004 the growth was more than 30%.

To sum up, the development of FDI in Mainland China can be divided into four stages:

(1) 1979–1982, the preliminary stage during which the total amount of realized FDI was about 1.7 billion USD;

(2) 1983–1991, the stage of stable development during which the total amount of realized FDI was about 23.289 billion USD;
(3) 1992–1996, the stage of rapid growth, due to Deng Xiaoping's encouragement during his inspection of southern China. Since 1993, Mainland China has attracted the most FDI among developing countries. From 1992 to 1996, Mainland China successfully reached a total amount of realized FDI of 151.537 billion USD. Mainland China took 13th place worldwide and third place among the developing countries that brought in FDI in 1991. In 1993, however, Mainland China became the second biggest country behind the United States in terms of FDI adoption, and also won first place among the developing countries. Hereafter, Mainland China kept its position of being the second largest country adopting FDI;
(4) After 1996, the application of foreign capital that entered Mainland China stabilized. But the Asian financial crisis as well as the huge decline of joint venture foreign capital led to the rise in number of government schemes aimed at slowing the decreasing amounts of foreign capital after 1997. Nevertheless, the number of FDI projects, the joint venture foreign capital, and the actual amount of realized foreign capital still displayed an overall decline. Mainland China lost its second place worldwide but still kept its first place among developing countries. Eventually, China was able to emerge from the Asian financial crisis, and the application of foreign capital rose again after 2000 (see Figure 2.1). The number of new FDI enterprises approved in 2000 was 22,347, an increase of 31.8% from the previous year. The total amount of joint venture foreign capital was 62.38 billion USD, which had grown by 51.3% from the year before. The total amount of application of foreign capital was 40.715 billion USD, which had risen 0.98% over the previous year. The rate of increase of FDI rose significantly in Mainland China in 2001. Both the total amount of joint venture foreign capital and application of foreign capital were increasing remarkably. The total amount of joint venture foreign capital was 69.19 billion USD and the total amount of application of foreign capital was

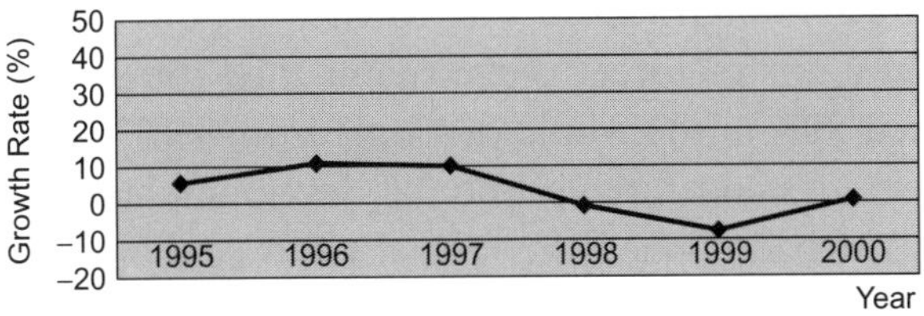

Figure 2.1. The Rate of Increase of Mainland China's Absorbing FDI

Source: UNCTAD: Analysis of Foreign Capital by Ministry of Commerce of PRC.

46.8776 billion USD, which had increased 10.43% and 14.9% respectively since 2000.

Looking at each area of allocation, it can be found that eastern China is the principle target of FDI, while the western areas are less concerned about bringing in foreign investment (see Table 2.2). This distribution shows the consistency between character gradation and progress of the open market in Mainland China. The process of bringing in foreign development begins in the special economic zones that

Table 2.2. FDI Distribution in Eastern, Central and Western China as of 31 December 2003 (100 million USD)

Area	Number of Projects	Percentage	Joint Venture FDI	Percentage	Realized FDI	Percentage
Eastern	381,527	82.00	8,191.64	86.86	4,326.07	86.27
Central	52,424	11.27	712.12	7.55	447.90	8.93
Western	31,326	6.73	527.54	5.59	240.74	4.80
Total	465,277	100	9,431.30	100	5,014.71	100

The eastern area: Beijing, Tianjin, Hebei, Liaoning, Shanghai, Jiangsu, Zhejiang, Fujian, Shandong, Guandong, Hainan.
The central area: Shanxi, Jilin, Heilongjiang, Anhui, Jiangxi, Henan, Hubei, Hunan.
The western area: Inner Mongolia, Guangxi, Sichuan, Chongqing, Guizhou, Yunnan, Shanxi, Gansu, Qinghai, Ningxia, Xinjiang, Tibet.

Source: Statistics of Foreign Capital by Ministry of Commerce of PRC.

strongly attract foreign capital due to their unique policy of preference. The amount of foreign investment into the special economic zones formed 70% in the whole of China from 1979 to 1982.

After 14 cities and three economic development zones along the southeast coast had applied the special open economic liberalization policy in 1984 and 1985, the foreign capital gradually expanded from the south to the north. Shanghai, Beijing, and Liaoning advanced greatly by bringing in foreign capital. It can be said that the major areas that attracted foreign capital were special economic zones, open cities along the coast and major cities with better economic foundation in the 1980s. After entering the 1990s, foreign capital started to extend from eastern to western China, which means the extension swept from the coastal areas to the inland areas. The coastal areas also showed an extension from major cities into smaller towns. Generally speaking, the FDI was focused on the southeastern area along the coast; however, the degree of realized foreign capital was still rather small. The gap between areas was still significant in the early 1990s — the FDI in the eastern area occupied 93.9%, while the percentages of the central and western areas were 3.8% and 2.2% respectively. This disparity was reduced slightly by the end of 2003. As shown in Table 2.2, the percentage of FDI in the east decreased from 94% to 82%, while in the central and western areas it rose to 11.27% and 6.73% respectively. Nevertheless, the statistics clearly show serious inequality in the FDI allocation in Mainland China, and the situation has not changed today.

1.2. Analysis of investment by industry

In terms of industrial distribution, foreign capital in Mainland China displays the basic pattern of "2-3-1" (see Table 2.3). FDI is located mainly in the second industry, especially for the industrial sectors. In the 1980s, the FDI in the second industry occupied 60.3%, and 2.9% and 36.8% in the first and third industries respectively. In the 1990s, the percentage of FDI in the first industry decreased to 1.7%, the third industry to 32.9%, while the second industry increased to 65.4%, an increase of more than 5% than what it was in the 1980s. The joint venture foreign capital applied by the first, second, and third industries were 1.76%,

Table 2.3. FDI Distribution by Type of Industry as of 31 December 2003 (100 million USD)

Industry Type	Number of Projects	Percentage	Joint Venture FDI	Percentage
First Industry	13,333	2.87	180.36	1.91
Second Industry	350,170	75.26	6,320.10	67.01
Third Industry	101,774	21.87	2,930.84	31.08
Total	465,277	100	9,431.30	100

Source: Statistics of Foreign Capital by Ministry of Commerce of PRC.

59.56%, and 38.67%, while the number of projects occupied 2.79%, 73.01% and 24.20% respectively. By the end of 2003, the projects' allocation of FDI composition in the first, second, and third industries was 2.87%, 75.26%, and 21.87%. The direct investment from foreign enterprises in the first industry was still small, and the percentage of investment in the third industry was relatively low. The essential reason might be that Mainland China always focused on production investment in the second industry when trying to attract FDI, which led to less investment in the first industry and lower percentage investment allocation in the third industry. Looking at FDI distribution by different industries (see Table 2.4), in the interior sector of the second industry, the FDI focused mainly on labor-intensive industries in the 1980s, and it shifted gradually into capital-intensive industries in the early 1990s and onwards. The technology-intensive industries have also attracted more and more FDI recently. In the interior sector of the third industry, the composition of FDI was also imbalanced — the percentage of real estate industry was so high that the development of the finance, telecommunication (which is one of the manufacturing industry), and logistics services ceased.

1.3. Analysis of sources of incoming investment

Looking at the original locations of FDI, a significant characteristic of territoriality can be observed (see Table 2.5). In terms of location, 50% of FDI realized by Mainland China came from Hong Kong and Taiwan. United States, Japan, and UK took the second most

Table 2.4. FDI Distribution by Composition of Industries as of 31 December 2003 (100 million USD)

Industry Category	Number of Projects	Percentage	Joint Venture FDI	Percentage
First Industry				
Agriculture, Forestry, Farming and Fishery Industries	13,333	2.87	180.36	1.91
Second Industry				
Mining Industry	524	0.11	16.81	0.18
Manufacturing Industry	338,952	72.85	6,003.98	63.66
Electricity, Gas, and Water Production and Supply Industries	654	0.14	56.82	0.60
Construction Industry	10,040	2.16	242.49	2.57
Third Industry				
Transportation, Logistics and Post Service Industries	5,235	1.13	238.13	2.52
Wholesale and Retail Industries	23,565	5.06	288.42	3.06
Finance Industries	48	0.01	8.65	0.09
Real Estate Industries	40,941	8.8	1,808.96	19.18
Lease and Commercial Service Industries	15,438	3.32	301.7	3.20
Scientific Research, Technological Service and Geological Prospecting Industries	3,528	0.76	41.59	0.44
Residential Service and Other Service Industries	10,333	2.22	163.19	173

(Continued)

Table 2.4. (***Continued***)

Industry Category	Number of Projects	Percentage	Joint Venture FDI	Percentage
Education	1,428	0.32	25.86	0.27
Sanitation, Social Security and Welfare Industries	1,204	0.26	54.33	0.58
Total	465,277	100	9,431.31	100

Source: Statistics of Foreign Capital by Ministry of Commerce of PRC.

important position. The currencies of Southeast Asian countries became devalued after the Asian financial crisis, which increased their cost of investing in Mainland China, thereby reducing their capability of investment. The enterprises in Europe and America, on the contrary, not only held a positive opinion of the investment environment in Mainland China, but also considered the possibility of increasing FDI in Mainland China because of its stable financial condition and high speed of economic development.

1.4. Analysis of investment methods and investment performance

Looking at the methods of investment, the major methods of foreign capital applied in Mainland China have been: Chinese-foreign equity joint venture enterprises, Sino-foreign joint managed enterprises, foreign exclusive venture enterprises, and joint developments (see Table 2.6). However, where the internal structure of FDI approved by Mainland China was concerned, no matter the number of projects, agreed amount, or the size of realized FDI, Chinese-foreign equity joint venture enterprises usually took the major position. At the end of 2003, the proportion of projects among Chinese-foreign equity joint venture enterprises, Sino-foreign joint managed enterprises, foreign exclusive venture enterprises and joint development enterprises was represented as 5:1:3.5:0.4. In terms of future development, joint venture enterprises are expected to retain their leading position for a

Table 2.5. Source of FDI in Mainland China by Particular Countries/Areas as of 31 December 2003 (100 million USD)

Country/Area	Number of Projects	Percentage	Joint Venture FDI	Percentage	Realized FDI	Percentage
Asia	**369,789**	**79.48**	**6,541.41**	**69.36**	**3,574.14**	**71.27**
Hong Kong	224,509	48.25	4,145.14	43.95	2,225.75	44.38
Indonesia	1,079	0.23	25.77	0.27	12.69	0.25
Japan	28,401	6.10	574.87	6.10	413.94	8.25
Macao	8,407	1.81	120.87	1.28	51.9	1.03
Malaysia	2,888	0.62	71.59	0.76	30.86	0.62
Philippines	1,945	0.42	37.43	0.40	16.45	0.33
Singapore	11,871	2.55	435.69	4.62	235.31	4.69
South Korea	27,128	5.83	366.53	3.89	196.88	3.93
Thailand	3,375	0.73	63.22	0.67	25.49	0.51
Taiwan	60,186	12.94	700.29	7.43	364.87	7.28
EU	**16,158**	**3.47**	**659.42**	**6.99**	**378.73**	**7.55**
Belgium	470	0.10	10.75	0.11	6.72	0.13
Denmark	267	0.06	14.12	0.15	5.2	0.10
UK	3,856	0.83	208.42	2.21	114.38	2.28
Germany	3,504	0.75	157.13	1.67	88.51	1.76
France	2,302	0.49	79.15	0.84	61.47	1.23
Ireland	61	0.01	1.29	0.01	0.46	0.01
Italy	2,137	0.46	38.14	0.40	25.46	0.51
Luxembourg	96	0.02	6.05	0.06	2.82	0.06

(*Continued*)

Table 2.5. ***(Continued)***

Country/Area	Number of Projects	Percentage	Joint Venture FDI	Percentage	Realized FDI	Percentage
Holland	1,254	0.27	99.25	1.05	50.63	1.01
Greece	47	0.01	1.10	0.01	0.31	0.01
Portugal	80	0.02	1.29	0.01	0.73	0.01
Spain	755	0.16	13.58	0.14	4.46	0.09
Austria	573	0.12	10.36	0.11	4.43	0.09
Finland	181	0.04	6.12	0.06	3.85	0.08
Sweden	575	0.12	12.67	0.13	9.30	0.19
North America	**48,281**	**10.38**	**984.30**	**10.44**	**480.09**	**9.57**
Canada	6,941	1.49	119.87	1.27	39.22	0.78
United States	41,340	8.89	864.43	9.17	440.88	8.79
Free Ports	**11,665**	**2.51**	**810.52**	**8.59**	**381.38**	**7.61**
Cayman Islands	923	0.20	111.76	1.18	46.69	0.93
Virgin Islands	8,877	1.91	620.12	6.58	301.65	6.02
Samoa	1,865	0.40	78.64	0.83	33.04	0.66
Total	465,277	100	9,431.31	100	5,014.71	100

Source: Statistics of Foreign Capital by Ministry of Commerce of PRC.

Table 2.6. Statistics of FDI Methods as of 31 December 2003 (100 million USD)

Method	Number of Projects	Percentage	Amount of Joint Foreign Capital	Percentage	Realized FDI	Percentage
Chinese-Foreign Equity Joint Venture Enterprises	238,367	51.23	3,518.37	37.31	2,060.2	741.08
Sino-Foreign Joint Managed Enterprises	54,512	11.72	1,707.99	18.11	866.1	917.27
Foreign Capital Enterprises	172,108	36.99	4,141.47	43.91	1,990	39.68
FDI Share	67	0.01	14.55	0.15	15.53	0.31
Joint Developments	191	0.04	47.4	0.5	73.98	1.48
Others	32	0.01	1.53	0.02	8.74	0.17
Total	465,277	100	9,431.3	100	5,014.71	100

Source: Statistics of Foreign Capital by Ministry of Commerce of PRC.

while, whereas Sino-foreign joint managed enterprises will eventually decline. While joint venture enterprises have been the major form among new FDI enterprises in recent times, yet the number of foreign exclusive venture enterprises and holding enterprises is steadily increasing.

Looking at the operating status of foreign capital, the overall application of foreign capital in Mainland China shows a tremendous performance (see Table 2.7). The growth percentage of foreign capital increased by 26% in each year. In the meantime, the sales status of products also show excellent results, with rates at more than 96% each year. This therefore shows that the investment from foreign enterprises in Mainland China has acquired enormous feedback.

1.5. Changes of foreign capital in Mainland China

International business economists have proposed the theory of "3 Centers" after studying the global strategic structure of FDI. The 3 centers included in the global network of foreign enterprises are the production and manufacturing center, the research and development (R&D) center, and the operation and management center. Before the mid-1990s, foreign enterprises viewed Mainland China as a nation of low technology skill in product processing and assembling. However, by the end of the 1990s when foreign enterprises had much investment in Mainland China, great changes took place. The role of Mainland China changed from a processing and assembling base into a manufacturing base. The changes are represented mainly as follows:

(1) Foreign capital has become the major provider of new products and technology in Mainland China. Technology levels in large multinational companies in Mainland China improved quickly in the late 1990s. According to the research, the reason why the high or new technological industries in Mainland China developed rapidly recently is that they have attracted more multinational investment. These investments have provided better or even the most advanced products and technology, which has well made up for the earlier inefficiency of technology in Mainland China.

Table 2.7. Main Economic Benefit Indices for "3-Capital" Industrial Enterprises

Year	Industrial Growth Rate (%)	Total Assets Contribution Ratio (%)	Debt-to-Asset Ratio (%)	Floating Assets Turnover Ratio (Frequency: Times/Year)	Cost Expense (% of Profit Margin)	Gross Labor (RMB/ Person-Year)	Product Sales Ratio (%)
1998	24.20	6.76	58.52	1.61	2.75	52,311	96.82
1999	25.59	7.93	57.73	1.69	4.39	61,260	97.18
2000	25.96	9.76	57.01	1.89	6.03	71,403	97.74
2001	26.19	9.83	54.87	1.89	5.85	75,913	97.64
2002	26.41	10.46	54.38	2.06	6.40	81,313	98.28
2003	26.15	11.46	55.43	2.20	6.83	92,158	98.17

Source: *2004 Mainland China Statistical Yearbook.*

(2) Foreign capital's assembly lines have been growing faster in Mainland China in recent years and they have focused on improving components and work processes that are more technology-related. In 1992, the growth rate of assembling lines of Japanese enterprises in Mainland China was only 20%, as compared to 47% in the countries of the Association of Southeast Asian Nations (ASEAN). In 1999, the rate in Mainland China rose to almost 50%, while the rate in ASEAN countries remained the same.

(3) The number of R&D organizations established by foreign companies in Mainland China has been increasing. At the end of August 2001, the independent R&D organizations established by large multinational companies numbered at least 100. They were focused mainly on the information communication, biopharmaceutical, precise chemical engineering, transportation facilities, and manufacturing industries. The multinational companies that engaged in R&D activities have had more than five years of investing experience in Mainland China. In addition, their investment was based on a long-term developing strategy in Mainland China. Although some foreign companies have established R&D organizations, their scope is still at an initial stage, at which investigations are conducted to assess the appropriateness of development in Mainland China. Recently, due to insufficient traffic and communication infrastructure, and especially because of the lack of knowledge-intensive service industries, Mainland China still has difficulty functioning as an operation center for foreign companies. The manufacturing base, R&D center, and operation center are usually interconnected. When Mainland China becomes a major manufacturing base and consumer market for foreign companies, establishing R&D centers and operation centers in Mainland China will become unavoidable.

1.6. Forecast of foreign capital investment trends in Mainland China in future years

In order to learn the investment trends of the industries in Mainland China, an investigation of the top 1000 global enterprises, which were listed in *Business Weekly*, was carried out. It included multinational enterprises in the following fields: information technology (IT), electronics, the auto industry, chemical engineering, and biopharmaceutical. The investigation, conducted by the Research Institute of the Ministry of Commerce of PRC, studied enterprises in Europe, America, Japan and Korea, in particular their tendency of investment from June to November in 2004.

Wang Zhi-le, the Director of the Multinational Companies Research Center in the Research Institute of the Ministry of Commerce of PRC, pointed out that the multinational companies within the preparatory stage have started to move into the "harvest period" stage as well as the period of adapting the development tendency of global industries. Multinational companies have to adjust their investment by observing the change in global markets and the market in Mainland China, in order to extend their reach into the Mainland China market. In addition, they need to enhance and complete their integrated strategy layout. "It shall bring a new chance of the investment of the industries in Mainland China as well as provide good opportunities for local governments. For those multinational companies within the preparatory stage, the investment will remain advantage of only technology and shift the basic industries to Mainland China in future three years," as pointed out by Wang Zhi-le. Experts indicated that after completing the shift from the assembly process, the production-based multinational enterprises in Mainland China will move into the more advanced stage of technology transfer. For example, in the IT industry, once the potential of technology development in Mainland China has been realized, multinational companies focusing on the software and system integration development will both enhance business extension and establish their R&D center and projects in Mainland China. This also helps develop the technological and management skills of the local people. The future trends of the foreign capital is represented as below:

1.6.1. Foreign capital investment in Mainland China will increase in future years

The statistics of the survey on above companies show that more than 80% of foreign companies will keep increasing their investment in

Figure 2.2. Investment Trend for 1,000 Global IT Enterprises in Mainland China

1% dropout stage, 16% entry stage, 22% development expansion stage, 26% investment management stage, 35% preparatory stage

Source: *2005–2007 Report of investment trend for multinational companies in Mainland China's IT and semiconductor industry, 2005–2007 Report of investment trend for multinational companies in Mainland China's automotive industry, 2005–2007 Report of investment trend for multinational companies in Mainland China's chemical engineering and biopharmaceutical industry* from the Ministry of Commerce of PRC.

Mainland China, with the range and speed of investment — regardless of production, sales, or technological development — increasing gradually. The results of the survey also indicate that about 35% of foreign companies were in the preparatory stage and planned to have new investment in Mainland China, as shown in Figure 2.2.

In entry stage, foreign companies usually use the "intervention investment", which means symbolically sharing the investment with domestic enterprises in Mainland China, whereby the proportion of shares will be lower and enterprises will not have leading management power. The aim of such investment is to seek trading rights. After entering the preparatory stage, the percentage of investment will be increased due to Mainland China being an investment target with comprehensive advantages, from funds, network and information. To ensure the benefit of management, proportion of sole proprietorship, and substantial investment in Mainland China, multinational companies should spend more time in the preparatory stage. As the multinational companies in that stage move to the "harvest period", they should also try to develop global industries at the same time. They have to increase and change their investment according to the change in global markets and the Mainland China market, in order to extend their market in Mainland China and enhance and complete the overall strategy layout. As a result, this will not only bring new developments to the industrial strategies, but also provide good opportunities for local governments in Mainland China.

The investment of multinational companies in the preparatory stage over the next three years is described as follows: the companies will initially retain technological advantages and accelerate the transfer of basic industries to Mainland China; that is, they keep the high-end products and technological developments within their own home countries and keep increasing the investment of production in Mainland China. Currently, the industries' clusters in investment locations are in the initial stages. Demand for the service and trading field has increased, therefore a sales channel has been established and branding strategies have been applied. The key point in investing in this type of company is to focus mainly on building up the sales and after-sales service. Some production-based multinational enterprises have completed the transfer of assembly lines to Mainland China, which will lead to the stage of further technology transfer.

1.6.2. Foreign companies will significantly increase investment in R&D, but mainly in research

In a survey conducted by the Ministry of Commerce of PRC, 61% of foreign multinational companies among 1000 enterprises indicated clearly that they will keep increasing their investment on R&D in Mainland China over the next three years. Some experts claim that the R&D investment of these multinational companies in Mainland China has started to change method from strategic alliance to the cross-border merger and acquisition.

Generally speaking, strategic alliance is usually in the capital-intensive technology field. Due to rapid developments in the revolution of IT, foreign companies have started to apply the method of cross-border merger and acquisition to obtain control of the R&D institutes which were set up in Mainland China, in order to establish their network of global R&D. These methods will be applied, modified, and used increasingly in the next three years.

According to Mr. Jin Po-sheng, the Director of the Department of Foreign Capital Research in the Research Institute of the Ministry of Commerce of PRC, the overseas R&D research institutes of foreign enterprises that are set up in Mainland China still focus on the study of applicable research, product development and localization, while basic research is limited. In terms of demand, the main purpose of overseas R&D investment by foreign companies is to support local production and sales. The development aspect goes beyond the demand of foreign companies. From the supplying aspect, the basic research usually receives support from the government, which means foreign companies are reluctant to move such R&D projects out of their home countries. Furthermore, basic research has important meaning for foreign companies' future development, hence confidentiality is a great necessity. Foreign companies with such concerns often situate their research in locations close to their headquarters for better management and control.

1.6.3. Future investment will tend towards sole proprietorship

"Multinational companies that plan to invest in production and research in Mainland China are more willing to invest by using sole proprietorship," as Mr. Jin Po-sheng stated, indicating that by using the method of sole proprietorship, multinational companies would achieve better self-management and intellectual property rights protection. In addition, they can adjust the direction of production and R&D in accordance with market conditions and the strategic decision of headquarters. According to the survey, 57% of multinational companies that made production investments tended towards sole proprietorship. 37% were willing to conduct joint ventures with the enterprises that owned specialized technology, production resources, or capabilities. 28% would prefer to cooperate with production factories by using the method of merger and acquisition (see Figure 2.3). In the study of R&D investment by multinational companies, 46% of them tended to establish independent R&D centers. 33% preferred to bring more advanced technology into Mainland China as they conducted R&D. At the same time, 25% planned to increase the number of R&D employees in Mainland China, and 24% chose to cooperate with Mainland China in research and development (see Figure 2.4).

The tendency of sole proprietorship of multinational companies in Mainland China has existed for a long time. From 2003 to 2004,

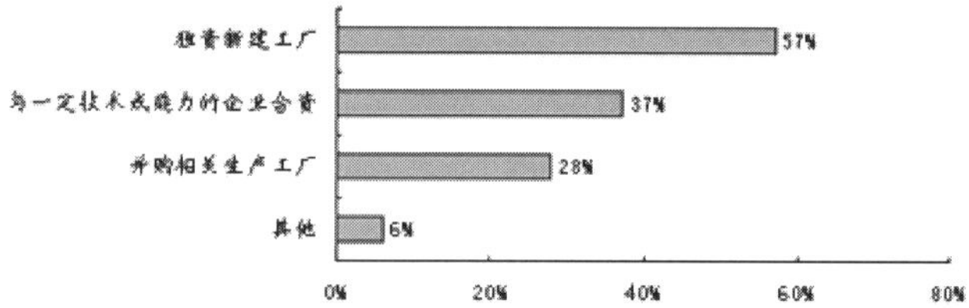

Figure 2.3. The Direction of 2005–2007 Production Investment Trend for Multinational Companies in Mainland China

57% sole proprietorship, 37% joint venture, 28% merger and acquisition with production factories, 6% others

Source: *2005–2007 Report of investment trend for multinational companies in Mainland China's IT and semiconductor industry, 2005–2007 Report of investment trend for multinational companies in Mainland China's automotive industry, 2005–2007 Report of investment trend for multinational companies in Mainland China's chemical engineering and biopharmaceutical industry* by the Ministry of Commerce of PRC.

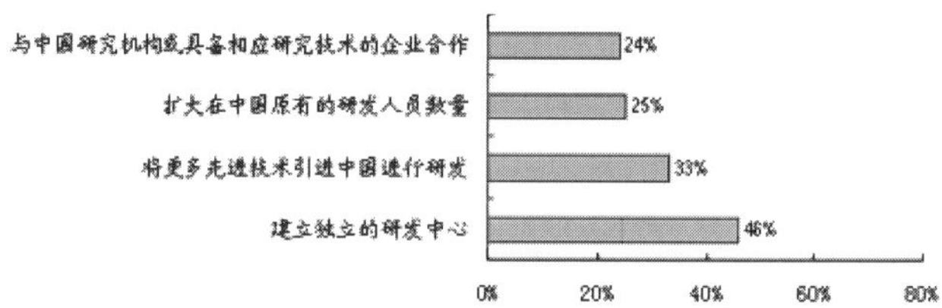

Figure 2.4. The Direction of 2005–2007 R&D Investment Trend for Multinational Companies in Mainland China

24% cooperate with research institute of Mainland China or enterprise with technique of R&D, 25% increase the number of R&D employees, 33% bring in more advanced technology to Mainland China to conduct R&D, 46% develop independent R&D centers

Source: *2005–2007 Report of investment trend for multinational companies in Mainland China's IT and semiconductor industry, 2005–2007 Report of investment trend for multinational companies in Mainland China's automotive industry, 2005–2007 Report of investment trend for multinational companies in Mainland China's chemical engineering and biopharmaceutical industry* by the Ministry of Commerce of PRC.

multinational companies in Mainland China, such as Panasonic Japan, Avon, P&G, Unilever, and Siemens were transforming their joint venture cooperation relationship into sole proprietorship, as stated by Lee Jain-Zhong, the market director of Shanghai Point.

1.6.4. Investment decisions of foreign multinational companies in Mainland China are mainly determined by overseas headquarters

In the survey conducted by the Ministry of Commerce of PRC, the decision of large-sized production investment in Mainland China by foreign companies are generally made by their headquarters in Asia-Pacific or elsewhere. Regarding the localized R&D of products and sales, however, the branches in Mainland China still retain decision-making power.

Investment in production-type enterprises: the investments in process production, renewal of equipment, and establishment of new factories in Mainland China are generally provided by the branch in Mainland China in order to decide the production plan and amount to invest. The headquarters of multinational companies will decide the product types to invest in and organize the layout of global strategies by referring to the global production plan and development of the Mainland China market. The headquarters of investing companies in Mainland China are responsible mainly for selecting investment locations and handling issues that occur in the process of investment, such as the relationship with governments or allocation of human resources.

Investment in R&D-type enterprises: the decision-making of a foreign company in R&D, such as software companies or biopharmaceutical research organizations, is handled by the headquarters in the home country. The investment decision-making ability of branches in Mainland China is poor because of weaknesses in the process of developing and investing in such fields as mentioned above. The branch in Mainland China will have stronger decision-making power only when the investment of R&D in overseas products becomes localized.

1.6.5. Foreign companies in Mainland China will invest mainly in the delta regions

In terms of investment locations, Yanzhijiang Delta Region, Bohai Rim Region, and Zhujiang Delta Region will be the initial locations of direct investment for multinational companies over the next three years. However, the investment in the western area is expected to increase gradually. The following is a breakdown of enterprises based on their choice of investment location: 47% based in Yanzhijiang Delta Region, 22% in Bohai Rim Region, 21% in Zhujiang Delta Region, 9% in the northeast areas, and 8% in the central-west areas (see Figure 2.5). Meanwhile, 92% of multinational companies have chosen to invest in the development zones.

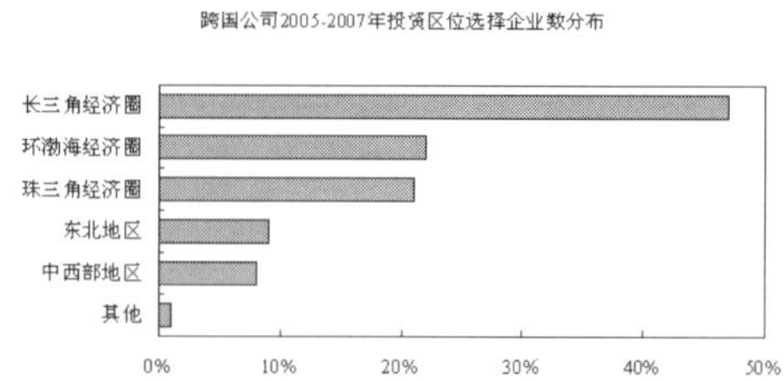

Figure 2.5. The Direction of 2005–2007 Investment Location Choice for Multinational Enterprises in Mainland China

Y: Area (from top to bottom): Yanzhijiang Delta Region, Bohai Rim Region, Zhujiang Delta Region, Northeast, Central-west, Others; X: Percentage (%)

Source: *2005–2007 Report of investment trend for multinational companies in Mainland China's IT and semiconductor industry, 2005–2007 Report of investment trend for multinational companies in Mainland China's automotive industry, 2005–2007 Report of investment trend for multinational companies in Mainland China's chemical engineering and biopharmaceutical industry* by the Ministry of Commerce of PRC.

2. Status Quo in Taiwan

The history of introducing foreign capital in Taiwan can be traced back to the 1950s. The Taiwanese government implemented many measures in order to attract investment from overseas Chinese and foreigners, and to encourage technological cooperation between domestic and foreign manufacturers due to the severe insufficiency of domestic capital at that time. These measures included the Statute for Investment by Foreign Nationals in 1954, the Statute for Overseas Investment in 1955, the Statute for Technological Collaboration in 1962, and the Statute for the Establishment and Administration of

Export-Processing Zone in 1965. The Statute for the Establishment and Administration of Export-Processing Zone is a significant example. The export-processing zone established by the government at that time was mainly to encourage investment activities. The exclusive characteristics such as exempted tariff and unlimited separate zones were the main incentives that attracted enormous foreign capital, and thus, almost all manufacturers in the export-processing zone were foreign enterprises.

After 1980, Taiwan accelerated its push for economic liberalization. Domestic enterprises entered the international market actively, and the domestic management environment was made more open by creating relevant policies and regulations. The project of Asia-Pacific Operation Center in 1995 even brought in significant investment from overseas Chinese. Opening the domestic market and removing restrictions allowed for liberalization and internationalization, which led to the diversification of overseas investment. In addition, these measures were also implemented in 1999 to remove restrictions on overseas investment in the finance and insurance industries, service industries, and telecommunication industries.

The reason the Taiwanese government encouraged overseas enterprises to invest was to increase employment opportunities, improve technological know-how, increase production, and advance the economic growth still in its early stage, at a time when capital was insufficient, foreign exchange was deficient, and production and management skill was weak and accompanied by a large labor force with low-wage standards. According to Hsue Chi (1978) and Wu Rong-Yi *et al.* (1980), the contribution of overseas Chinese investment included the formation of domestic capital, the introduction of foreign technological know-how, the development of the manufacturing industry and export trading. Hsue Chi *et al.* (1986) conducted an evaluation of the influence on Taiwan's economy from overseas Chinese investment. They discovered that overseas Chinese investment has greatly contributed to export trading, the introduction of technological know-how, increasing employment opportunities and improved domestic purchasing. Thus, there is no doubt about the contribution to economic growth from foreign investment.

2.1. Analysis of overall national investment trends

Table 2.8 indicates the change of overseas investment in Taiwan over the years. From 1952 until the end of 2004, a total of 17,066 investment projects by overseas Chinese and foreigners have been approved, with a total amount of 60.495 billion USD. The total number of investment projects brought in by overseas Chinese was 2,834 with 3.93 billion USD, and 14,232 projects with 56.565 billion USD was invested by foreigners. The proportion of both was 1:5 and the percentage of investment amount was 1:14. It can be seen that overseas investment came mainly from foreigners, not overseas Chinese, which helps explain why economic development accelerated in Taiwan during the past four decades.

Figure 2.6 shows the trend of overall overseas investment in Taiwan by year. Taiwan has been encouraging overseas investment for a long time, and with the challenge of economic aid termination from the United States and the promotional efforts of the Taiwanese government, overseas investments began to increase significantly after the late 1960s. Foreign capital kept growing after the 1970s due to many beneficial measures adopted by the government and the comparable advantages of Taiwan. In 1986, Plaza Accord compelled the exchange rate to appreciate from 250 Yen to 1 USD, and this influenced the currency appreciation of the New Taiwan Dollar as well. As a result, the amount of overseas investment almost doubled the following year. Investment increased greatly from 770 million USD in 1986 to 1.42 billion USD in 1987. The amount of foreign capital peaked in 1989, reaching 2.4 billion USD that year.

Because of economic changes in both domestic and overseas markets in 1989, overseas investment in Taiwan gradually dropped every following year and did not grow again until 1994 (see Table 2.8). During 1995 to 2000, the economic status of Taiwan, especially the development of the electronics industry, grew remarkably, which brought upon a growth of overseas investment. After Taiwan achieved a historic peak of 7.6 billion USD in investments in 2000, there was a global capital shrink due to the events of 11 September 2001, and this worried capital providers, who feared a worldwide terrorist threat. In the meantime, Taiwan was experiencing the disputation about its 4th nuclear power plant and an

Table 2.8. Statistics of Approved Investment Projects by Overseas Chinese and Foreigners by Year (1,000 USD)

Year	Overseas Chinese		Foreigners		Total	
	Number of Projects	Amount	Number of Projects	Amount	Number of Projects	Amount
1952–1964	142	36,150	88	56,956	230	93,106
1965	30	6,470	36	35,140	66	41,610
1966	51	8,377	52	20,904	103	29,281
1967	105	18,340	107	38,666	212	57,006
1968	198	36,449	123	53,445	321	89,894
1969	90	27,499	111	81,938	201	109,437
1970	80	29,731	71	109,165	151	138,896
1971	86	37,807	46	125,147	132	162,954
1972	113	26,466	52	100,190	165	126,656
1973	201	55,166	150	193,688	351	248,854
1974	85	80,640	83	108,736	168	189,376
1975	44	47,235	43	70,940	87	118,175
1976	53	39,487	45	102,032	98	141,519
1977	52	68,723	50	95,186	102	163,909
1978	50	76,210	66	136,719	116	212,929
1979	50	147,352	73	181,483	123	328,835
1980	39	222,584	71	243,380	110	465,964
1981	32	39,463	73	356,294	105	395,757
1982	50	59,720	82	320,286	132	380,006
1983	49	29,086	100	375,382	149	404,468
1984	74	39,770	101	518,971	175	558,741
1985	67	41,757	107	660,702	174	702,459
1986	80	64,806	206	705,574	286	770,380
1987	117	195,727	363	1,223,069	480	1,418,796
1988	89	121,377	438	1,061,161	527	1,182,538
1989	70	177,273	477	2,241,026	547	2,418,299
1990	85	220,115	376	2,081,657	461	2,301,772
1991	65	219,462	324	1,558,957	389	1,778,419
1992	73	312,146	338	1,149,228	411	1,461,374
1993	62	123,50	261	1,089,975	323	1,213,476
1994	57	106,790	332	1,523,927	389	1,630,717
1995	43	168,554	370	2,756,786	413	2,925,340
1996	52	170,451	448	2,290,385	500	2,460,836
1997	44	387,463	639	3,879,166	683	4,266,629

(*Continued*)

Table 2.8. (*Continued*)

Year	Overseas Chinese		Foreigners		Total	
	Number of Projects	Amount	Number of Projects	Amount	Number of Projects	Amount
1998	81	184,721	1,059	3,554,037	1,140	3,738,758
1999	36	132,380	1,053	4,099,024	1,089	4,231,404
2000	40	503,84	1,370	7,557,355	1,410	7,607,739
2001	33	47,223	1,145	5,081,306	1,178	5,128,529
2002	25	44,960	1,117	3,226,787	1,142	3,271,747
2003	22	14,917	1,056	3,560,739	1,078	3,575,656
2004	19	13,740	1,130	3,939,048	1,149	3,952,788
Total	2,834	3,930,472	14,232	56,564,557	17,066	60,495,029

Source: *2004 Annual Report of Investment Statistics for Overseas Chinese and Foreign Investment, FDI, and Mainland China Investment*, Investment Commission, MOEA.

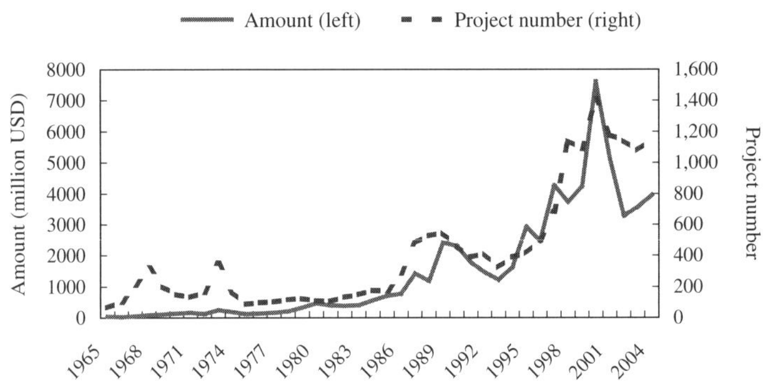

Figure 2.6. Trend of Overseas Investment by Year
(X: Year; Y: Amount, Project number)

unstable political situation stemming from an argument between two parties. This led to a drop in international capital investment. The influence of the above factors led to a reduction in overseas investment by about 2.5 billion USD to a total amount of 5.12 billion USD in 2001, Taiwan's greatest reduction in recent years. With the lack of confidence in global capital providers, overseas investments dropped to 3.27 billion USD in 2002. In 2003, Taiwan made an economic recovery with various measures such as Taiwan First Investment Incentive Program and the 2003 Taiwanese Business Alliance Conference, which promoted government businesses, overseas investment, and Chinese investment.

2.2. Analysis of investment by industry

As seen in Table 2.9, which lists foreign investment by industry, foreign investment was focused mainly on the electronic and electric appliance manufacturing industry in the past 50 years. The total amount accumulated was 14.53 billion USD, which occupied 24.01% in the total amount of foreign investment. The electronic and electric appliance manufacturing industry was the major industry in the past 50 years, and so became the most likely industry to bring foreign investment into Taiwan.

With the Taiwanese government's effort in pushing the liberalization and internationalization of service industries,[3] the finance and insurance industry, other service industries and the wholesale and retail industry have gradually become the major industries to attract foreign investment. The percentages are 17.71%, 7.67%, and 7.66% respectively, placed in the second, third, and fourth place in Table 2.9. The chemicals manufacturing industry, whose growth has almost stopped in recent years, fell to fifth place. Together with the international trade industry, basic metal and products manufacturing industry, telecommunications industry, mechanical manufacturing industry, and food and beverage manufacturing industry, they retain the top 10 rankings; these investments have occupied more than 80% (83.04%) of the overseas investment in Taiwan in the past years.

In 1999, the government lifted restrictions on overseas investment in finance and insurance industries, service industries, and telecommunication industries, bringing significant growth in terms of overseas investment. Telecommunication industries (fixed network and mobile communication categorized into the service industries), in particular, reflected the fastest growth in the amount of investment. Its annual investment was 294 million USD, and its growth reached 501.13% compared to the same period in 1998. Since then, the

[3] Service Industries in this research covered the wholesale and retail industries, international trade industry, restaurant industry, transportation industry, logistics industry, finance insurance industry and other service industries.

Table 2.9. Overseas Investment in Taiwan by Industry (1952–2004) (1,000 USD)

Type of Industry	Number of Projects	Occupied Rate	Amount Approved	Occupied Rate
Electronic and Electric Appliance Manufacturing Industry	2,326	13.63%	14,525,168	24.01%
Finance and Insurance Industry	1,161	6.80%	10,715,202	17.71%
Other Service Industries	1,523	8.92%	4,636,955	7.67%
Wholesale and Retail Industry	3,053	17.89%	4,633,224	7.66%
Chemicals Manufacturing Industry	685	4.01%	4,478,566	7.40%
International Trade Industry	2,378	13.93%	2,740,693	4.53%
Basic Metal and Products Manufacturing Industry	697	4.08%	2,565,262	4.24%
Telecommunications Industry	75	0.44%	2,443,885	4.04%
Mechanical Manufacturing Industry	644	3.77%	2,086,964	3.45%
Food and Beverage Manufacturing Industry	358	2.10%	1,408,279	2.33%
Transportation Industry	364	2.13%	1,117,771	1.85%
Information Service Industry	451	2.64%	1,055,670	1.75%
Consultant Service Industry	537	3.15%	1,006,016	1.66%
Transportation Equipment Manufacturing Industry	122	0.71%	964,394	1.59%
Building and Construction Industry	566	3.32%	938,529	1.55%
Nonmetal and Products Manufacturing Industry	263	1.54%	908,493	1.50%
Textile Industry	159	0.93%	607,662	1.00%
Precise Instruments Manufacturing Industry	169	0.99%	605,565	1.00%
Rubber Products Manufacturing Industry	318	1.86%	536,002	0.89%
Lease Industry	57	0.33%	247,513	0.41%

(*Continued*)

Table 2.9. (*Continued*)

Type of Industry	Number of Projects	Occupied Rate	Amount Approved	Occupied Rate
Plastic Products Manufacturing Industry	72	0.42%	234,751	0.39%
Garments and Accessories Industry	231	1.35%	225,118	0.37%
Papermaking and Printing Industry	106	0.62%	221,907	0.37%
Leatherwear Manufacturing Industry	75	0.44%	204,808	0.34%
Wood, Bamboo, Vines, and Willow Products Manufacturing Industry	124	0.73%	173,884	0.29%
Restaurant Industry	58	0.34%	170,828	0.28%
Mineral, Soil, and Stone Mining Industry	16	0.09%	67,358	0.11%
Fishery and Forestry Industry	68	0.40%	53,756	0.09%
Logistics Industry	17	0.10%	45,662	0.08%
Agriculture and Forestry Industry	42	0.25%	38,850	0.06%
Others	351	2.06%	836,294	1.38%
Total	17,066	100.00%	6,0495,029	100.00%

Source: Same as Table 2.8.

amount accumulated by the end of 2004 has reached 2.44 billion USD, which formed 4.04% of the total amount of overseas investment.

Figure 2.7 displays the proportion of foreign investment in the manufacturing industry in Taiwan in past years, which remained at more than 50% of the total investment amount in the 1960s and 1970s. This increased to more than 80% within a few years. However, after investment reached 89.9% in 1984, the rate dropped gradually each year. It did not rise again until after it hit a low of 22.8%, the historic lowest point, in 2000. The proportion started to grow slightly afterwards. In Figure 2.8, the foreign investment in service industries shows a reverse trend from the manufacturing industry. The proportion of foreign

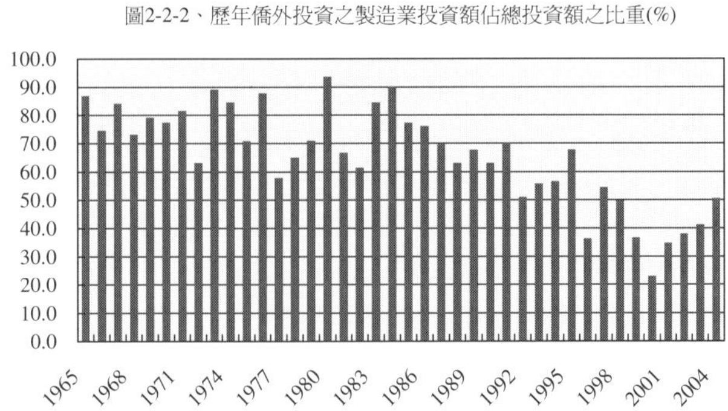

Figure 2.7. Proportion of Foreign Capital Invested in the Manufacturing Industry in Past Years (%)

(X: Year, Y: Percentage)

investment in service industries was 7.3% in 1984, but then it started to grow. After reaching its peak of 58.7% in 2001, it started to drop slightly over the next three years but still retained its high levels.

This trend shows that foreign investment in Taiwan initially leaned towards the low labor cost of the manufacturing industry, but has gradually shifted to other industries such as the financial service industry. This transfer implies that the major industry type in Taiwan has shifted from the manufacturing to the service industry.

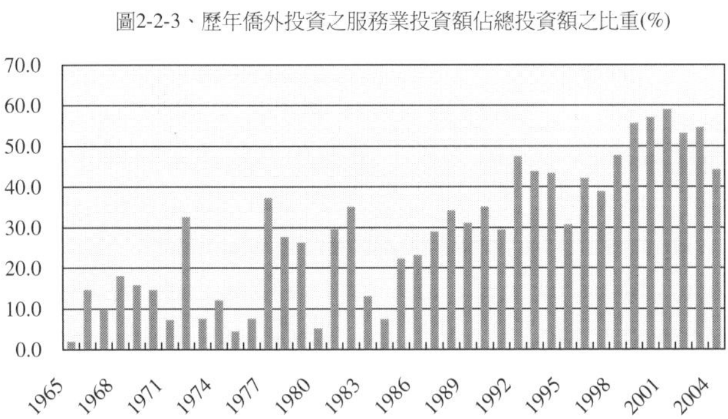

Figure 2.8. Proportion of Foreign Capital Invested in the Service Industry in Past Years (%)

(X: Year, Y: Percentage)

Tables 2.10, 2.11 and 2.12 show the statistics of approved overseas investment in Taiwan by year and industry. It can be seen that the electronic and electric appliance manufacturing industry was always the major industry that foreigners invested in over the past years (see Figure 2.9 to study its historical trends). Investment in the finance and insurance industry rose rapidly after 1997, during which foreign investment almost doubled from 545 million USD in 1997 to 1.03 billion USD in 1998. As mentioned previously, from 1999 to 2000, the government lifted restrictions on overseas investment in finance and insurance industries, service industries and telecommunication industries. Therefore, except for the electronic and electric appliance manufacturing industry, all other types of service-related industries seem to be the most popular industries for overseas investment.

By studying the tables of statistics of foreign investment by year and industry, it can be found that the foreign investment in each industry was not stable. For example, the main business of the rubber products manufacturing industry was to produce tyres for the automobile industry. Thus, when the domestic market became saturated, the foreign investment dropped gradually since 1992. The amount did not even reach 400,000 USD in the years 1993, 1998, and 2001, all of which differ from the 1980s, when the amount of investment was often more than 10 million USD. In 2004, the manufacturing industries that obtained more than 100 million USD of foreign investment were the electronic and electric appliance manufacturing industry, chemical products manufacturing industry, and mechanical manufacturing industry. The industries that gained less than 10 million USD of investment were the textile industry, leather and fur products manufacturing industry, and papermaking and printing industry. Foreign investment in the manufacturing industry focuses on particular industries, which are worth further study.

2.3. Analysis of sources of incoming investment

Table 2.13 shows the breakdown of overseas investment in Taiwan by source locations. Overseas investment in Taiwan came

Table 2.10. Statistics of Approved Overseas Chinese and Foreign Investment by Year and Industry (1,000 USD)

Year	Agriculture and Forestry Industry	Fishery and Forestry Industry	Mineral, Soil, and Stone Mining Industry	Food and Beverage Manufacturing Industry	Textile Industry	Garments and Accessories Industry	Leather and Fur Products Manufacturing Industry	Wood, Bamboo, Vines, and Willow Products Manufacturing Industry	Papermaking and Printing Industry
1952–1964	742	1,264	73	13,448	6,612	791	131	527,842	842
1965	—	1,105	—	753	503	1,706	461	254	13
1966	—	294	—	533	1,832	58	272	618	237
1967	200	670	—	2,039	1,541	6,040	146	282	1,143
1968	137	350	—	2,144	3,133	5,250	104	1,000	775
1969	405	953	10	1,913	4,353	1,536	224	125	593
1970	455	201	—	222	4,177	2,529	817	250	254
1971	6	82	—	145	6,752	2,258	771	223	620
1972	87	1,318	—	726	2,898	1,853	108	1,314	597
1973	—	1,596	49	1,652	16,026	4,537	2,962	3,289	1,814
1974	20	225	—	3,796	10,720	1,100	798	2,265	5,382
1975	—	365	—	1,763	7,329	1,641	2,508	21	1,310
1976	895	187	—	3,259	3,972	1,526	487	2,743	25
1977	300	53	—	1,663	1,493	814	339	915	940
1978	80	2,000	—	673	5,557	910	213	629	94

(Continued)

Table 2.10. (*Continued*)

Year	Agriculture and Forestry Industry	Fishery and Forestry Industry	Mineral, Soil, and Stone Mining Industry	Food and Beverage Manufacturing Industry	Textile Industry	Garments and Accessories Industry	Leather and Fur Products Manufacturing Industry	Wood, Bamboo, Vines, and Willow Products Manufacturing Industry	Papermaking and Printing Industry
1979	—	1,150	—	8,701	5,262	3,410	72	4,480	955
1980	—	—	142	14,004	8,201	1,928	—	3,865	587
1981	—	—	—	18,367	6,582	2,172	—	1,554	82
1982	386	842	—	20,375	4,130	4,062	—	614	7,252
1983	201	1,925	—	10,773	1,426	913	—	1,375	3,826
1984	—	2,225	—	9,275	3,533	1,993	2,023	1,525	479
1985	63	1,238	—	21,058	7749	2,006	2,782	1,565	555
1986	427	68	100	9,442	5,725	3,028	1,026	1,356	2,658
1987	469	—	12,399	77,073	16,218	4,541	2,431	7,823	9,806
1988	1,474	6,819	—	56,916	25,778	3,805	885	3,587	13,017
1989	—	—	—	219,206	52,400	5,296	772	11,934	33,157
1990	368	—	—	110,179	31,403	3,287	2,543	22,977	8,786
1991	—	219	88	38,410	46,631	13,332	478	13,320	1,245
1992	—	—	—	55,014	15,218	1,392	1,767	8,613	12,355
1993	86	—	—	55,354	9,279	13,437	2,899	5,801	1,008

(*Continued*)

Table 2.10. ***(Continued)***

Year	Agriculture and Forestry Industry	Fishery and Forestry Industry	Mineral, Soil, and Stone Mining Industry	Food and Beverage Manufacturing Industry	Textile Industry	Garments and Accessories Industry	Leather and Fur Products Manufacturing Industry	Wood, Bamboo, Vines, and Willow Products Manufacturing Industry	Papermaking and Printing Industry
1994	424	1,628	1,235	93,118	53,595	16,407	10,248	1,897	3,419
1995	—	—	—	43,187	39,935	3,448	7,476	1,681	5,704
1996	—	—	1,488	108,285	30,200	1,251	17,898	3,502	10,571
1997	—	—	626	47,718	46,959	33,969	34,296	1,844	6,604
1998	—	—	2,530	59,752	54,763	4,439	41,094	1,566	4,900
1999	—	—	115	17,728	21,795	1,749	13,190	2,441	7,377
2000	171	32	353	81,501	4,907	16,050	16,906	2,832	46,798
2001	5,416	13,221	39,447	75,838	5,072	18,326	6,712	1,714	3,365
2002	1,415	355	—	52,796	11,440	2,175	11,721	3,687	3,971
2003	1,696	—	32,349	26,100	15,146	2,185	9,729	219	10,092
2004	8,073	7,729	6,054	43,380	7,417	27,968	7,519	47,657	8,699
Total	38,850	53,756	67,358	1,408,279	607,662	225,118	204,808	173,881	221,907

Source: Same as Table 2.8.

Table 2.11. Statistics of Approved Overseas Chinese and Foreign Investment by Year and Industry (1,000 USD)

Year	Chemicals Manufacturing Industry	Rubber Products Manufacturing Industry	Plastic Products Manufacturing Industry	Nonmetal and Products Manufacturing Industry	Basic Metal and Products Manufacturing Industry	Mechanical Manufacturing Industry	Electronic and Electric Appliance Manufacturing Industry	Transportation Equipment Manufacturing Industry	Precise Instruments Manufacturing Industry
1952–1964	35,156	558	—	1,802	1,851	242	14,446	—	—
1965	10,061	463	—	195	344	—	21,377	—	—
1966	585	178	—	626	1,061	387	15,381	—	—
1967	6,595	2,078	—	2,958	5,873	4,642	14,559	—	—
1968	1,443	43,980	—	3,330	2,727	1,650	26,993	—	—
1969	3,490	3,288	—	717	2,004	3,349	64,766	—	—
1670	10,638	1,891	—	5,920	3,516	3,090	73,832	—	—
1971	11,903	7,442	—	3,176	7,500	1,592	32,970	—	—
1972	3,883	4,982	—	156	3,327	40,058	19,720	—	—
1973	28,864	6,995	—	24,557	11,029	43,818	75,748	—	—
1974	12,633	2,557	—	36,210	6,587	8,174	69,652	—	—
1975	30,568	2,201	—	1,802	3,816	5,661	24,701	—	—
1976	12,433	1,594	—	12,330	5,108	8,177	72,395	—	—
1977	14,010	2,719	—	4,151	6,991	7,627	52,801	—	—
1978	35,170	9,806	—	3,246	8,593	9,417	63,495	—	—
1979	22,866	6,246	—	44,045	16,936	8,887	110,408	—	—
1980	57,258	9,965	—	186,740	29,325	14,519	108,602	—	—
1981	42,982	18,700	—	5,729	44,099	38,462	83,948	—	—
1982	35,994	5,395	—	6,922	46,383	30,727	70,797	—	—
1983	36,934	12,418	—	10,435	13,922	141,642	10,7267	—	—
1984	132,507	23,430	—	895	17,412	41,739	26,7652	—	—
1985	214,259	6,927	—	6,644	51,425	86,471	13,9304	—	—
1986	139,896	21,662	—	11,566	45,637	110,565	231,741	—	—

(*Continued*)

Table 2.11. ***(Continued)***

Year	Chemicals Manufacturing Industry	Rubber Products Manufacturing Industry	Plastic Products Manufacturing Industry	Nonmetal and Products Manufacturing Industry	Basic Metal and Products Manufacturing Industry	Mechanical Manufacturing Industry	Electronic and Electric Appliance Manufacturing Industry	Transportation Equipment Manufacturing Industry	Precise Instruments Manufacturing Industry
1987	171,625	72,427	—	47,776	121,163	75,946	377,375	—	—
1988	104,237	56,701	—	34,737	68,419	137,098	237,329	—	—
1989	519,996	83,079	—	32,667	175,561	102,885	391,042	—	—
1990	506,678	33,176	—	33,089	187,232	128,384	377,039	—	—
1991	200,963	18,409	—	30,797	132,960	175,049	570,099	—	—
1992	106,517	70,088	—	10,490	40,192	97,529	323,308	—	—
1993	107,682	364	22,158	44,386	72,897	36,923	226,651	65,197	8,873
1994	189,567	4,015	8,608	14,953	42,935	45,882	296,088	115,609	22,272
1995	374,874	460	6,798	13,572	22,578	51,772	1,241,040	146,220	16,984
1996	64,526	1,991	19,014	7,247	63,696	45,422	443,147	49,645	23,795
1997	188,689	953	11,173	8,972	763,444	62,595	952,463	99,335	50,888
1998	171,008	334	29,618	25,815	67,634	46,576	1,205,776	93,401	47,473
1999	163,015	954	44,387	13,019	62,661	57,733	1,034,058	68,687	34,546
2000	146,122	609	10,644	83,117	59,027	87,078	1,092,871	26,451	62,690
2001	136,062	397	20,622	47,636	118,938	102,380	1,055,614	48,496	129,284
2002	98,750	2,403	20,309	32,018	6,3993	40,491	662,994	169,874	57,700
2003	168,771	4,039	19,195	22,400	30,261	66,188	963,620	37,738	94,590
2004	146,365	30,128	22,225	34,509	76,205	113,958	1,322,099	43,741	56,200
Total	4,478,566	536,002	234,751	908,493	2,562,262	2,086,964	14,525,168	964,394	605,565

Source: Same as Table 2.8.

Table 2.12. Statistics of Approved Overseas Chinese and Foreign Investment by Year and Industry (1,000 USD)

Year	Building and Construction Industry	Wholesale and Retail Industry	International Trade Industry	Accommodation and Eating–Drinking Restaurant Industry	Transportation Industry	Logistics Industry	Finance and Insurance Industry	Other Service Industries	Others
1952–1964	10,187	—	742	—	1,091	—	242,079	6,409	1,593
1965	2,467	—	522	—	128	—	79	50	1,129
1966	93	—	100	—	274	—	535	3,363	2,854
1967	1,062	—	481	—	4,249	—	—	850	1,598
1968	6,450	—	414	—	3,470	—	3,916	8,327	1,310
1969	1,818	—	198	—	1,125	—	1,187	14,668	3,138
1670	1,336	—	362	—	3,137	—	2,251	14,402	9,616
1971	16,732	—	524	—	54	—	9,254	1,792	2,017
1972	1,866	—	690	—	1,508	—	31,429	7,544	2,592
1973	3,717	—	2,504	—	2,914	—	1,698	11,360	3,725
1974	4,976	—	332	—	4,050	—	10,420	7,874	1,605
1975	27,250	—	169	—	2,510	—	544	1,964	2,052
1976	4,199	—	244	—	8,309	—	982	1,002	1,652
1977	4,432	—	39	—	4,842	—	4,783	51,268	3,729
1978	10,032	—	—	—	3,115	—	15,892	39,733	4,274
1979	3,063	—	172	—	2,669	—	11,577	71,350	6,586
1980	1,595	—	724	—	3,440	—	4,138	15,580	5,351
1981	1,669	—	115	—	26	—	7,546	108,599	15,125
1982	2,016	—	50	—	2,674	—	12,037	118,179	5,519
1983	—	—	1,211	—	2,503	—	6,466	42,094	9,137
1984	—	—	10,077	—	4,146	—	6,850	19,960	13,020
1985	525	—	4,746	—	4,049	—	97,739	49,360	3,994

(*Continued*)

Table 2.12. (*Continued*)

Year	Building and Construction Industry	Wholesale and Retail Industry	International Trade Industry	Accommodation and Eating–Drinking Restaurant Industry	Transportation Industry	Logistics Industry	Finance and Insurance Industry	Other Service Industries	Others
1986	—	—	4,776	—	46,650	—	80,332	45,478	8,247
1987	3,761	—	40,742	—	43,182	—	15,050	310,800	8,189
1988	11,250	—	109,655	—	63,576	—	52,096	177,060	18,099
1989	17,144	—	222,094	—	65,306	—	150,128	312,097	23,535
1990	12,528	—	284,080	—	39,100	—	315,206	167,686	38,031
1991	7,656	—	231,845	—	29,973	—	122,099	136,315	8,531
1992	21,301	—	221,462	—	22,655	—	196,431	250,859	6,183
1993	10,238	105,516	139,668	4,744	66,094	7,617	100,915	104,984	405
1994	4,183	213,543	134,243	41,918	28,862	5,281	168,558	112,229	—
1995	8,246	95,747	173,814	9,036	169,933	259	259,110	188,061	45,405
1996	34,802	208,945	159,351	26,392	68,870	62	318,551	249,199	502,986
1997	14,730	248,979	265,189	24,958	107,644	5,954	545,159	454,293	289,195
1998	37,386	191,538	284,642	20,346	49,699	3,473	1,030,040	197,477	67,748
1999	22,168	344,323	310,275	43,434	64,185	799	1,029,831	555,606	317,328
2000	254,148	990,645	36,415	—	52,557	10,028	2,182,475	1,504,097	1,288,945
2001	106,049	790,157	68,720	—	13,151	2,271	1,474,490	664,206	180,945
2002	94,625	390,917	11,053	—	89,970	5,194	879,242	356,557	208,097
2003	39,545	450,674	5,982	—	21,066	4,169	754,446	710,402	99,787
2004	133,284	602,240	12,271	—	15,015	555	809,330	303,020	67,177
Total	938,529	4,633,224	2,740,693	170,828	1,117,771	45,662	10,715,202	6,946,154	3,280,179

Source: Same as Table 2.8.

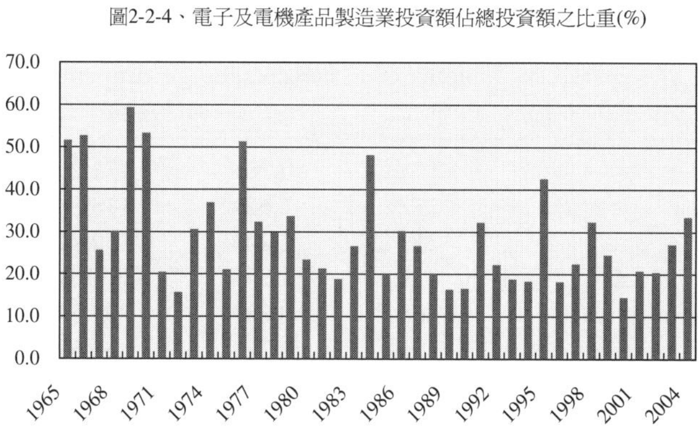

Figure 2.9. Proportion of Foreign Capital Invested in the Electronic and Electric Appliance Manufacturing Industry in Past Years (%)

(X: Year, Y: Percentage)

Table 2.13. Source of FDI in Taiwan by Particular Countries/Areas (1952–2004) (1,000 USD)

Area	Number of Projects	Percentage	Amount Approved	Percentage
US	2,897	16.98%	13,342,388	22.06%
Japan	4,659	27.30%	12,054,885	19.93%
The Caribbean	2,037	11.94%	10,087,105	16.67%
Singapore	901	5.28%	4,368,189	7.22%
Hong Kong	2,434	14.26%	4,002,463	6.62%
Holland	251	1.47%	3,060,068	5.06%
UK	334	1.96%	2,440,797	4.03%
Other areas in America	225	1.32%	1,677,137	2.77%
Germany	304	1.78%	1,206,043	1.99%
Philippines	227	1.33%	1,135,917	1.88%
Other areas in Europe	373	2.19%	1,083,306	1.79%
Australia	139	0.81%	950,177	1.57%
Liberia	35	0.21%	822,442	1.36%
Malaysia	521	3.05%	779,686	1.29%
Panama	50	0.29%	723,129	1.20%
Other areas in Oceania	208	1.22%	529,394	0.88%
Switzerland	147	0.86%	477,324	0.79%

(*Continued*)

Table 2.13. ***(Continued)***

Area	Number of Projects	Percentage	Amount Approved	Percentage
France	128	0.75%	472,207	0.78%
South Korea	277	1.62%	443,362	0.73%
Canada	211	1.24%	400,424	0.66%
Other areas in Asia	319	1.87%	111,016	0.18%
Other areas in Africa	65	0.38%	109,570	0.18%
Thailand	123	0.72%	68,597	0.11%
Indonesia	114	0.67%	68,567	0.11%
New Zealand	23	0.13%	28,162	0.05%
Vietnam	27	0.16%	4,569	0.01%
South Africa	27	0.16%	3,925	0.01%
Mexico	1	0.01%	24	0.00%
Others	9	0.05%	44,156	0.07%
Total	17,066	100.00%	60,495,029	100.00%

Source: Same as Table 2.8.

mainly from three areas. The first was America (the investment amount was 26.231 billion USD, which occupied 43.36% of the total amount and 31.78% by project number), followed by Asia (the investment amount was 23.037 billion USD, which occupied 37.72% of the total and 56.26% by project number), and lastly, Europe (the investment amount was 8.735 billion USD, which occupied 14.44% of the total and 9.01% by project number). The proportion of the total amount of foreign investment and number of projects from these three regions were 95.52% and 97.05% respectively. The proportion from Oceania and Africa were not large.

The United States and Japan took the first and second position among other countries. The total amount they have invested engaged about half (41.99%) of the total amount of foreign investment. American and Japanese enterprises can be viewed as two major investment sources in Taiwan, as well as sources of major foreign capital and technological skills. This makes Taiwan dependent on the US and Japan for their technology and raw materials. The British Territories in the Caribbean region (British Virgin Islands and British Cayman

Islands, also known as tax-free havens) took third place as a major overseas country to invest in Taiwan. Foreign companies from that area earned tax benefits, which would also be available if the companies invested in domestic industries in Taiwan again. Singapore and Hong Kong took the fourth and fifth place; they were also given many tax-free benefits because of their role as headquarters of many multinational companies. Many foreign enterprises also have investment in Taiwan by using capital from Singapore and Hong Kong in order to avoid unnecessary difficulty while transferring capital. From Europe, Holland took first place and sixth place overall because of the investment from Philips Group. The seventh, eighth, ninth and tenth places were occupied by UK, Germany, Philippines and Australia respectively.

In conclusion, the 10 major investment countries mentioned above contributed almost 90% (87.03%) of overseas investment to Taiwan. The countries whose investment amount was over 100 million USD in past years include seven countries: Liberia (822 million USD), Malaysia (779 million USD), Panama (723 million USD), Switzerland (477 million USD), France (472 million USD), South Korea (443 million USD), and Canada (400 million USD).

Chapter 3
Investment Environment Index

The economy of a country depends on basic production factors like labor power, capital, land, water and electricity, oil, coal, mine, other natural resources, and technology to develop smoothly. In particular, the development of technology is an essential factor in the adjustment and improvement of the structure of industries, and upgrading and increasing the competitiveness of a country. The investment from foreign capital brings capital and technology, but first, the foreign enterprise needs to evaluate the investment environment of the location in order to determine whether to invest or not. The factors to evaluate the investment environment are listed in Appendix A: knowledge capital, innovative ability, utilization of information technology, and social infrastructure. These factors may be quantified in order to conduct comparisons (see Appendix B).

By the 18th century, the flourishing era of Emperor Chienlung of Ching Dynasty, Mainland China had remained a nation of agriculture for a thousand years. In addition, the country was conservative about overseas trading. Thus, the Ching Dynasty knew of the Industrial Revolution in Europe, but the leader chose to retain its economic structure. The economic production in Mainland China at that time depended mainly on labor power and land, that is, manpower played an important role in the development of the economy. Today, machinery has replaced labor and plays a crucial role in the economic growth due to technology sweeping all over Europe after the Industrial Revolution occurred in the UK in the 18th Century.

Robert M. Solow, the winner of the Nobel Prize Laureate in Economics, proposed the neoclassical growth theory in 1956, which explains that the economic growth of a country will eventually lead to the stationary state of balanced growth and the growth rate of long-term balanced average national income will be influenced by the growth rate of the development of technology. However, Solow also pointed out that the economic growth rate would be greatly affected by population growth rate, the capital accumulation rate and the rate of technological development in a short period. The endogenous growth theory or new growth theory developed by Nobel Prize Laureates in Economics, Robert E. Lucas, Jr. (1988) and Paul Romer (1986, 1990), has become one of the major economic theories since 1986. The theory brings up the idea that the source is related to the development of technology. Basically, the development of technology can be created by combining human resources and research on innovation. The human resource and research on innovation can be controlled by human beings, thus strong education policies and R&D strategies can encourage or induce private enterprises to achieve the target of technology development.

The popular growing model of a knowledge-based economy in 2000 further quantified the productivity factor of the development of technology. The theoretical model of a knowledge-based economy can be seen in Figure 3.1. According to the experience of OECD (Organization for Economic Cooperation and Development) countries, the influences on the aspects of input, output, and process can be applied while studying the major factors in the knowledge-based economy model.

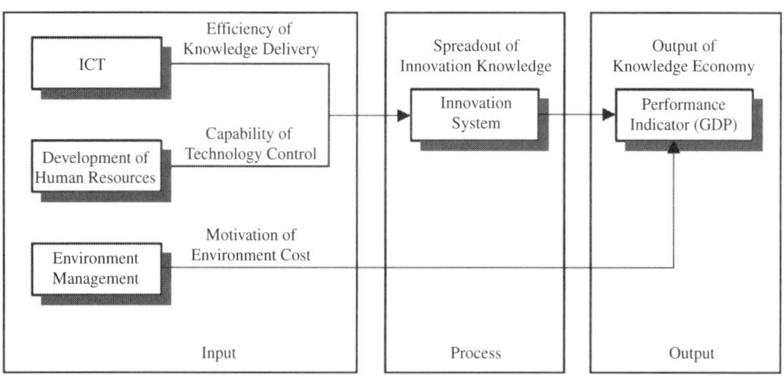

Figure 3.1. Theoretical Model of a Knowledge-Based Economy

1. R&D and Innovative Mechanisms

Knowledge has already replaced factors of production such as land, labor, and capital, and has become the most important production input factor in the knowledge-based economy system. It is able to bring higher economic profits and permanent competitive advantages. Therefore, each country should be actively enriching its own knowledge capital.

Knowledge capital is a crucial production factor in the knowledge-based economy. It has a distinguishing characteristic of a high spillover effect, that is, it is easily attainable. Having the courage to take risks, innovate continuously, and keep creating products or services to meet customers' demands is the only way to ensure the sustainable success of a country's economy, which is also the essential factor to evaluate the society of a knowledge-based economy. In other words, if knowledge capital is the "stock", then national innovative capability is the "flow". A country should have enough "creative capability" or the ability to design innovative mechanisms to encourage creations, nurture an adventurous spirit, and integrate the "creative realization", all of which are necessary in evaluating national innovation capability.

Knowledge capital and innovative capability are the two factors in the investment environment essential to the foundation of developing technology in a country. The knowledge accumulated will represent the performance of human capital and R&D innovation. In a modern society, which is led by a knowledge-based economy, the investment in R&D and human capital are principle factors in maintaining the competitiveness of an enterprise and avoiding elimination by others. The Austrian economist Joseph Schumpeter proposed an important viewpoint in 1942. He indicated that entrepreneurs could seek an enormous margin through innovation, then create new technology to replace the old one by innovating further. As a result, elimination in a market can not only be avoided, but also be carried out against other companies, hence "creative destruction". However, the success of research on innovation relies on the high quality of human resources. Moreover, enormous profit will only be obtained when a specific innovation has received a patent right. Not every innovation can earn a patent right successfully, but innovation is still an essential condition in increasing productivity. The improvement and partial modification of products can be applied to raise productivity too.

Research in development was not common in Taiwan before 2000, as the expense of R&D only occupied a low percentage of GDP. 50% of R&D was focused on governmental sectors. Recently, however, the incentive policy from the government and the competitive international environment have led to growing R&D in the electronic technological companies, and the expenditures of R&D keep rising. Many beneficial measures of tax exemption and allowance have been created by the Taiwanese government including tax deduction on investment, tax deduction on R&D expenses, and governmental allowance for projects of technological R&D.

R&D can be divided into innovative development and process development. Innovative development involves developing new products, new markets and new management skills in companies. Process development seeks to reduce the production and management costs of enterprises. The purpose of R&D is to increase the competitive advantages for enterprises, but the success of R&D is very uncertain. SMEs will find it difficult to invest in R&D because of their lack of capital. It might not be achieved if they do not acquire assistance from external sources. The beneficial policies of government is not enough; they will still need support for technology development research from venture investment organizations and technological research institutes in universities. With the transfer of technology, operational support, and capital, certain results can be achieved. Apart from providing tax benefits and allowance, governments should also protect the intellectual property rights of inventions. R&D press announcements, the departure of R&D staff, or disclosure from those terminated employees put such intellectual information at risk, discouraging enterprises from engaging in research and development.

The evaluation indicator of knowledge capital is "quantity and quality of invention and design patent", with six sub-indicators. The five indicators of innovative capability are "quality and quantity of research institutes", "quality and quantity of an enterprise's R&D", "encouraged level of new established enterprises and innovative activities", "knowledge circulation among industrial, academic and research circles", and "innovative R&D and international cooperation level".

2. Human Resources and Human Literacy

Labor is an essential factor in production and the quality of human resources will influence production quality and the value of products. High-quality products have higher-added value, which requires production with a higher standard of manpower. The quality of manpower is directly connected to education. Both education quality and content will influence the quality of manpower in a country. The objective indicator of the quality of manpower is surely the country's supply of education; however, the importance of demand is also undeniable. The demand for education means the desire for education from individuals and family. The demand for education from individuals and family is determined by the willingness to receive education, the learning capability, time availability, and income. Therefore, the demand for education can be observed from the investments in education.

The remarkable research on economic growth accelerated by human capital comes from the study of two Nobel Prize Laureates in Economics from the University of Chicago: Gary Becker and Robert E. Lucas, Jr. They theorize that in the knowledge-based economy, economic growth will occur with investments in human capital by individuals and families, which depend on the demand of such an investment. It can be assumed that if governments can provide allowances or lower education fees, it may enhance the willingness of the population to invest in education.

The disadvantage of these economic theories is that investment in education will reduce families' income levels. Individuals and families should be aware of this when they decide to spend more time and money on education. However, in the long run, higher incomes depend on whether the future economy can provide better job opportunities;

thus, the future reward of personal educational investment is directly linked to the overall economic growth. For example, it will be difficult to persuade parents to allow their children to more education if the future development of the economy is vague, and if the economic prospects and living standards are unable to provide a better vision for individuals and families. Such circumstance often occur, especially in the aboriginal society.

Let us examine a case study of one of the aboriginal tribes, the Yami Tribe in Lanyu. One of the authors, Chen-Min Hsu, interviewed the tribe in Lanyu in 1985. The teachers from junior high schools in Lanyu pointed out that parents think their children should start work once they graduate from elementary schools, but the teachers were encouraging parents to allow their children to go to junior high schools. It can be seen from this example that if future success cannot be guaranteed, parents will not invest in further education as they have to consider the family economic situation. The demand for educational investment and the educational investment environment have a strong relationship of causation, and are therefore inter-connected. In order to consider educational investment, individuals and families must also consider the environment of education. This includes the future vision, quality of education and intention, and supply of educational investment, which involves macroeconomics and human capital investment from the government.

The supply of human capital investment determines the quality and quantity of education. The importance of education is recognized in both Taiwan and Mainland China, and both governments have created a low-tuition of education policy by using the government budget to subsidize education. Although the quantity of education might be increased by increasing schools, the quality of education cannot be improved in the short-term. The improvement of education quality depends on the continued and improved R&D efforts of the educational staff in schools, which also involves innovative capability, as the quality of manpower is the original source for innovative capability. The school's quality of academic research, especially its technological science such as the publication of SCI (Science Citation Index) and ECI (Engineering Citation Index), is the principal indicator. The result of research can be made commercial by obtaining the patent right, then applying for development. Schools can also rely on the industrial technology research institutes to promote the innovation or commercialization of their technology research.

According to Professors Philippe Aghion and Peter Howitt (1992), the innovation of technology can be obtained by two ways — internal R&D of enterprises, or basic research carried out by schools with the cooperation of industrial technology research institutes and venture organizations. The educational training of human resources can be completed through either the regular general education system, or the vocational education system. The former refers to the general university education, and the latter refers to the education is a technological institute or college. While specialized technological education is crucial for innovation, education in humanities and social sciences are still important in order to develop the overall innovative capability of the individual.

The evaluation of the index of human capital is based on two indicators in the range of knowledge capital. They are the "quantity and quality of human resources" and "public basic technological and humanity quality and development", with five sub-indicators. The indicator for their innovative capability is "innovative capability and entrepreneurship performance", which includes eight sub-indicators.

3. Information Technology

Information technology is the activator of a knowledge-based economy. The tools, technological development, and application of information technology have accelerated the access, accumulation, and recombination of knowledge, further accelerating the increase of knowledge capital and the speed of innovation. In fact, the application of knowledge capital, innovative capability and information technology has created a significant positive feedback, which can be described as a causal relationship.

Sociologist Daniel Bell expressed that information is a strategic resource for today's society, and forms the means to step into the post-industrial society, as well as the principal key point when entering the new social structure of a society based on telecommunication. He also indicated that the three major characteristics of the post-industrial information society are: (1) the shift from currency and finance production to the services economies, (2) the increasing dependence on theoretical knowledge, and (3) the application of computers and other machines to create new intellectual technology. People are increasingly relying on information, knowledge, and integration of relevant telecommunication in order to automate daily activities and business processes. Therefore, whether each country can supply enough input and output for IT infrastructure will become a vital issue when to comes to winning a position within the competition of a knowledge-based economy.

The three indicators to evaluate IT are "industrial electronic level", "abundance of digitalization contents", and "e-learning popularization (family/school/others)". "Industrial electronic level" has four sub-indicators, "abundance of digitalization contents" has three sub-indicators, and "e-learning popularization (family/school/others)" has two sub-indicators.

4. Social Infrastructure

Given that the operating characteristics of the knowledge-based economy are different from those of the traditional industrial economy, the social infrastructure has to be different from what it was in the past. In Taiwan, for example, high-tech industries are now confronted with a fiercer competitive environment than before; so apart from prioritizing the issue of internationalization, the government should establish a production system that integrates work, leisure, and consideration of the ecological environment in order to create a most appropriate living place that will attract those international experts.

The ten chosen indicators to evaluate social infrastructure are "encouraged measures of FDI, financial environment and capital market", "social internationalization and degree of liberalization", "living medical care environment and social security", "governmental administration transparency and administrative efficiency", "integrity and executive level of intellectual property and relevant scientific legal system", "level of social trust and mutual respect", "positiveness of innovative teaching and innovative education", "convenience and quality of communication and transportation facilities", "convenience and quality of telecommunication facilities", and "public environment and facilities and environmental protection", all of which are within fifteen sub-indicators respectively.

5. Relevant Indices for the Introduction of a Foreign Knowledge-Based Economy

5.1. IDC

Index name: Information Society Index (ISI)

Announcing organization: International Data Corporation (IDC)

Background introduction: Information Society Index is the result of cooperation between IDC and World Times Inc., and is based on the Society, Economy and Information Technology Relevant Index issued by *World Times* in 1996. After many discussions and statistically analyses, the variables were reduced from 40 to 23. This index has been issued regularly since 1997 and currently has 55 countries participating in the evaluation.

The ISI consists of the arithmetic sum of the scores evaluated from four major elements. The score of each major element is the sum of composed variables. The evaluation of each variable has been appropriately adjusted to avoid the innate inequality between small and big countries and to reflect the different degree of influence in the future. It uses the simple mathematical mode with the record of the past three years to calculate the annual compound growth rate (CAGR) in order to learn future estimations, and then makes adjustments by considering the different stage of the CAGR curve of each country. Countries whose total score is over 6,000 points are "Skaters"; those with scores between 4,000–5,999 points are "Striders"; those whose scores lie between 2,000–3,999 are "Sprinters"; and the "Strollers" are the countries whose scores are less than 1,999.

Index content:

(1) Computers

(i) The average number of computers for teachers and students at schools
(ii) Quantity of computers for government or commercial use
(iii) Quantity of computers for family use
(iv) Quantity of Internet computers for non-home use
(v) Quantity of personal computers
(vi) Price comparison between software and hardware

(2) Information

(i) The average number of TVs per person
(ii) The popularization of mobile phones
(iii) The cost of mobile phones
(iv) The popularization of fax machines
(v) The popularization of broadcasting
(vi) The rate of wrong number setting of phone line
(vii) The average number of phone lines per household
(viii) The popularization of cable TV

(3) Internet

(i) The connection rate of non-agricultural commercial work
(ii) The average cost of e-Commerce
(iii) The rate of Internet service at home
(iv) The rate of Internet service at schools

(4) Society

(i) The degree of civil liberalization
(ii) The popularization of newspapers
(iii) The liberalization of publishing
(iv) The percentage of population with high school degrees
(v) The percentage of population with university or college degrees

5.2. EIU (The Economist Intelligence Unit)

Index name: E-readiness Index

Announcing organization: EIU/Pyramid Research

Background introduction: E-readiness Index is issued by e-Business Form, the relevant research institute of the Economist Intelligence Unit in the UK, to help the governments of every country, companies and individuals understand the weaknesses and strengths of the developing environment of an Internet economy. This index has been issued the annual evaluation of the "Industrial Electronic Development Maturity", and includes six elements with 70 indices carried out regularly on 60 major countries worldwide since 2000.

The score of each index ranges from one to ten, summed up later according to the statistical weight of each element, and then each country is ranked. The statistical weight of each element will be adjusted appropriately according to the actual situation in each year.

Index content:

Connectivity and technology infrastructure — 25%

To evaluate opinion regarding the access of traditional and mobile phone services, from both individuals and enterprises (both include voice mail and data transmission of broadband and narrowband). It also evaluates whether or not the price of service is affordable to the user and if this service can be accessed easily.

Business environment — 20%

Almost 70 indices are used to evaluate the attractiveness of a country's general business environment within a five-year period. These indices include economic strength, political stability, legal environment, tax and the opening of trade and investment.

Consumer and business adoption — 20%

Payment and logistic support systems is the main axle to evaluate in this element. The percentage of credit card holders and the security, reliability, and efficiency of the current e-payment system is evaluated here. In particular, it evaluates the opinion of goods suppliers regarding their timing ability, reliability in delivering goods to buyers and the degree of website development by local companies.

Legal and policy environment — 15%

The legal structure leads the development of e-commerce or e-business, which can be seems as an advantage but might also be a disadvantage. The support from the law for the virtual business and electronic signature is evaluated here. The process of authorizing simple and easy but effective regulation is another standard to evaluate.

Supporting e-services — 5%

No enterprise or industry can be managed effectively without support and help from agents or supporting companies. In the e-business market, the relevant support services include: portal sites, online companies, web-hosting firms, application service providers (ASP), website construction companies, and enterprise electrization consultant service providers. The access of local companies to these services will be evaluated in this element.

Social and cultural infrastructure — 15%

Unrestricted Internet service to the local population, together with education and reading and writing abilities, are the priority conditions for the development of the Internet. In addition, the spirit and risk adoption of entrepreneurship also plays an important role in the model of e-commerce; thus, website contents and the reformation of enterprises by a country will be evaluated. Both of these control the direction of social and cultural infrastructure when preparing electrization.

5.3. IMD

Index name: World Competitiveness Index

Announcing organization: Switzerland International Institute for Management Development (IMD)

Background introduction: IMD World Competitiveness Index is executed and investigated by the Switzerland International Institute for Management Development. They publish the annual report, *The World Competitiveness Yearbook*, which lists rankings by the evaluation results of the local and global competitive environment, provided by countries to the companies and enterprises in their territory.

The IMD World Competitiveness Index has integrated the original eight major indices into four since 2001, and the number of sub-indices has increased from 286 in 2001 to 314 in 2002.

Index content (the number inside the brackets is the quantity of sub-indices):

(1) Economic performance	(i)	Domestic economy (33)
	(ii)	Export trading (20)
	(iii)	International investment (10)
	(iv)	Employment (7)
	(v)	Prices of commodities (4)
(2) Governmental efficiency	(i)	Financial balance (11)
	(ii)	Financial policies (14)
	(iii)	Systematic structure (22)
	(iv)	Business laws and regulations (24)
	(v)	Education (13)
(3) Enterprise efficiency	(i)	Productivity (11)
	(ii)	Labor market (20)
	(iii)	Finance market (19)
	(iv)	Management capability (11)
	(v)	Impact of globalization (5)
(4) Infrastructure	(i)	Infrastructure (20)
	(ii)	Technological facilities (20)
	(iii)	Scientific facilities (22)
	(iv)	Health and environment (18)
	(v)	Value system (10)

5.4. Harvard University and WEF

Index name: Networked Readiness Index (NRI)

Announcing organization: Harvard University and World Economic Forum (WEF)

Background introduction: NRI is planned and implemented jointly by Harvard University and World Economic Forum. The purpose of publishing *The Global Information Technology Report* is to understand the major role played by the economic development of IT in order to evaluate the application of information and communication technology (ICT) of each country, as well as the degree of preparation of an information-based society for each country. 2000 marked the first year of evaluating the 75 countries worldwide with the Network Readiness Index, which includes two principal indices and 65 sub-indices.

Index content:

A. Degree of overall network use

(1) Percentage of computers with Internet connection
(2) Internet users per host
(3) Internet users per 100 inhabitants
(4) Cellular subscribers per 100 inhabitants
(5) Availability of public Internet access*

B. Information enabling factors

(i) Network Access	A. Information infrastructure (6) Teledensity (7) Years taken to first adopt mobile telephones (8) Waiting list for telephone lines (9) Telecommunication staff per 1,000 mainlines (10) Telephone faults per 100 mainlines (11) Availability of telephone lines for business* (12) Perceptions of broadband Internet access*

	(13)	Price and quality of Internet connection*
	(14)	Availability and costs of mobile telephones*
	B.	Hardware, software, and support
	(15)	PCs per 100 inhabitants
	(16)	Software piracy
	(17)	Availability of specialized IT services*
	(18)	Software products fitting local needs*
	(19)	Competition and the domestic software market*
(ii) Network Policy	A.	ICT policy
	(20)	Internet access cost
	(21)	Perceived effect of telecommunications competition on quality and price*
	(22)	Perceived effect of ISP (Internet Service Provider) competition on quality and price*
	(23)	Legal framework supporting IT business*
	(24)	ICTs as overall priority for the government*
	B.	Business and economic environment
	(25)	Income per capita
	(26)	Rule of law*
	(27)	Government effectiveness*
	(28)	Regulatory burden*
	(29)	Number of days taken to start a new firm*
	(30)	Women's participation in the economy*

	(31)	Minority groups' participation in the economy*
	(32)	Country's relative position in technology*
	(33)	New government's respect for previous government's commitments*
	(34)	Trust in public postal system*
(iii) Network Society	A.	Network learning
	(35)	Investment in employees' IT skills development
	(36)	Quality of IT training and educational programs
	(37)	Internet access in schools
	(38)	Brain drain of IT-skilled workforce*
	(39)	Brain drain of scientists and engineers*
	B.	Social capital
	(40)	No schooling in total population
	(41)	Average years of schooling in total population
	(42)	Literacy
	(43)	Political rights*
	(44)	Quality of public schools*
	(45)	Difference in quality of schooling between rich and poor children*
(iv) Network Economy	A.	E-commerce
	(46)	B2C e-commerce transactions*
	(47)	B2B e-commerce transactions*
	(48)	Business Intranet sophistication*
	(49)	Commercial websites*
	(50)	Domestic venture capital investment in e-commerce*
	(51)	Competition in dotcom market*

	(52) Prevalence of Internet start-ups*
	(53) Use of Internet-based payment systems*
	(54) Sophistication of online marketing*
	B. E-government
	(55) Government effectiveness in promoting the use of ICTs*
	(56) Availability of online government services*
	(57) Extent of government websites*
	(58) Business Internet-based interactions with government*
	C. General infrastructure
	(59) Electricity consumption
	(60) Electric power transmission and distribution losses
	(61) Percentage of paved roads
	(62) Television penetration
	(63) Typical driving speed between cities*
	(64) Quality of ports, facilities and waterways*
	(65) Quality of air transport*

*Survey by questionnaire.

5.5. Brown University

Index name: Global E-Government Evaluation

Announcing organization: Brown University

Background introduction: The evaluation of global e-Government, which evaluates the effect of e-Government in each country, is executed by Brown University Taubman Center for Public Policy in the United States.

The evaluation object of government websites are: national policy-making administration (such as the Presidential Office, Executive Yuan), legislatives, judiciaries, the Cabinet, and the major institutions that provide essential function services for government such as medical treatment, tax, education, interior affairs, economics, natural resources, diplomacy, overseas Chinese and foreign investment, communication and transportation, military affairs, tourism, commercial laws and regulations, etc.

Websites of those subordinate organs or organs of regional government are excluded in this evaluation.

The items of evaluation for government websites are mainly divided into integrity of information content, abundance of services items, and convenience of public access. The score is given according to the following items (4 points maximum if the item is present): office phone, address, online information, online database service, outside link, video and audio services, foreign language services, disable service, privacy protection policy, information security, government portal site, application of digital signature to conduct trade, online payment by credit card, e-mail address, search engine, message board or forum, message broadcast, personal webpage, no commercial advertisement, no extra charge and no user-restricted services, etc. One point will be given if the item is present, with a maximum of 4 points. The handling of email replies on the government's website is evaluated by conducting a real test.

5.6. WEF

Index name: Global Competitiveness Report (GCR)

Announcing organization: World Economic Forum (WEF)

Background introduction: *WEF Global Competitiveness Report* is carried out by Michael E. Porter, the expert in strategy and competitiveness from Harvard Business School, Jeffrey D. Sachs, a supervisor of Harvard International Development Center, *et al.*

It has two important competitive indices — Growth Competitiveness Index (GCI) and Current Growth Index (CGI) (or jointly named as the Microeconomics Competitiveness Index [MCI]). The former is the growing competitiveness prospects of economic growth in the next five years; the latter means the index of current competitiveness, which describes current productivity and economic performance by viewing the economic fundamentals. The GCI can be divided into three principal sub-indices, which are technology index, public sectors, and macroeconomic environment. Current Growth Index, which was introduced in 2000, can be separated into two principal sub-indices, namely enterprise operational and strategic maturity, and operation environment quality. Each sub-index also includes detailed indices which approximate 161 items in total. The calculation of the CGI is based on a questionnaire. This questionnaire, conducted by 5,000 high-level supervisors and officials from 80 countries, applies GNP in 2000 (which has been adjusted by using the price level as the dependent variable), and then conducts dual-variable regression.

WEF has used two different methods to calculate the technology index since 2001. They divide the evaluating countries into two groups — the core country with innovation orientation, and a non-core country. The core country focuses on technology as a major factor to advance the growth of the country's competitiveness. Thus the technology index in this country would be the innovation sub-index and average value of sub-index of ICT. The technology index of the non-core country is the total value of weights among the innovation index, technology transfer index and ICT index, weighted as 1/8, 3/8 and 1/2 respectively.

The first step in acquiring each index is to apply the actual data by referring to the relationship between all countries and conducting standardized calculation using the same formula. The data gained from the above formula will then be applied to obtain the average of weights with the result of survey afterwards. The end result will be the index, which is obtained from the combination of quantitative and qualitative indices.

Chapter 4

Analysis of the Information Technology and Electronics Industry Market in Mainland China

1. Current Status of Electronics Information Industries in Mainland China

The continued high-speed economic development of Mainland China is truly an economic miracle and also a powerful engine to advance global economic growth. Of all the major industries in Mainland China, the electronics industry is developing the fastest, and has already become the pillar of Mainland China's economy. According to statistics from the Ministry of Information, in 2002, Mainland China's electronics information industry reached a total output value of 1536.167 billion RMB, with an added value of 226.357 billion RMB and sales income of 1120.347 billion RMB, placing Mainland China behind USA and Japan as the third major producer of IT products worldwide. Mainland China's enormous consumption of electronic products and its potential market drove the global electronics industry to accelerate their transfer to Mainland China. This allowed Mainland China's electronics industry up/downstream products supply chain to be completed gradually, and continue with better high-speed loop operations. Up to March 2003, the total realized foreign capital for Mainland China's electronics industry was more than 70 billion USD, with more than 10,000 foreign-funded enterprises established. In addition, the continuously increasing foreign capital has become an important factor in advancing the development of Mainland China's electronics industry.

1.1. Mainland China's IC design industry

Currently, Mainland China's IC (Integrated Circuit) design industry is increasing progressively with an annual average of 30%, and its output value is 1.5 billion RMB, an increase of 36% from the previous year. According to statistics from the China Semiconductor Industry Association (CSIA), there are currently about 390 IC design companies in Mainland China in total, four times higher than the number in 2001. Among these companies, the sales volume of these six local IC design companies — Hangzhou Silan, CEC Huada Electronic Design Co. Ltd., Datang Microelectronics Technology Co. Ltd., Shanghai Hua Hong NFC, Wuxi SEMICO, and Jiangsu IdeaBank Microelectronics Technology Co. Ltd. — exceeded 100 million RMB. At present, most multinational and Taiwan-based design firms have entered the Mainland China market by establishing sales branches and organizations. Only a few design companies will be equipped with market technology support.

At present, existence is still the priority for Mainland China's local design companies. Thus, most of their product lines focus on circuit fields that are used in smart cards and mid/low-end consumptive electronic products. As for the high-end products, they mainly consist of the CPU from Beijing's ARCA Technology and BLXCPU IC design, and the DSP of Shanghai Jiao Tung University's Shanghai Chinese Core Semiconductor Science Technologies Co. Ltd. These design projects are sponsored mainly by government resources and have already reached the industrialization readiness phase. However, it will still be some time before large-scale commercialization is reached. Other projects, such as the EVD standard proposed by HOMAA, the TD-CDMA chip jointly developed by Datang Mobile and foreign-funded companies, and the audio/video processing chipset from VIMICRO, etc., are achievements of the Mainland China IC design level.

Formation of alliances is the main way for industrial operations of current major design projects to be conducted. For example, the Mainland China C*Core industrial alliance, initiated by Suzhou C*Core Technology and joined by more than 60 Mainland China enterprises, tries to use the 32bit RISC Embedded CPU Technological Development Platform which is built on the basis of Motorola Technology, as well as the commercial mode of marketization, to upgrade the level of Mainland China's SoC design and application. ARCA Technology and BLXCPU also had similar activities prior to this. The commercial mode of IC Design OEM is also currently rising in Mainland China, and it will reduce the time needed for IC design company products to enter the market and achieve effective promotion.

1.2. Mainland China's semiconductor equipment industry

Mainland China's semiconductor equipment is almost totally dependent on imports, and this is the weakest part of the electronics industrial chain in Mainland China. However, due to the recent rising fever of IC-making in Mainland China, the enormous business opportunities have attracted lots of semiconductor equipment manufacturers to cluster in Mainland China. This situation has also resulted in the relaxing of restrictions on semiconductor manufacturing and advanced equipment imports from other countries, including USA and Japan. In 2000, Semiconductor Manufacturing International Corporation purchased large quantities of equipment from Holland and Sweden, putting a lot of pressure on US equipment manufacturers. US authorities were forced to send relevant personnel to visit Mainland China to examine the actual situation. Thus, we expect that in the second half of this year, the US will relax restrictions on the semiconductor equipment exported to Mainland China.

In general, the international semiconductor equipment suppliers that have already entered Mainland China have chosen to establish their sales branches in the Greater Shanghai Area, except for Applied Materials Inc. which entered Mainland China very early. Japan TOWA set up its Mainland China headquarters and the R&D and Services Center in Shanghai. The TOWA Semiconductor Equipment (Suzhou) Inc., Japan ANELVA Corporation and Taiwan GPT jointly invested and established Anelva (Shanghai) Semiconductor Equipment Inc. in Shanghai, and planned to directly proceed with equipment manufacturing in Mainland China. Other semiconductor equipment companies like South Korea's Jusung Engineering, K.C. Tech, PSK Tech, and Samsung Techwin have also accelerated their operations in Mainland China, and Taiwanese Gallant Precision Machining and HT semiconductor equipment manufacturers have already set up contact branches in Mainland China.

Faced with the embarrassing seizure of the market and restrictions by foreign businesses, the Mainland Chinese government has led the industry to make a bold move during the 10th Five-year Plan by investing a total of 20 billion RMB. 12 major technological specialities were officially initiated in 2002, among which the developing bottleneck for equipment manufacturing of Mainland China's semiconductor industry has been listed as one of the key fields in which there needs to be a breakthrough. Currently, Beijing is the major R&D and manufacturing location for Mainland China's semiconductor special equipment. Using this as the foundation, Beijing will possibly take the lead in breaking through the bottleneck for semiconductor manufacturing equipment. In addition, it has also brought new developing opportunities for Beijing's IC industry.

1.3. Mainland China's IC manufacturing industry

The microelectronics industry is a fundamental industry. The reason it has developed so fast in Mainland China is that apart from the enormous contribution to the national economy with its technology, it is also related to the IC manufacturing industry's penetration.

According to the developing trend of international microelectronics, the characteristic sizes of IC will become smaller and smaller, and IC will be developed as the System-on-a-Chip (SoC). The microelectronic technology and other sciences will be integrated to produce many scientific growing points, and in addition, combined with other industries, will become a key economic growing point. In 1999, the total IC consumption in Mainland China amounted to 43.6 billion RMB, of which the total amount of IC made in Mainland China came to 8.38 billion RMB, or 0.6% of the total global IC output. However, even though Mainland China made significant development in the microelectronics industry, it still fell behind greatly compared to the developed countries. In general, there is a difference of about two generations between their production technologies. The self-supply rate for the domestic demand for IC is rather low, especially for products with higher technological content, which basically depend on imports.

In 2000, Shanghai initially formed the most standardized and complete structural system of IC industrial chain in Mainland China. In addition, the Putung-based IC-making Highland led the Greater Shanghai Area and formed the biggest IC manufacturing center in Mainland China. After Fab 1 of the Semiconductor Manufacturing International Corporation began mass production in 2001, its Fab 2 and Fab 3 (Mainland China's first domestic IC OEM Foundry providing the copper process) stages were completed and already put into production in 2002. Currently, monthly output has reached 30,000 pieces (see Table 4.1). The 8-inch line of Grace Semiconductor Manufacturing and Advanced Semiconductor Manufacturing Corporation of Shanghai are also completed and in the process of test runs. Even though the scheduled time of mass production for Shanghai Belling's 8-inch line has been postponed, its fabrication plant will be installed and put into production within 2003 without a problem. As a result, at the end of 2003, Shanghai would have a productivity of 100,000 8-inch wafers monthly.

UMC-related Suzhou HeJian Technology has already started production. This is the first 8-inch line not located within Shanghai in the

Table 4.1. Current Production Lines of 8-inch Wafers in Mainland China

Manufacturer	Location	Manufacturing Technology	Productivity (Sheet/Month)
Hua Hong NFC NEC	Putung Jinqiao	0.24–0.35 μm	20,000
Semiconductor Manufacturing International Corporation	Putung Zhangjian	0.13–0.25 μm	30,000
Motorola	Tientsin	0.25 μm	20,000

Source: EDN Mainland China.

Yanzhijiang Delta Region. In addition, the global No. 1 Foundry, TSMC's Shanghai Songjiang 8-inch line project, has also gone into operation. TSMC President Morris Chang stated that, in the second half of 2004, the Songjiang Foundry would start production and its production schedule would not be changed. With the entrance of these two major wafer OEM foundries to Suzhou, Shanghai, the contest for IC investment between Beijing, Shanghai and Shenzhen has now ended, with Shanghai as the winner. The cluster effect of IC manufacturing is also appearing in the Yanzhijiang Delta Region, in Jiangsu Province's Wuxi, Taicang, and Zhangzhou; and in Zhejiang Province's Hangzhou, Ningpo, and Huzhou. Construction and contractual projects are widespread in this region (see Table 4.2).

In April of 2003, the news disseminated that the global major player Intel is making a decision to select the location for its new 12-inch wafer production line, and Mainland China has been selected as a candidate. The effect on the residence of TSMC and UMC in the Yanzhijiang Delta Region for those packaging and testing companies shall not be neglected. In fact, Yanzhijiang Delta Region was originally the base of Mainland China's packaging industry, and its output has accounted for more than 40% of national packaging productivity. Two of the four domestic packaging companies have a total annual packaging output of more than 500 million pieces stationed in the Yanzhijiang Delta Region (Jiangsu Changjiang Electronics Technology and Nantong Fujitsu). In addition, Jiangsu Changjiang Electronics Technology has become the biggest local manufacturer in Mainland China's semiconductor packaging industry. Except for productivity, Yanzhijiang Delta Region is the packaging technology leader nationwide. Moreover, Intel's Pentium 4 microprocessor, which is packaged in Shanghai Jinqiao, has entered the market. ACE Semiconductor has cooperated with APIA (Advanced Packaging Industrial Alliance) to establish the first complete 8-inch IC packaging production line in Mainland China's Zhangjian high-tech park, and will be in production within this year. US-based National Semiconductor and Fairchild are also establishing their IC packaging and testing foundries in Suzhou. Thus, a relatively complete IC industrial chain has been formed in Shanghai and its peripheral region.

In September 2000, the total investment of 1.25 billion USD for Semiconductor Manufacturing International Corporation's 8-inch/0.18 micron and advanced IC production line project went to the operation in Beijing, and this was the first 8-inch IC production line established in the city. From 2001 to 2003, Mainland China's north region IC investment project has not gone well, like with the failure in progression of the HSMC 8-inch Line project caused by the turning of the market. Motorola's 8-inch Line in Tianjin was

Table 4.2. New Investment Items of Wafer Manufacturing since 2002

Name	Specifications	Location	Investment	Output Production Time
SMIC	8-inch	Beijing	1.25 billion USD	End of 2003
TSMC	8-inch	Shanghai Songjian	890 million USD	2nd half of 2004
Zhuhai Nanke	8-inch	Guangdong Zhuhai	255 million USD	Construct is the 2nd half of 2003
Dongshan Phnix	6-inch	Suzhuan Dongshan	231 million USD	Simulating circuit, completed this year
Nanfang KS	6-inch	Jiangsu Nanjing	360 million USD	60,000 sheet/ month 2 Lines
Semiwel	6-inch	Jiangsu Zhangzhou	50 million USD	12,000 sheet/ month (6-inch)
	5-inch			24,000 sheet/ month (5-inch)
HK Kefeng	6-inch	Liaonin Shenyang	300 million USD	5,000 sheet/ month (6-inch)
	8-inch			10,000 sheet/ month (8-inch)
Yuntai Tek	6-inch	Zhejian Huzhou	600 million USD	Put into production in 2003
Shanhua	6-inch	Jiangsu Wuxi	150 million USD	10,000 sheet/ month

Source: EDN Mainland China.

equipped two years ago, but it has not been put into production because of the depreciated market. Currently, with Semiconductor Manufacturing International Corporation headed north, Motorola's Tianjin Fab has started limited production. This move will provide a strong motivation to the north region's microelectronic up/downstream industries, as well as attract more investors, as seen in the selective contest for the IC investment project of Hong Kong Ke-Feng which has finally ended in Shenyang.

1.4. Mainland China's electronic machinery industry

Mainland China's electronic consumer market for mobile phones, PCs, notebook computers, high-end color TVs, digital cameras, etc., has greatly increased. The international market is also gradually recovering, which means offshore investors' productive capability has been realized in Mainland China. TFT-LCD displays, high-end digital cameras, notebook computers, color TVs, etc., have already become the most increased export products for Mainland China.

In order to achieve low-cost manufacturing, capture Mainland China's market share and upgrade international market competitiveness, offshore investors are continuously accelerating the speed of transfer of their production lines to Mainland China. All Japanese-branded digital camera companies have entered Mainland China to conduct their production layout. The productivity of Taiwan's top 10 notebook computer manufacturers operating in Mainland China have accounted for more than 60% of their total productivity, and the world-class, top six EMS service companies have also upgraded their productive capability three times through methods of purchase and new construction in Mainland China. During the first quarter of 2003, foreign capital and Hong Kong and Taiwanese enterprises accounted for about 69.2% of Mainland China's electronic industry production, which is a 21% to 30.3% increase of the whole industrial production.

While Mainland China's local brand-name companies have not only defended their conventional products market, but also striven to expand their export share, they have also actively moved into the high-end products field. Changhong's DLP Color TV, TCL and Hisense's PDP TV, etc., have achieved a six to seven-fold increase, which also created an impact on those multinational companies that have traditionally monopolized the high-end color TV market. In addition, Mainland China's own mobile phone manufacturers have taken about 45% of the domestic market share from international brand-name companies.

As for the emerging electronic products market, in order to increase the supply capability rapidly, Mainland China's local brand-name companies have contracted production internationally. Currently, domestic notebook computers are mostly manufactured by Taiwanese manufacturers. With the exception of Lenovo, there is no company that owns production lines. Apart from digital camera production in Mainland China, there is only one production line which has been established by Founder Electronics. Recently, Soutec and Solectron established a joint venture company, and outsourced its mobile phones manufacturing business.

At present, the most significant productivity of Mainland China's entire electronic machinery manufacturers is in these three major product fields — mobile phones, PCs, and color TVs. Even though DVD production is also one of the output advantages of Mainland China companies, its high patent fees have weakened the competitive capability for Mainland China's DVD manufacturers. Therefore, within this product field, many Mainland China manufacturers are hoping that DVD technology will be commercialized as soon as possible.

1.5. Mainland China's mobile phone industry

Currently, the mobile phone industry is the biggest hotspot for Mainland China's electronic manufacturing industry. According to data published by the Ministry of Information, Mainland China's mobile phone manufacturing has formed three major bases: one is the northern base in Beijing and Tianjin with nine enterprises, which counts for 40.7% of the total amount nationwide; the second is the southern base, mainly in the Zhujiang Delta Region with nine production enterprises counting for 27.8% of the total amount nationwide; and the third is Huadong in the Yanzhijiang Delta Region, which has six production enterprises counting for 25.6% of the total amount nationwide. These three major bases' total production counts for 94.1% of the total amount nationwide.

Mainland China's present mobile phone productivity reaches 250 million units annually. Manufacturers consist of multinational companies, local enterprises, Taiwanese manufacturers and EMS contractual manufacturers. Local companies have shown that they are putting more investment into expanding production. TCL has increased its capital of 10 million USD and increased its annual output to 15 million units; Bird and France Sagem established a joint venture company, with its mobile phone productivity expanding to 15 million units, and it will hit the mark of 20 million units within the next two years; Panda mobile phones has announced that its productivity will be expanded to 10 million units within 3 years. The total investment of phase two construction in the Soutec production base is 200 million RMB, and its annual output will reach 10 million units; as for the Kejian Samsung CDMA production base that Mainland China Kejian and Samsung Electronics invested 120 million RMB in, after phase one construction is completed, its annual output will be 10 million units.

1.6. Mainland China's PC industry

The expected mature capacity of Mainland China's computer market is about 30 million units, which compared to the current total amount of 10 million units, still leaves a lot of ground to cover. In contrast to the expectation of European and US markets that users will be replacing computers, our domestic computer market is still not saturated, so even more first-time users will be added. Some experts expect that, according to the current trend, the increase in the Asian market will take the lead globally this year, and Mainland China's computer shipments will be more than Japan's making it the second major computer market next to the USA.

IDC analyst Kitty Fok stated in 2002 that in the second half of that year, Mainland China's PC shipment volume would reach 6.3 million units, thereby exceeding Japan's 5.7 million units and becoming the second major PC maker after USA. She also expected that in 2003,

Mainland China's and Japan's PC shipment volume would increase by 18% and 4% respectively. Fok said that education and government-related organizations were the major demanding factors for Mainland China PCs, and such consumption would help increase this year's PC shipment volume. In fact, since 2000, Mainland China's computer-related products have generally developed at a stable rate, with product manufacturing being undertaken by Mainland Chinese and Taiwanese manufacturers. After the Zhujiang Delta Region, the Wujian and Kunshan regions of Suzhou form the main cluster for Taiwanese manufacturers and is another emerging computer products manufacturing base in Mainland China (see Table 4.3).

Most of Mainland China's PC suppliers are adopting the self-production method to occupy more than 80% of the domestic market share, but the contracted production pattern has also been used by domestic manufacturers. According to the report, many foreign manufacturers have requested Lenovo to produce their products. There are more than 20 Taiwan OEM manufacturers which also have the supply-chain relationship with Lenovo. The second major local computer company in Mainland China, Founder Technology, owns the biggest PC production

Table 4.3. Output Status of Mainland China's Major PC Manufacturers

Manufacturer	Annual Capacity (10,000 units)	Production Location
Lenovo	500	Beijing Shanghai Guangdong Huiyang
DELL China	100	Xiamen
Founder Tech	400	Guangdong Dongguan
TCL Computer	100	Guangdong Huizhou
Great Wall	100	Shenzheng
Langchao Group	100	Guangdong Dongguan
Shida Computer	100	Anhui Wuhu
THTF	100	Beijing
Hisense	100	Xiandong Qingdao
Haier	100	Xiandong Qingdao

Source: EDN Mainland China.

lines in Mainland China, but it still needs to outsource its production of PC products to Taiwanese manufacturers due to cost issues. Haier, with its interest in the PC market, has also selected production outsourcing and focused on building marketing channels instead (see Table 4.3).

On 8 December 2004 at 9:32 am, Mainland China's PC industry leader Lenovo Group and IBM called for a press conference in Beijing and officially announced that Lenovo would cooperate with IBM to build the third major global PC enterprise. Mr. Liu Chuanzhi, who represented the Board of Directors of Lenovo Group, announced to the world that Lenovo Group had purchased the IBM PC Business for 1.25 billion USD. The business they acquired from this cooperation was IBM's global desktop and notebook computer business, which included the R&D, purchasing, production, and marketing departments. Lenovo Group thus became the third major global PC manufacturer with more than 10 billion USD in annual income.

2. Economics of Electronics Information Industry's Operations

2.1. The industrial structure of electronics information industry

The industrial structure's modulation is obviously faster than before, with electronic specialty equipment and computer products production rapidly increasing. In particular, the ratio of computer products production and marketing to the entire industrial ratio has climbed very quickly, from 27.3% in 2000 to 33.6% in 2003, which is an increase of about 6.3%. It ranked as first place within the entire industry. In 2003, Mainland China's computer market reached a sales volume of 332.69 billion RMB, which is an increase of 15.4% compared to 2000. Digital products have performed well in the computer market, and its sales volume was 12.23 billion RMB, which is an increase of 77.4% compared to the same ratio. Computer systems and peripheral equipment increased at a relatively slower pace; the ratios increased 9.2% and 8.8% respectively.

The electronic information industry's main products' production and marketing volume for 2003 is shown in Table 4.4.

Table 4.4. Electronic Information Industry's Main Products' Production and Marketing Volume

Product Names	Production Output		Sales Volume		Export Amount	
	Accumulation (10,000 units)	In/Decrease (%)	Accumulation (10,000 units)	In/Decrease (%)	Accumulation (100 million USD)	In/Decrease (%)
Mobile phones	18,644	54.5	18,321	56.1	73.8	39.4
Program control switchboards	5807	39	5387	34.6	4.5	81.8
Color TV	6521	30.3	6500	23.8	25.6	16.6
Microcomputers	3216	98	3083	98.9	22	55.5
Monitors	7326	56.2	7373	55.2	95.7	62.4
Color picture tube	9051	16.1	8906	15.7	7.5	17.7

Source: China's Ministry of Information.

2.2. The high-end products ratio is continuously increasing

According to the data for 2004, electronic information enterprises are continuously increasing technological innovative capability to enhance the products' technological level. The entire income for industrial completion of telecommunication product marketing has reached 438.1 billion RMB, and the ratio has increased 28.6%. Notebook computers and high-end color TVs have increased most significantly, and the increased margin is more than 70%. Notebook computers accounted for around 12 million units, which in turn has accounted for more than 50% of the sales volume of computers. The ratio of high-end color TV's (DLP, LCD, and PDP) realized product marketing income to all color TV's sales income is up from 15% in 2003 to 25.9% in 2004. New electronic components production and marketing have also increased with a wide margin, especially for the micro monitor and IC products; their production and marketing ratios have increased more than 50%.

2.3. Software products market is rapidly increasing

According to data from 2004, Mainland China's software industry followed the continuous improvement of the country's policy and market environment, and its industrial scale clearly expanded. In addition, it achieved an integrated income of 242.4 billion RMB for software products and systems, and its ratio increased 48.4%, which is higher than the 7.2% of the entire industrial increase. From this amount, the software products' (including software services) income was 149 billion RMB, with a ratio increase of 35.3%, which counted for 61% of the software industry ratio; system integration income was 93.4 billion RMB, with a ratio of 75.7%, which is an increase of more than 20% from the previous year. In 2004, the entire industry's exports amounted to 2.8 billion USD, which is an increase of 40% from the previous year.

2.4. Benefit increasing rate is higher than scale increasing rate for the first time

According to the 2004 data, the economic benefits of Mainland China's electronics information industry have significantly improved.

The growth rate of economic benefits for the entire industry has exceeded the growth rate of its scale for the first time. The economic benefits composite index also increased. In 2004, the entire industry value-added increased by 41.3%. The realized profit tax increased 44.6%, considerably higher than the entire industrial product and sales revenue by 0.1% and 3.4% respectively. The total profit growth rate is higher than the sales revenue by 8.1%.

2.5. Internationalization level is gradually increasing

The effects of three-capital enterprises still drive the entire industrial economic growth. Up to 2004, the electronics information industry's realized foreign capital was more than 100 billion USD. In addition, among the Fortune Global Top 500 IT enterprises, about 90% of them have invested in Mainland China. Electronics and IT products exports have accounted for a sales income of about 62.5%. Foreign-funded companies in Mainland China's sales income, industrial value-added profits, and other major economic indices have accounted for an entire industrial total volume of over 75%. In the new round of international industrial transfer, the transfer of multinational companies' R&D design centers toward Mainland China is gaining speed. The most famous international electronic information enterprises have already established R&D centers in Mainland China, and have gradually been upgraded from regional R&D centers to global R&D centers.

2.6. Industrial concentration is increasing further

Increasing industrial concentration has resulted in industrial bases. Currently, Mainland China has formed the Yanzhijiang Delta Region, Zhujiang Delta Region and Bohai Rim Region. These three major electronics information industry regions, with the entire machinery production as the leading components and parts industry, form a complete set development. In 2004, these three regions' labor force, sales income, industrial value-added and total profit amount accounted for an entire industrial ratio of about 83.8%, 89.3%, 86.7%, and 88.7% respectively, which is an increase of 4.7%, 1.5%, 5.4%, and 6.8% on the previous year. After a series of operations, Mainland China's first batch of nine national electronics information industrial bases with bigger industrial scale, relatively concentrated layout, stronger integrated strength, significant divergent effects, and completed foundation of the industrial chain, have already gone into operation. In addition, according to statistics from 2004, these nine industrial bases have jointly completed the product marketing income mark of 1920.5 billion RMB, and the ratio has increased by about 39%. The industrial value-added is 399.95 billion RMB, and the ratio has increased by about 51.4%; the total realized interest and profits amount to 99.4 billion RMB, and the ratio has increased by about 48.5%, which accounts for the entire industrial ratio of 79.6%, 77%, and 75.8% respectively.

3. Analysis of Mainland China's IC Industry Development

Mainland China is becoming a key player in the global electronics manufacturing industry. In order to secure clients, many well-known IC suppliers are also joining Mainland China's IC industry, especially Taiwanese IC companies which form the largest group. They have either moved their design centers to Mainland China, established independent companies, or even invested in and taught local IC companies. Mainland China has decided to adopt the motto of "using information to lead industrialization", and has set the goal of "doubling GDP in 2020 from that of 2000 on the basis of optimized structure and increasing high-efficiency". With this prerequisite, the question of how to evaluate the developing room for Mainland China's IC industry is really a considerable issue. To answer this question, one has to begin with evaluating the status quo of Mainland China's IC industry. We start with the relationship between the IC industry's development and relevant economics index in order to make a primary exploration of this issue.

3.1. Status quo of Mainland China's IC industry

What does the status quo of Mainland China's IC industry look like? Existing research can be categorized as follows: (a) Mainland China's IC industry has been through decades of development, and it has already formed a good industrial foundation; (b) at present, Mainland China's IC industry has started a new phase of accelerating its development, and is in the early stages of launch; and (c) the current issues for Mainland China's IC industry are mainly its small industrial scale, low technological level and innovative capability, weak support of industrial foundation, and insufficiently smooth marketing channels.

Our analytical and research results, which we obtained with plenty of data, support the following viewpoints:

(1) For many years, the population of Mainland China has been about 20% of the global population, but Mainland China's IC sales volume counts for only 2% of the global IC market sales volume. As for the ratio of IC sales volume to electronics industry sales volume, the US is at about 13%, Japan about 15%, South Korea about 30%, but Mainland China is at only 2%. In terms of the ratio of IC sales volume to IC demand, the US is at about 80%, Japan about 125%, South Korea about 140%, but Mainland China is at only 12%.
(2) Within the past eight years before 2004, the average marketing price of global IC products has been about 2 USD per piece, but only 2 RMB in Mainland China. Even though this price increased after 2000, it still remained at around 3 RMB.
(3) During these years, most domestic IC production lines are accepting outsourcing or have authorized the operations to foreign-funded companies. Since 1994, Mainland China's total IC export amount has always been greater than the domestic sales volume, and the total IC import amount is greater than the total domestic demand.
(4) Since 2000, there have been more than 10 newly constructed and planned IC chip production lines in Mainland China, and the construction capital that has already been invested is about 3 billion USD. In addition, it is expected that more than 3 billion USD will

be invested in the short-term future. This 6 billion USD is double the amount of investment accumulated in Mainland China's IC industry over the last three decades. In addition, the investment of the 6 billion USD construction capital indicates that the demand of IC specialty equipment will amount to more than 4 billion USD. However, in 2001, the total sales volume of domestic semiconductor equipment was only 200 million RMB.

(5) From 1996 to 2000, global IC annual average growth rate was 4.3%, with 10% in the US, 3.3% in Japan, 2.6% in South Korea, and Mainland China as high as 40%. In 2001, global IC sales volume decreased by 24% compared to 2000. Japan's sales volume decreased by 25.7%, but Mainland China's increased by 1.1%.

(6) In 2000, Mainland China's electronics and IT products manufacturing industry size jumped to third place globally. The total number of domestic telephone users clinched first place globally, the number of Internet users remained in second place, and the increase in imports and exports for Mainland China's electronics and telecommunications equipment directly promoted the increase of IC demand, and also greatly advanced IC production. Currently, Mainland China's IC market has become the third major market globally, and in 2001, Mainland China's IC sales volume accounted for 2.1% of total global IC sales, with the total sales volume equal to 16% of the domestic IC market demand. The bulk of Mainland China's production technology is the production of the 5-6-inch line and 0.5-0.8 micron width. At present, there are 651 domestic IC enterprises, mainly distributed in the Yanzhijiang Delta Region, Beijing and Tianjin region, and Zhujiang Delta Region, among which the Yanzhijiang Delta Region's IC output value counts for 70% of national output.

3.2. Important ratios in Mainland China's IC industry development

Our findings in this section is based on the survey and research of EBN-Mainland China. The main focus of its research is the USA, Japan, South Korea and Mainland China, and it spans from 1981 to 2000. The main research objective is the economic data relating to the IC industry, and its methodology entails extracting the more regular developing trace and ratio relations from comparative research. According to their data, we can find the following primary results:

(1) The ratio of the electronics industry's sales volume to Mainland China's GDP possesses a relative stability within a longer developing phase, but this ratio is different for different countries within different developing phases. In the US since 1984, it has maintained a level of about 5%; it is mostly unchanged, except for 1983 to 1991 due to Japan's rapid development of its electronics industry which made the ratio remain at 5.5% for those eight years. As for other years, the ratio basically stayed at 4.5%. South Korea's ratio is unstable, and the value is higher, with an average of about 12%. This was due to South Korea's outward-based economy and its third industry being not well-developed yet.

(2) The ratio of the electronics industry sales volume to the total value of the second industry is in the stairstep-type increase status during the procedure of informatization. In the past 20 years, USA's development has been the most outstanding. It continuously increased four stairsteps, and the ratio increased from 11% to 21%; South Korea has also been more active during these past two decades; its ratio jumped from 14% to more than 33%. It briefly showed a relapse during 1990 to 1992, but the general trend still climbed quickly. In Japan, the ratio reached the stairstep of 14% from 10% during 1981 to 1984. It later recorded a small margin of adjustment, but basically fluctuated between 13% and 14%. Mainland China, on the other hand, has shown the characteristics of a later-developing country: in the early stages of the 1980s, its ratio had not yet reached 4%, and even though it climbed two small stairsteps later, it still stayed at the 7% level for 11 years. It started to accelerate its step from 1998, however, with its progressive force, and reached the mark of 17% in 2001. China's sped-up situation may keep going for a couple of years in the future.

(3) In the developing process of information technology, IC demand and electronic sales volume keep on increasing. This situation reflects the growing ratio of IC to electronic industrial product manufacturing cost. In our research, IC sales amount is contained

in electronic industrial sales amount. Therefore, the changes of IC values included in electronic industrial product cost cannot be precisely shown simply by the ratio of IC demand amount to electronic industrial sales amount. As a result, we also study the relations between IC demand amount and the electronic industrial sales amount after deducting IC sales amount. In the past twenty years, the latter ratio has kept on growing in America, Japan, Korea, and Mainland China. From 1981 to 1990, America achieved about 7%. After a continuous increase for five consecutive years, the ratio reached 18% in 1995. The ratio in Japan increased four times in the twenty years from 5% to 16%. Korea experienced a great jump from 5% to 20%. Mainland China started experiencing a great leap from 3% to 9% in 1992 and then continued growing to over 16% in 2000. The above figures reflect that the ratio of IC value to electronic industrial products is dramatically increasing. This is also an important requirement for Mainland China in its entry into the international market for electronic industrial products.

(4) The ratio of IC sales amount to electronic industrial sales amount reflects the basic development condition of the IC industry in a country. The ratio in America before 1990 stayed at around 5%. After growing for the next five years, the ratio rose to about 13% in 1996. A smaller growth rate is shown afterwards. Japan is an IC product export country, and its general export ratio is around 20%. In the past twenty years, Japan has experienced three transitions — the ratio increased from 6% to 10%, then rose to 12.5, and finally reached 16%. Korea has a greater IC product export ratio which hovers at around 50%. Therefore, Korea may be more greatly affected by international factors which cause difficulty for Korea to form a relatively independent and more stable ratio relationship.

(5) The final aspect that we specifically need to point out is that the growth rate of GDP has slowed down in America, Japan, Korea and Mainland China in the past twenty years. Among the above countries, Korea has experienced the most obvious slowed growth. The linear trend dropped from 16% to 7%, followed by Japan which has a decrease from 7% to 1%. The figures show that both Japan and Korea have gone through a much slower economic growth rate. The linear trend shown by America has decreased

from 8% to 6%, while that shown by Mainland China has decreased from 20% to 14%. Nevertheless, the economies of Mainland China and America are under stable development. In addition, the national economic growth rate of Mainland China is far higher than that of America.

Furthermore, in the last 20 years, the trend of the sales growth rates in the electronic and IC industries and the growth rate of IC demand has declined. The reason is obvious: the growth rate increases only when the country is under an acceleratively developing stage, while a stable and slower growth rate is one of the basic characteristics under normal development. The exceptions during this period are:

- The growth rates of sales amount in the electronic and IC industries have shown an increasing trend for the past twenty years. The increasing trend indicates that the sales amounts for the electronic and IC industries in Mainland China are going through an ascending period. In addition, the slope of growth rate of IC sales amount is greater than that of electronic sales amount in Mainland China. Therefore, IC industry enjoys greater ascending development.
- The growth rate of electronic industrial sales amount in America during the past twenty years shows a slight descending slope while the growth rate of IC sales amount still maintains a slight ascending slope. This shows that the IC industry in America keeps on growing. Meanwhile, it also indicates that the self-sufficiency rate of IC supply in America is progressing further.

4. Analysis of Mainland China's Automotive Electronics Market

4.1. Mainland China's automotive electronics market profile

The automotive electronics industry has become the most attractive new expansion point for Mainland China. Statistics show that in 2003, the global auto industry output increased less than 3%, but 80% of this 3% increase was from Mainland China. Mainland China's automotive market's rapid development has greatly advanced the penetration and development of Mainland China's electronics information industry to the automotive electronics field. According to research data, demand for the whole vehicle can be a great influence on the automotive electronics market development — the related coefficient between them is as high as 0.997. From this, we can conclude that Mainland China's increase of complete vehicle requirement will lead to a wide-margin increase in Mainland China's automotive electronics requirement.

In 2004's National Conference on Information Technology and Application Works, the information industry sector's personnel indicated that the auto industry's fuel control system and the vehicle driving adjustment system should be realized through IT. With the deep application in the auto industry field, through revolutionary changes in the automotive industry, automotive electronics will become one of the new economic growth points, and it will also greatly expand the room for developing the information industry.

Up to 2005, Mainland China's automotive market sales volume may have been able to reach 1000 billion RMB, with the ratio of every global new vehicle's average automotive electronics expenditures to the whole vehicle's value of around 25% to 30%. However, from our calculations, Mainland China's automotive electronics actual market was only worth about 50 billion RMB, a very significant difference. Nevertheless, difference means opportunity, and if Mainland China's automotive electronics enterprises can exclusively explore this potential demand for consumers, then the increased potential is very good for the future of Mainland China's automotive electronics industry.

Generally speaking, the automotive electronic facilities refer to controlling, driving, safety, displaying, communications, and entertainment fields. It involves software, computers, system integration, telecommunications, and electronics and basic components consumption. Almost all electronic IT products can be applied to a car. The electronic technologies applied in cars not only enhance the performance and levels of traditional cars, but also expand the applications of electronic technologies, thus establishing the rapid development of the automotive electronic industry. It is estimated that 24.88 million sets of automotive electronics at 30.04 billion dollars were sold in Mainland China in 2003, which is at an increase of 33.7% and 41.5% respectively compared with the same period last year. Such an increase rate was five times the average global rate. Such an increase rate was also higher than 29.2%, the growth rate of national automotive sales during the same period.

As Mainland China develops its automotive industry and vehicle technology, the standardization of vehicles in Mainland China is realized from obligatory standards to a combination of obligatory standards and design. The stricter requirements placed on safety, environmental protection, and energy saving in the automotive industry further promote the research and applications in automotive electronics technologies. Consumers' requirements regarding automobiles are constantly changing. They not only request for an excellent driving experience and performance, but also for in-car entertainment and office facilities. For example, customers may request for an in-car multimedia system that includes information processing, access to the Internet, communications, a GPS system, anti-theft accessories, a language identification system, image display and entertainment. A driver can input the destination to an in-car computer which provides the driver with the best route to the destination. The application of wireless access to the Internet makes e-mail transmission and business handling possible in a car. The automotive industry will take measures to fulfill the changeable needs of consumers. However, the fulfillment of these needs will be carried out through the automotive electronics industry. Consequently, the development of the automotive electronics industry will be activated accordingly.

According to the statistics of CCID, the average annual growth rate of the domestic automotive electronics industry in Mainland China is 35%. It is estimated that the market demand will reach 54 million sets at 78 billion dollars in 2006. The increasing market demand and higher profits will attract foreign investments in the automotive electronics industry. Recently, the leading companies worldwide engaged in household electronics and IT have announced their participation in the automotive electronics industry. For example, Neusoft and Alpine cooperated to establish Neusoft Automotive Electronic R&D Center to produce related in-car electronic products for BMW and Mercedes-Benz on 16 June 2004. Soon after this collaboration, the Desay Group, which specializes in disk players, visual and audio facilities, energy and optical displays, announced on 17 June that it was going to make a large investment in the automotive acoustic industry.

It plans to spend two to three years establishing itself as a dominant brand in the automotive electronics market. The management of TCL also suggested not long ago that TCL was going to enter the automotive electronics field.

4.2. The development directions and dilemmas faced by the automotive electronics industry in Mainland China

So far, the manufacturers engaged in in-car electronic systems have developed a certain economic scale in Mainland China (discussed later). Some are also ready to enter the market for international high-end cars. The development of in-car electronic facilities is one of the important directions guiding the automotive electronics industry in Mainland China. Compared with other countries, Mainland China still has a long way to go regarding the industrialized level and R&D ability of its automotive electronics control system.

With increasing market competition ahead, technical personnel in Mainland China cannot afford to stop their progress. As the international atmosphere for energy resources tightens, the "863" project is positively promoted for the research in automatic automobiles and the fuel cell electrical control system in Mainland China. The research and industrialization of automotive electronics focus on:

- Information: speed meter, tachometer, and fuel meter will be replaced by touch-type LCD with Internet functions, site diagnosis, and digital display. The functions of site diagnosis, automatic drive, e-map, mobile display, track replay and so on can be put into practice through the "expert think-tank" of the in-car mobile information system.
- Energy saving: an extra 2% of fuel is consumed when more effort is needed to turn the steering wheel at low speed and more ease in turning the steering wheel at high speed. The newly developed automatic auxiliary steering system aims to improve the situation in order to enhance safety and save energy.
- Comfort: the air-conditioning system operating on a multifunctional motor is manually controlled by the driver. R&D focuses on an automatic air-conditioning system, which collects the information of interior and exterior temperature, moisture, sun, and

speed. The air-conditioning system will adjust and automatically calculate an ideal setting based on a specific mathematical model.

These three research results will first be applied to fuel cell automobiles. Traditional cars will then be improved in accordance with individual conditions. Furthermore, the Ministry of IT in Mainland China has stated that it will support six automotive electronic products: uniform automotive calculation platform, ECU (Electronic Control Unit), driving recorder, automotive visual and audio navigation system, automotive semiconductor components and special chip, and intelligent traffic system.

One of the major difficulties of the newly developed automotive electronics industry in the process of research and industrialization of products is that the industry does not know much about the international standard system. Also, an industrial chain has not been formed by the automobile assembly manufacturers and newly developed automotive electronic manufacturers. The business model for developing automotive electronics has not been established either. Automative electronic products are highly associated with the total assembly of a car. Therefore, without guidance in the requirements of a complete automobile, the electronic manufacturers have to carry out research by themselves and make sample units of related products. Then, they send the sample units to undergo tests by the automobile assembly manufacturers. After passing the tests, they have to seek out cooperative alliances. A senior engineer of the automotive R&D center in Mainland China suggests that these products have less stability and safety because of fewer tests and high test expenses. As a result, the automotive electronic industry in Mainland China is unable to fulfill the needs of the market and expand in the short-term. The newly developed automotive electronic industry is unable to move away from the joint venture model (i.e., with automobile assembly manufacturers) if it hopes to expand its scale to a certain extent.

In the R&D process, the upstream suppliers of automotive electronic products are controlled by joint ventures, which create an unhealthy automotive electronic packaging market. Such an unhealthy market acts negatively on the automotive electronic R&D industry. Common problems such as a longer cycle of purchase, inconvenient procurement, and complicated specifications persist. Manager Ma Hsi Ping of Hsin Tai Vehicle Technology Corporation in Shanghai states that "the automotive electronics had been in difficulty of technology in the high power and high current requirement of stricter standards on cables and components. But we have difficulties in packaging the whole thing".

So far, the electronic industry in Mainland China has started promoting products such as the automatic support system, safety control system for high voltage power supply, vehicle frame electrical control, digital meters, low-speed control system, tyre pressure monitoring system, and alarm system. The execution of National Automotive Calculation Platform Engineering in Mainland China will hopefully produce the core technologies including automotive control, steering, information processing, intelligent traffic sensors, transmission, execution, etc., and provide further important opportunities for the applications of 3G mobile communications, high resolution TV, GPS (Global Positioning System), and mobile Internet. In addition, there is much hope for positive mutual interactions between the automotive industry and IT industry. As the top automotive semiconductor service providers join the market in Mainland China, the automotive electronic industry is expected to grow faster.

4.3. *The market and developing opportunities for major automotive electronics in Mainland China*

In 2004, CCID Consulting Co. predicted that the market scale of automotive electronics products in Mainland China will keep growing faster in future years. The automotive electronics technology will develop according with the needs for accessing the internet, intelligence, and integration. The continuous expansion of the automotive electronics market is derived from the dramatic development of the automotive market in recent years. In 2003, the sales of automobiles reached 4.44 million cars and 4.39 million cars respectively, an increase of 35% and 34% compared with the same period the previous year. The rapid growth of the automotive market facilitates the growth of the in-car entertainment market in Mainland China. Therefore, there is huge potential in the automotive electronic market in Mainland China. Market surveys show that a total of 12–16 million automobiles was manufactured from 2003 to 2006. The market for automotive entertainment systems is approximately 5–8 million sets. OEM packaging service is provided. The in-car computer is used as a platform to integrate in-car communications, navigation, video and audio entertainment and internet controls into a multimedia information system. This is the trend for future development. The potential automotive electronic products in Mainland China are shown in Table 4.5.

4.3.1. Rapid development of the market for in-car navigation systems

Mainland China plans to make Beijing, Tianjin, Shanghai, and Guangzhou into four centers for the production of navigation e-maps. General Manager Sun Yu Kuo of NavInfo Beijing, which provides the navigating e-map databank, says, "The automotive navigation e-map products which meet current automotive industrial standards and laws in Mainland China have been introduced in Beijing, Tianjin, Shanghai, and Guangzhou." It is estimated that navigation e-map products which display areas in Mainland China will be promoted in the market in 2005. These navigation e-map products will fulfill the needs generated from the 2008 Beijing Olympic Games and 2010 Shanghai World Expo. The in-car navigation system takes advantage of the in-car GPS to receive the current position of a car and stores the information in a disc or built-in memory, like a hard disk. Subscribers can get information

Table 4.5. Potential Automotive Electronic Products in China

Automotive Electronic Products	Market Prospect	Compound Annual Growth Rate of 2002–2004 Market Demand
GPS	28.35%	82.6%
E-map	23.61%	–
Vehicle body electronic control	23.15%	29.2%
ABS	9.02%	46.1%
In-car DVD	4.56%	25.8%
Automatic system	7.20%	–
Air bag	2.73%	34.2%

Source: Market analysis derived from *Automotive Electronic World* and *ESM-Mainland China*.

on destinations from the e-map and the system will calculate the route and provide the route and driving time to the driver.

The market development of the in-car e-map will create a huge demand in GPS, e-map storage facilities, and display facilities. Strategy analysts predict that the annual demand for global navigation systems will increase from 4.3 million units in 2002 to 12.2 million units in 2010. The Telematics Research Group predicts that the demand for GPS receivers will increase from 9.4 million units in 2004 to 30 million units in 2010.

Technically speaking, old cars face space limits when being equipped with GPS navigation systems. This problem can be alleviated by taking advantage of the in-car DVD system with its large storage capacity, which would enable the e-map information to be displayed on the DVD player. Therefore, the in-car DVD market is expected to grow fast. The Telematics Research Group made a forecast in July 2004, stating that the global demand of the in-car DVD market will increase from 4.5 million sets in 2004 to 33 million sets in 2010.

The growth of the global automotive market comes mainly from Mainland China. Therefore, the rapidly growing market for the in-car navigation system in Mainland China will dominate the global market. According to the Japanese press, the demand for DVD/HDD navigation systems is growing and is expected to exceed 10 million units in 2004. The estimated sales will exceed 10 million units in 2008 in Mainland China, even though the market for navigation systems there started only in 2003. E-map information suppliers will obtain high profits in such a huge market. However, the common use of e-maps may create problems for national security.

Automotive wireless receiving technology is developed from digital systems. The digital signal processor that applies soft wireless technology will lead to the development of the in-car digital wireless receiver. Consequently, AM/FM, TV receiver, satellite digital radio receiver, GPS and communications may be integrated together. The solution provided by Philips CarDSP can help strengthen the rectification of multiple paths and adaptive functions of antennas. Such development means a lot to the improvement of in-car visual and audio facilities such as AM/FM or digital wireless radio.

4.3.2. Development of vehicle body electronics

In the future, cars will be equipped with more and more electronic facilities. To reduce the number of connecting leads and vehicle body weight, there has been a breakthrough development with the Internet and hybrid data bus technology. It aims at solving problems such as security and reliability of information transmission and greater weight of vehicle body with much longer in-car cables, by using an electronic system and satisfying the needs of wire control. The international leading automobile companies and top automotive semiconductor manufacturers have promoted a series of automotive hybrid data bus standards including IDB-1394 and MOST (for multimedia), FlexRAY and TTCAN (for wire control), CAN, IDB-C and GMLAN SWC (for power assembly control), J1850 and LIN (for vehicle body control), and Keyword 2000 and ISO 9141 (for diagnosis). Figure 4.1 shows the current hybrid data bus standards, purposes and rates of automobiles in the market.

Based on the above, some experts think that the development of the vehicle body net system will bring excellent opportunities to the automotive electronics industry in Mainland China. There are two reasons, one of which is that such development lowers the threshold for enterprises. By taking advantage of the single chip project with LIN or CAN hybrid data bus interface, the cycle for studying vehicle body electronics is thus shortened. The threshold for enterprises entering the automotive electronic packaging market is lowered accordingly. The second reason is the need to enhance the comfort in a car and to expand value-added space. The manual controlled mechanical units such as the air-conditioner, locks, windows, seat, internal and external lighting, rear mirror, rain brush, etc., can be connected with vehicle body net system through CAN or LIN hybrid data bus. The manual controlled units can be changed to electrical controlled units. It becomes possible to make middle and low-end automobiles and light trucks equipped with a mechanically and electrically integrated control unit.

Nancy Dang, senior analyst of iSuppli for the automotive market, says, "There have been over 50 automotive electronic units in a foreign high-end car, while there are about six to seven units in a middle-end car in Mainland China. Therefore, there is still a great space for the automotive electronic industry to develop in the future in connection with

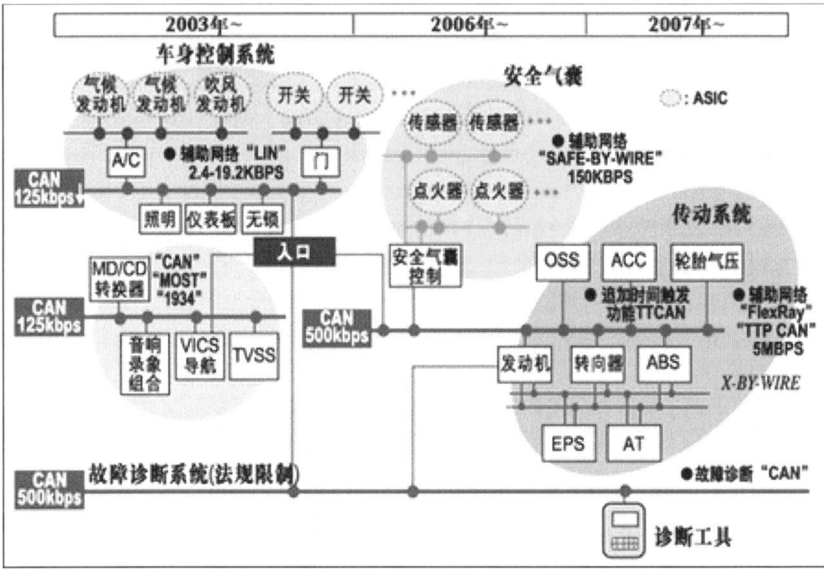

2003~					2006~			2007~
Self-control system					Air Bag			
Climate Engine	Climate Engine	Blowing Engine	Develop ment	Developm ent	Sensor	Sensor	Transmission Network "SAFE-BY-WIRE" 150KBPS	Transmission system
CAN 125 kbps	A/C Transmission Network "LIN" Door 2.4 – 19.2 KBPS				Igniter	Igniter		
	Lighting	Dashboard	Opening Lock					
CAN 125 kbps	MD/CD Changer	"CAN" "MOST" "1934"	**ENTRANCE**		Air Bag Control	COSS	ACC / Tire Pressure	Auxiliary Network "Flex Ray" "TTP CAN" 5MBPS
						Additional Igniting Time Function TTCAN		
	Stereo & Video Combina tion	VICS Navigation	TVSS	CAN	500 kbps	Engine	Steerer ABS	X-BY-WIRE
CAN 500 kbps	Failure Diagnose System (Limited by Regulations)				E	PS	AT	Failure Diagnose CAN
								Diagnostic Device

Figure 4.1. Current Automotive Hybrid Data Bus Standards, Purposes and Rates

Source: Ruei Sa Technology.

the vehicle body electronics." According to the Economic Operations Department of IT Ministry, the vehicle body electronics in Mainland China will increase from 1.032 billion RMB in 2003 to 5.033 billion RMB in 2007, which is 3.8 times the increase rate of the in-car information communication system (4.2 times compared between 2003 and 2007). As a result, there will be a great demand for vehicle body electronic products such as MCU with LIN/CAN hybrid data bus interface, related power driven components, electrically controlled MCU and products executing electrical mechanical components.

So far, the manufacturers supporting LIN alliance include BMW, Volvo, Audi, VW, Daimler Chrysler, etc. Such a trend indicates the developing direction in the future.

4.3.3. Automotive safety system — the fastest growing field

According to the market survey by Strategy Analytics, the automotive safety system is one of the greatest demands in the field of automotive electronics. The average annual growth rate reaches 25%. The latest survey by Dupon Automobile indicates that most users consider safety as a top priority compared with performance, entertainment, and fuel consumption. Airbags and ABS (Anti-lock Brake System) will become standard equipment. The study by Visteon also suggests that consumers are most concerned about safety. Renesas conducted research on middle-end automobiles and indicates that airbags have been treated as standard unit equipment in cars since 1996. Starting from 2002, the international society has started making two sensor passenger seats, pretensioner seat belt, sensor system, and airbags as standard units equipped in cars. It is estimated that the number of pretensioner seat belt and sensor systems will be increased to four in 2006. The sensor system dramatically enhances the stability and impact ability and therefore may become a part of the airbag net.

In the market of international automotive safety systems, there are automatic and passive systems being developed respectively, which may be integrated in the future. Figure 4.2 shows the development routes of the automotive safety system. The newly developed automotive electronics industry in Mainland China has been concentrating on automotive safety. For example, the study on ABS and airbags has just begun in Mainland China. The TPMS is a radar system that keeps the vehicle from impact coming at the front, side, and back.

The TPMS promoted in the market can be categorized into two parts:

(i) *Direct TPMS.* This system allows a sensor and a chip with an RF transmitter to be equipped near the valve of a tyre. When the tyre

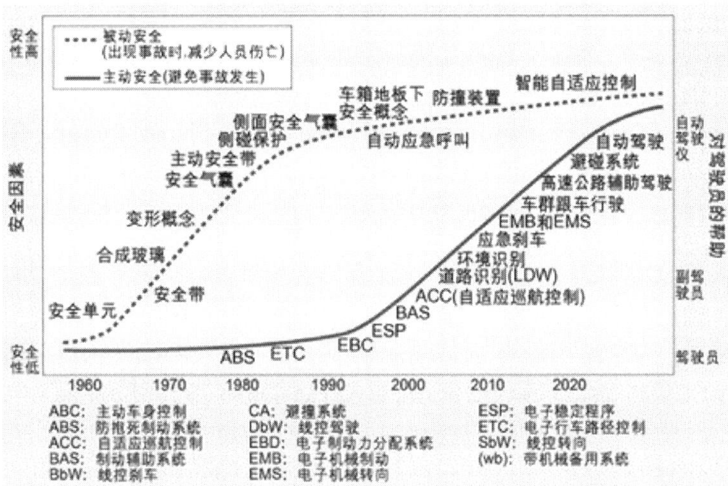

<table>
<tr><td rowspan="3">Safety Factors</td><td rowspan="2">High safety</td><td>….Passive safety (activated while occurring accident to reduce the casualty)

------- Active Safety (Prevent from Accident Occurrence)</td><td></td><td>Auto Driving Device</td><td rowspan="3">Help for Drivers</td></tr>
<tr><td colspan="2">Safety Unit
Seat Belt
Compound Glass
Deforming Concept
Air Bag
Active Seat Belt
Side Impact Guarantee
Side Air Bag
Auto Emergency Warning
Safety Concept that under the floor of car cabinet
Anti-collide System
Intelligent Auto Control</td><td>Deputy Driver</td></tr>
<tr><td>Low safety</td><td colspan="2">ABS
ETC
EBC
ESP
BAS
ACC(Auto Cruise Control)
LDW
Environment Distinguish
Emergency Brake
EMB and EMS
Cars Trial Driving
Highway Auxiliary Driving system
Anti-collide System
Auto Driving</td><td>Driver</td></tr>
</table>

ABC: Active Body Control ABS: Anti-brake System ACC: Auto Cruise Control BAS: Brake Assist System BbW: Brake by Wire	CA: Collision Avoidance DbW: Driving by Wire EBD: Electronic Brake Distribution EMB: Electronic Mechanic Brake EMS: Electronic Mechanic Steering	ESP: Electronic Stability Program ETC: Electronic Transportation Control SbW: Steering by Wire (wb): With Backup Mechanic System

Figure 4.2. Development Routes of Automotive Safety Performance Led by Automotive Electronic Technology

Source: Ruei Sa Technology.

pressure is too high or too low, experiences a slow leak or abnormal temperature changes, alarms will be transmitted to that in-car wireless receiver to effectively prevent a flat tyre. The difficulty of this system lies in creating a high reliability. Battery life, wider range of temperature changes, and vibrations can compromise the reliability of the system. Another difficulty is that there is no standard for wireless receiving and transmitting frequency used for direct TPMS to follow at the moment. Some manufacturers adopt the same frequency (433 MHz) as a remote control for opening doors or for the anti-theft unit, which is made by the FSK method. However, it is difficult to avoid co-channel interference. Finally, the wireless sensor equipped on a tyre has a design problem in low power consumption which has bothered the industry for a long time. The development of wireless sensors without any cells will soon dominate the market.

Direct TPMS is divided into automatic and passive types. The automatic system provided by Infineon allows a pressure sensor to be equipped on each wheel disk. The various existing remote receiving units including RKE frequency components can be used to receive measurements of tyre pressure. The strength of the automatic TPMS system is its mature technology with modules applicable to tyres of different brands. The weakness is that the modules need batteries to supply power and therefore affects the length of battery life. Passive technology is free of power supply by batteries, but it needs to be integrated with the transmitter into a tyre. The technology involves establishing common standards among tyre manufacturers. So far, only Siemens and Michelin have developed such technology. No other mature products have been developed yet.

(ii) *Indirect TPMS* uses ABS tyre speed sensor to measure the rotating speed of each tyre in order to detect the tyre pressure. This system has the advantages of endurance and reliability without being influenced by batteries and other wireless waves. It is not necessary to reassemble a tyre. The cost of the passive system is much lower than that of direct TMPS. The indirect TPMS is especially suitable for private cars and light trucks. The system can cut directly into the market by packaging it with automobile assembly manufacturers. The scale of indirect TPMS market is determined by the automobile market equipped with ABS.

The safety of heavy-load vehicles has become widely acknowledged so far. The Chinese government is considering a policy that all heavy-load vehicles are obliged to install ABS. Direct TPMS can be used to monitor the tyre pressure for long-wheel base trucks, trailers, and heavy-load vehicles so as to prevent accidents caused by flat tyres.

Manager Chen Chiang of Burnon Technology says, "The demand in direct tyre pressure monitoring system in the global market reached about 2.3 million units in 2001. Over two million cars provided by GM and Ford are using the direct TPMS at the moment. It is estimated that the number will reach 29.3 million units in 2008. The CAGR is 44%." The market for the tyre monitoring system in Mainland China is still at a preliminary stage. The unit price is as high as thousands of dollars (four tyres of a private car). The industrial packaging market has not been activated yet. However, the attention the Chinese government and automotive users are paying to safety will bring the tyre pressure monitoring system into one of the fastest growing automotive electronic application systems.

4.3.4. Development of automotive power supply control

According to the forecast made by Delphi, the demand in automotive electronic systems will grow at a rate of 5–8% from now to 2030. In the transition period, the 42V/12V power system may become the standard vehicle power system. Currently, a foreign high-end automobile needs over 2,000 leads with over 4,000 touch points for various controlling components and units. The total length of leads exceeds 4,000 meters at 40 kg. There are over 100 electrical units and MCUs. After adopting the 42V power system, the current lead sectional area will be reduced from 0.5–1.5 mm2 to 0.35 mm2. The advantages are: the consumption of ohm is reduced with lower current through lead at the same power; the fuel consumption is reduced by 10%–20%; and the space saved from lead can be used for new electronic systems. The aluminum wire replaces copper wire.

More soft lead is used to make smaller electrical units. These advantages can meet the need of hybrid fuel automobiles.

The latest news suggests that Mainland China will start mass production of hybrid automobiles within one to two years due to energy and environmental problems. The automotive industry in Mainland China will gradually increase the voltage (12V, 16V, 24V, 36V) of the power supply system based on the need to fulfill the fast-growing demand for power supply. The interactions between the above two factors may create greater demand for power transformers (DC-DC transformer) and high voltage/high current components required by various automobiles.

According to the forecast made by International Rectifier, the ratio of power supply IC demand to the automotive electronics market will increase at an average annual rate of 25% in the following three years. The requirements of users to power the components, including the overall need for MOSFET and IGBT, are to increase current density, lower component cost, reduce space, etc. Many manufacturers including Freescale, Toshiba, NEC, Philips, On Semiconductor, Vishay Siliconix, Linear Technology, Fairchild Semiconductor and IRF are working hard on these. The characteristics shown in temperature control for power supply components will directly influence energy saving and result in environmental protection. Lowering Ron of power supply components is one of the most competitive advantages in the automotive electronics industry. The automotive semiconductor service providers and their market share in 2003 are shown in Table 4.6. Figure 4.3 shows the market demand for automotive power supply control components.

The power device specially produced for automobiles accounts for a major part of automotive power cost. The industry is forced to continuously develop new manufacturing crafts of power device in order to further lower the cost. SiC and GaN devices which do not consume too much of the power cooling system possess higher transferable efficiency and therefore may become the potential technologies aimed at strengthening the performance of automotive power devices. It is anticipated that SiC and GaN devices may be widely applied in automotive power after five to ten years. SiC and GaN devices are important development trends worth paying attention to. The progress in automotive electronics has just begun. Therefore, the line width of chips required by the

Table 4.6. Auto Industry's Main Semiconductor Solution Suppliers and 2003 Market Share

Company	Market Share
Freescale	12.90%
Infineon	8.70%
ST Microelectronics	7.60%
Renesas	7.00%
Bosch	6.40%
Toshiba	6.10%
NEC	6.00%
Philips	5.70%
On Semiconductor	1.50%
Texas Instruments	3.50%
Intel	2.20%
Delphi	2.10%
Fujitsu	2.00%
Vishay Siliconix	1.30%
Maxim/Dallas Microchip Technology	26.70%
Fairchild Semiconductor	
International Rectifier Company	
National Semiconductor Corporation	
Atmel Frontier Silicon	

Source: Freescale.

automotive semiconductor power device is not as strict as IC. Such fields may become the best tool for assisting IC design companies and power module design companies in Mainland China in entering the automotive electronics market. The participation of the top international suppliers for semiconductor devices is expected to accelerate the progress.

5. Analysis of Mainland China's Consumer Electronics Market

Promoted by pan-digital waves, the consumer electronics market in Mainland China, which includes consumer electronics, cell phones and digital products, is turning into the most energetic segment market in the IT field. The huge market potential has turned Mainland China into an important place of growth in the global consumer electronics industries. Statistics indicate that Mainland China is the

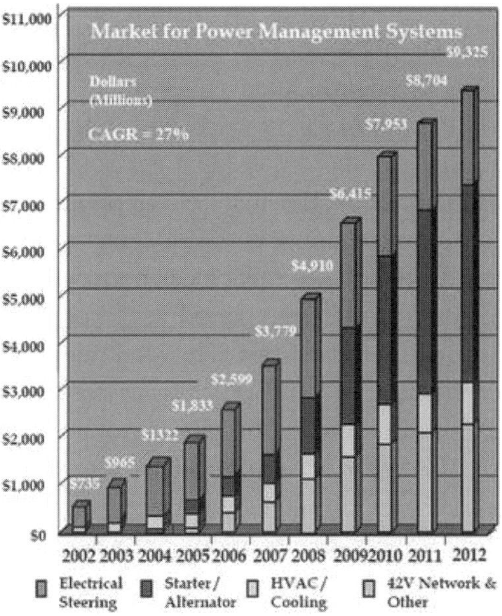

Figure 4.3. Forecast of Market Demand for Power Management Systems by Automotive Electronics

Source: IR.

biggest base for producing consumer electronics and the second-largest consumer market next to America.

In 2004, 44 million color TV sets and 78.7 million cell phones were sold in Mainland China. There were also dramatic increases of sales in digital products such as digital cameras, digital video cameras, MP3 players and CD players. According to the forecast made by the famous US market analysis company IDC, the consumer electronics market including consumer electronics, cell phones, and digital products will keep growing steadily in Mainland China in the coming years. By 2008, the consumer electronics market in Mainland China is expected to reach 100 billion USD. Mainland China has greeted the new consumer electronics age.

Lo Wen, the Deputy Director of the R&D Institute of IT Industry in Mainland China, says that many consumer electronics fields including cell phones and color TV sets have become dominant in the global industries. More and more foreign investors also hold positive views on the potential of the market in Mainland China, and are planning to make investments and developments in the consumer electronics field in Mainland China.

According to the statistics provided by CCID Consulting Co. in Mainland China, the number of cell phones sold in Mainland China reached 78.7 million in 2004. The sales of cell phones exceeded 120 billion USD. The rate of upgrading and replacement of products is obviously accelerating. A total of 226 million cell phones, which is an increase of 31.3% compared with the same period in the previous year, are produced by 37 domestic manufacturers. The total number of domestic cell phones exported reaches 146 million, further consolidating Mainland China's leading position among the global cell phone manufacturers. It has been forecasted that a compound annual growth rate of 7.9% in the cell phone market should be maintained in the next five years in Mainland China. The number of cell phones sold will exceed 100 million in 2008.

In 2004, 1.7507 million high-end TV sets (inclusive of projection TVs, LCD TVs and plasma TVs) at sales of 20.53 billion USD were sold in Mainland China, which is an increase of 54.1% compared with the same period in the previous year. According to the forecast made by CCID, the compound growth rate of the high-end TV sets market will reach 47.1% in the following five years.

The sales volume of digital cameras and digital video cameras in Mainland China in 2004 increased by 94.4% and 116.2% respectively compared with the previous year. The annual sales volume of MP3 players totals 4.284 million units, an increase of 141.6%, at the sales amount of nearly 3.5 billion dollars, an increase of 123.7%. A rapid growth in sales of these digital products will continue in the next five years.

According to a report, the number of analog cable TV subscribers has reached 0.13 billion and digital cable TV subscribers, 773,000, which is an increase of 36% compared with the previous year. The worth of the digital TV industry has reached 29.5 billion USD, and the industrial worth of digital TV sets is expected to exceed 500 billion dollars in the following five years.

Chung His Ho, Secretary General of the Digitalization Industry Union (3C products) of the Association of Information Industry, says the consumer electronics market keeps on growing at a high speed. In addition to the high-speed market expansion, innovative technologies and bright spots of products continuously emerge. The concept of digitalization has indeed entered into the lives of the Chinese.

In a recent symposium discussing innovative technologies and business opportunities of newly risen industries, Wan Ning, Vice CEO of IDC China, claimed that the demand for high-end consumer electronics is getting stronger after the GDP exceeds 1,000 USD in Mainland China. The consumer electronics market is gradually reaching maturity in Mainland China. In the past, the major consumer electronic products that Mainland China consumed were telephones, TV sets, and computers. Now, the consumer electronics include game accessories, PDAs, digital TV sets, digital home multimedia centers, etc.

Another famous company specializing in market analysis Global Resource, suggests in a report that the sales amount of consumer electronics in Mainland China in 2005 is expected to reach 49.6 billion USD, with an increasing annual growth rate of 20%. By 2007, the sales amount of consumer electronics in Mainland China will reach 94 billion USD. The report also indicates that the sale of traditional electronic products will gradually slow down in Mainland China. Instead, Chinese manufacturers are increasingly turning towards products with advanced technology. The production technology for products such as TV sets, DVD players, automotive audios, MP3 players, and other consumer electronics is being rapidly upgraded. Notebooks, regarded as one of the consumer electronic products, are dramatically produced in Mainland China. In 2003, the market development of notebooks in the Asia-Pacific region increased drastically, and maintains a growth rate of 25% to 30%. Notebooks are rapidly replacing traditional PCs and are becoming the dominant product in the market. Mainland China is the biggest market for notebooks in the Asia-Pacific region.

Wan Ning says that the rapid development of the market for digital products in Mainland China has begun. He anticipates that digital cameras in Mainland China will exceed a delivery growth rate of 35% in 2008. The average growth rate of sales will exceed 30%.

The enormous potential of the consumer electronics market in Mainland China has attracted the global leading electronic manufacturers' attention. Samsung in Korea has already announced the continuous expansion of its market share in Mainland China. Samsung predicts that it will reach 25 billion USD in 2010. Samsung also indicates that all electronic products including plasma TV sets, cell phones, and chips in Mainland China will account for a very essential position in the buyer's market. Yi Chung Lung, Vice CEO of Samsung, also holds that the market in Mainland China will determine the future global marketing direction of electronics. He states, "Mainland China will become the biggest market of consumer electronics in the world. We predict that the demand for high-end electronic products in Mainland China will keep on growing."

6. Analysis of Mainland China's Mobile Storage Industry

Shuei Mu Ching Hua Research Center conducted a detailed analysis on seven leading mobile storage chip manufacturers, eleven flash memory card manufacturers, and the current five micro drive manufacturers in Mainland China. The results regarding future development trends and the competition between flash memory card and micro drive are as follows:

(1) The difference between the flash memory card and micro drive is divided by 2GB. Micro drives dominate the market for products with a storage capacity over 2GB, while flash memory card has a stronger presence in the market for products with a storage capacity lower than 2GB. The competitive advantage of micro drives over flash memory products lies in the ratio of capacity to price. Looking at the current conditions in the market, the price of flash memory chips keeps reducing while its capacity increases. In the market for the capacity of 1-2 GB storage products and below, micro drives lack the advantage in capacity/price ratio. Therefore, the future development trend for micro drives should focus on higher capacity. From the current market, it can be seen that competition increases when the capacity of the product is higher than 2GB. This phenomenon can be proven by the Apple iPod mini. The capacity designed for this kind of MP3 with the Hitachi 1-inch micro drive may be 4GB or 6GB. MP3 players with micro drives below 2GB have become rare in the market. Toshiba is also feeling the pressure brought on by the lowering price of flash

memory and has given up products with capacity of 2GB at 0.85-inch. Instead, Toshiba has started focusing on 4GB products.

(2) The future of the flash memory card market lies in SD and MMC. SM became rare in the 2004 market due to its limitations on storage capacity and compatibility. We could hardly find SM among the consumer electronics promoted in 2005. SM is being replaced by the xD card, which is only used in Olympus and Fuji digital cameras at the moment. The MS series is used in Sony products. xD and MS maintain a certain share in the market, but they lack potential development. Though the CF card once became the leading product in the market, it has a size limitation which made it impossible to use in cell phones. In addition, the CF card has to compete with micro drives and SD cards in the market above and below 2GB. The SD card has become a leading product by replacing CF cards at the moment. From a long-term point of view, the CF card cannot compete with the micro drive either in price or in capacity. Therefore, the future for the CF card is not optimistic.

The SD card accounts for the biggest share in the mobile storage market, followed by the MMC card. Both cards have their respective advantages. There are two reasons why the SD card has a greater market share. Firstly, advanced technology and successful promotion; and secondly, its manufacturers monopolize the consumer electronics market. The SD card is used widely in global digital products. However, the MMC card has an advantage in opening standards. The competitiveness with lower cost becomes more favorable in the market than SD cards since there is no charge on patent fee. Furthermore, the manufacturers of the MMC card have gained support from major cell phone manufacturers such as Nokia, Samsung, LG and Siemens. MMCA has made more effort in exploring the market in Mainland China. Some cell phone manufacturers including governmental units and operators in Mainland China have great interest and recognition in the MMC card. Konka was the first manufacturer in Mainland China to join the MMCA system. TCL, OKWAP, CECT, Lenovo and Dopod have also promoted cell phones with built-in MMC cards. SDA has valued copyright protection from the beginning. MMCA introduced its 2.0 version with a complete sealing function in 2005. In the future, there will be much competition between MMCA and SDA in providing lower costs, security, and high speed in the mobile storage market. The manufacturers in Mainland China will put more emphasis on cost and be open to adopting flexible standards. It is anticipated that MMC cards will be more popular in Mainland China.

(3) The T-flash card is booming in the mobile storage card market. This market will be shared by small storage cards including T-flash, RS-MMC, and miniSD in the future.

At the moment, the T-Flash card promoted by SanDisk in 2004 is the smallest storage card in terms of physical size. It received much support from Motorola and Samsung when it was first promoted, and the T-flash card rapidly became dominant in the market. In a survey of mobile phones equipped with storage cards released in 2005, T-flash accounts for 29.5%, miniSD accounts for 27.9%, and RS-MMC accounts for 18.0%. The total market share of these three types of cards amounts to 75.4% (see Figure 4.4). The memory stick produced by Sony is used mainly by Sony Ericsson phones, which gradually took parts for middle- and low-end products. Therefore, the percentage of products using expensive memory sticks would not increase. Memory sticks may not have much potential in the future cell phone market because lighter, thinner and smaller cell phones are in demand, and smaller

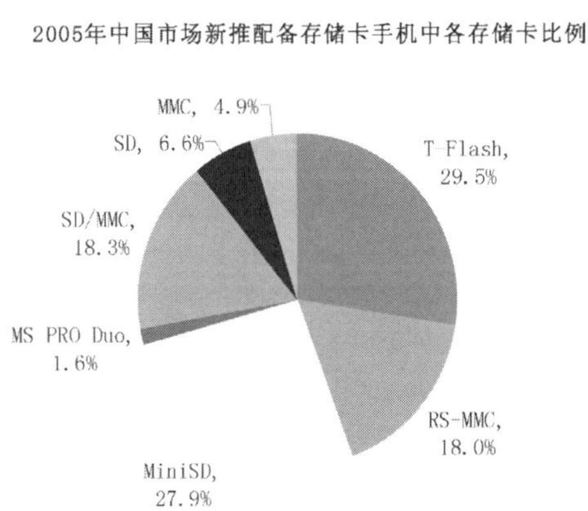

Figure 4.4. Percentages of Different Types of Storage Cards Equipped in Newly Promoted Mobiles in Mainland China in 2005

Source: PDAY Research.

storage cards are requested accordingly. It is foreseen that small types of storage cards are the future trend in the mobile storage card market, thus leading to the decline of standard SD and MMC cards.

(4) The 1-inch hard disk is more successful than the 0.85-inch hard disk. The 0.85-inch hard disk is higher in price and of a smaller capacity, while the flash memory card is continuously falling in price and increasing its capacity. Therefore, the 0.85-inch hard disk is under great pressure. The advantage of the 0.85-inch hard disk lies in its volume. As a result, its target is mobile phones. However, the demand for smaller volume in mobile phones has not appeared yet. Portable multimedia players like Hitachi Global Storage and Seagate also prefer 1-inch hard disks rather than 0.85-inch ones, as the 1-inch hard disk has a higher capacity/price ratio. On 8 June 2005, Seagate promoted the 1-inch 8GB hard disk which further distinguishes the capacity/price ratios between hard disk and flash memory. In addition, this hard disk can provide a faster rotating speed and stronger performance which are stronger advantages compared to those of flash memory. It is hard to increase the capacity of the 0.85-inch hard disk, however, and it will soon be overwhelmed by the popularity of flash memory.

(5) Micro drive mobile phones will become as popular as the iPod. The development of the micro drive in cell phones depends on whether the micro drive can successfully imitate the pattern of MP3s in iPod hard disks (see Table 4.7). Ching Hua Research Center thinks that the coexistence of micro drive and flash memory products for the

Table 4.7. Distribution of iPod Products by Apple

Product	iPod	iPod Mini	iPod Photo	iPod Shuffle
Storage media	Micro drive	Micro drive	Micro drive	Nand Flash
Capacity	20GB	4GB; 6GB	30GB; 60GB	512MB; 1GB
Number of songs stored	5,000	1,000; 2,000	7,500; 15,000	120; 240
Length of battery	12 hours	18 hours	15 hours	12 hours
Size (cm)	10.4*6.1*1.4	9.1*5.1*1.3	10.4*6.1*1.6	8.4*2.5*0.8

Source: PDay Research.

cell phone storage market is the same as that for the MP3 market. The application of micro drives in cell phones should be on a larger capacity. So far, Samsung and Nokia have both introduced built-in micro drive cell phones. The model numbers are V5400 and N91 respectively. Their capacities are 1.5GB and 4GB. The vertical recording technology applied to micro drives by Hitachi and Toshiba will make the capacity of 1-inch micro drive boom. Within one to three years, there will be 10GB and 20GB micro drives appearing in the market, which will, in turn, create further advantages for micro drive cell phones.

7. Analysis of Mainland China's IC Design Industry

There is a saying in the industry: "Choose IC, choose hope." However, the road leading to such hope is not that easy. Mainland China has become one of the major manufacturing countries of world electronic products. In 2001, the demand in the IC market reached 24.3 USD, but the IC products produced by domestic manufacturers in Mainland China accounted for less than 8% of the 24.3 USD. Such a gap reveals both huge business opportunities and reflects the enormous difference in IC design industries between Mainland China and the rest of the world. According to the data in connection with IC design companies in Mainland China gathered by ESM-Mainland China and *Electronic Engineering Times* from 2001 to 2004, there have been great increases in revenue, design levels, and industrial chains in the IC design industry in Mainland China.

7.1. Current situation of IC design industry in Mainland China

In four short years, the IC design industry in Mainland China has undergone many changes. The year 2004, in particular, marked the first year the domestic IC products became high-end. The following five factors contributed to the changes: a ground-breaking journey began in the core chip market of cell phones many years before; the companies first entering the IC design field moved towards high-end products such as the reformatting processor, RFID chips, and MPU products from smart cards and low-end consumer electronic products; multimedia IC companies began booming and becoming competitive; an increasing number of new companies that can compete with international companies are being set up in Mainland China in the new and rising IC market of fingerprint sensors and MEMS; and lastly, domestic IC companies have entered the home multimedia entertainment market such as LCD TV, PC TV, and Set-top Box by visual control IC.

The above mentioned changes show the overall upgrade of design capabilities. According to the statistics, 47.8% of companies adopt technology below 0.18 μm for design. 4.9% of companies adopt 0.13 μm technology for design; one example is Commit and Chung Hsing Micro Electronics Institute. On the other hand, the changes of these design companies are also present in versatile products. A survey suggests that 28.7% of companies own at least ten products at the moment. This shows that these companies have entered the stage of providing a series of products and solutions for customers, compared to a single product in the initial period.

The performance of the IC design industry in Mainland China shows an upgrade in analysis and design capability and versatile products. The trend of providing versatile products is closely related to the maturity of the industries in upper and lower streams. Li Wei, the Chief Technological Executive of Fu Dan Micro-Electronics, says, "2004 is the year that the IC industrial chain began to develop rapidly in Mainland China."

According to the statistics, the OEM industry led by SMIC in Mainland China entered a period of maturity in 2004. Most IC design companies show that the level of domestic OEM can satisfy their needs either in technology or IP bases. The domestic OEM not only provides a convenient manufacturing environment, but also lowers cost and shortens the industrial chain. Meanwhile, domestic OEM's strategy is to give a large amount of support to IC companies in Mainland China: "SMIC supports domestic IC design companies very much. Basically, SMIC does not limit its production. Their strategy is to fully support domestic IC companies and provide many IPs in Mainland China," says Chang Lung Hua, General Manager of

Chipsbank MicroElectronics. Therefore, more and more domestic IC companies are choosing OEM in Mainland China. A survey shows that over 60% of IC companies choose domestic OEM. Chang Huei, Vice CEO of Vimicro, indicates that there is obvious progress in the interactions between chip OEM and IC design companies.

However, they must pay attention to the design technology entering into 0.18 μm. In 2004, the domestic high-end technological line reached maturity. Relative IP service quality has also improved a lot. But further progress is needed for high-end wafer packaging and tests.

Currently, most tests and packaging factories are located in Beijing, while chip OEMs are located in eastern Mainland China. This is not a convenient location for many companies. Liu Wei Ping, General Manager of HED, suggests that the chips be sent to the chip factory in Shanghai for processing after design completion in Beijing. Then, the chips would be sent to BHTC in Beijing for tests and delivered to Shanghai for slicing. After that, the processed products would be sent to Beijing again for packaging. However, such sequences may cause problems in security, transportation and time.

The problems related to IP intellectual property right protection are still serious in Mainland China. Some IP service providers are concerned when their products are delivered to domestic design companies because it confines the development of high-end design to a certain degree. However, it is believed that when intellectual property rights are gradually taken into serious consideration and successful IC companies keep growing, the related problems will improve the following year.

The greatest change in the cooperation between the design industry and downstream customers lies in the former being able to provide customers with more and more comprehensive solutions. In addition, for strengthening the competitiveness of solutions, the design industry invests more on the R&D efforts of system engineers than chip engineers. Many companies suggest that the number of system engineers and test engineers should be twice that of chip engineers.

The above condition has become a key selling point for domestic IC manufacturers. For example, not only Anyka, Spreadtrum, Jade Tech and Actions Semiconductor, which are engaged in platform solutions, but also Mainland Chinaasic, Risong and MC Devices, which are engaged in peripheral appliances, spend more energy on the development of package solutions. They also provide free solutions or solutions at a very low cost to their customers. The idea is for the domestic design industry to make money on chips by providing solutions at a sacrificing cost. This is also an advantage when competing with international IC companies. The chip R&D of international companies is not based in Mainland China. Therefore, their chip designers are unable to show thorough understanding of the market while developing solutions. Such a problem does not exist in the IC companies in Mainland China.

Apart from enhancing their abilities in solution design, IC companies have more opportunities to cooperate with distributors and third-party design service providers as product lines expand. Most domestic IC companies like to adopt this strategy because of its benefits. Yang Kang Nung suggests that "we cooperate with different design service providers for different product lines so as to develop some competitive solutions for the market in Mainland China. This strategy establishes a foundation for our sustainable development." As distributors, Chipsbank, Actions Semiconductor and Symwave are the three leading companies in the field. Basically, they use distribution channels to sell their products. "Distributors can help us to explore the market so that we can focus on the development and technology and applied solutions. Different parties play different roles so as to achieve the best results," says Yeh Chun, General Manager of Symwave. The above companies effectively control the bulk of distribution channels including international distribution. They have also gained success internationally.

In short, dramatic changes have occurred to IC design companies in Mainland China in terms of design capabilities, product range and business ability. A great number of students studying abroad and experienced experts have returned to Mainland China and started businesses and facilitated the changes. Diligent, intelligent and hardworking domestic engineers in Mainland China worked together to achieve such results. For example, they created the world record of developing cell phone core chips within one year.

7.2. Operational characteristics of IC design industry in Mainland China

7.2.1. More precise market positioning and focus on consumer electronics

The positioning of IC products is determined by technological capabilities of IC companies, market demand, and the maturity of the industrial chain. The IC design industry in Mainland China no longer follows the trend blindly. They have become more rational in positioning IC products. According to a survey in 2004, IC companies engaged in consumer IC products account for 59% of the total number of IC companies in Mainland China. In 2003, IC companies engaged in telecommunications IC products accounted for the most at 42%. These figures show that the views of IC design companies in Mainland China are changing. They have sufficient reasons to support such change. Shih Lai, General Manager of Fudan Microelectronics, says, "Looking at the global IC design market, the PC is dominated by American suppliers (micro processors, etc.) and Taiwanese suppliers (chipsets, peripheral chips, etc.). High-end telecommunication products (telecommunication chips, digital signal processors, etc.) are dominated by Europe and America. The opportunities for IC design industry in Mainland China is at the phase where they should lie — in consumer electronics." International IC experts also raise a similar suggestion to the IC design companies in Mainland China. Yuan Chi Fang, Assistant General Manager of ISSI, says, "The market demand in Mainland China shows that consumer IC products is the first priority we should consider. Consumer electronics are not high-end, which do not need established technological standards in innovative products. The IC companies in Mainland China can start from this point. As for those products in need of establishing new standards and planning a brand new marketing project such as high-end telecommunication IC, it is suitable to do that abroad."

7.2.2. Overall enhancement of design

The design level of IC companies in Mainland China has dramatically enhanced in the process of accumulating experience. Looking at digital IC technology, a 2004 survey indicates that 31% of companies say 0.18 μm line-width technology is mainly applied in their products, while 22% of companies say they use 0.5 μm line-width technology in their products. An exciting trend emerges when comparing the above figures with those surveyed in 2003. The data for 2003 shows that only 20% of the companies used 0.25 μm line-width and below, while 46% of the companies used 0.5 μm line-width. Similarly, the scale in design has also largely increased in Mainland China. In the development of ASIC products in 2004, the companies having a gate count of one million accounted for 34% of the total number of IC companies, five million for 5%. In 2003, the companies having a gate count of over one million only accounted for 15% of the total number of IC companies. The following reasons explain the fast growth of technological levels of IC design companies in Mainland China. First, the mainstream technology of international semiconductor companies is inclined to 0.13 μm. In the meantime, many companies have already adopted 0.09 μm technology. Second, the IC companies from Taiwan have moved their design centers to Mainland China on a large scale, thereby improving the technology of domestic IC design companies. Third, many IC companies led by overseas Chinese experts, who bring the advanced technology from Silicon Valley back to Mainland China, have emerged in recent years. Fourth, the development of chip OEM industry grows very fast. The advanced OEMs like SMIC and Grace Semiconductor have upgraded themselves to a commercialized stage. Their technology in 0.25 μm and 0.18 μm activates and enhances the design capability of domestic IC design companies in Mainland China.

7.2.3. Incomplete industrial chain

As the technology improves in Mainland China, the insufficiency of the IC industrial chain becomes more obvious. A 2004 survey indicates that 56% of companies say they need to search for offshore OEM. Yet, the situation has improved compared with that shown in the previous year. In 2003, 63% of IC companies chose offshore OEM. Generally speaking, there are still many companies searching for offshore OEM. This is because the domestic OEMs are either insufficient in technology, or unstable in production with a low yield rate despite having sufficient technology. For example, some earlier OEMs like HJMicro, CSMC, Shanghai Belling and ASMC own mature production lines based on the 0.5 μm line-width technology. Other OEMs such as the recently established Grace Semiconductor and SMIC have advanced technology but lack IP base and mix-signal technology due to their position in the growth stage. Therefore, those advanced OEMs do not form a complete design and service-corresponding chain. Such conditions lead to IC companies seeking offshore OEMs, while domestic IC OEMs still have a long way to go before achieving full operations. Hsiu Chun Liang, General Manager of Acer Labs, China Region, says that the production lines will not be moved to Mainland China even though the IC design centers are moving there. He also says, "We prefer working with chip factories to raise the yield rate, for example, from 85% to 98%, to lower the cost. It takes time for the newly established chip factories in Mainland China to improve the yield rate to a certain level. And we won't easily move our production lines to Mainland China during this period." The domestic IC companies in Mainland China are expecting to work with OEMs with abundant IP bases because they can get IP at a more preferential price. As for back-end packaging and tests, 57% of IC companies indicate that they have already conducted their packaging in Mainland China.

The packaging industry is developing very quickly in Mainland China. According to the statistics provided by iSuppli, Mainland China accounts for 10% of the total sales of global packaging and test markets. Even so, the packaging and test industry still needs to make some adjustments. ISSI Assistant General Manager Yuan Chi Fang suggests that "the product range provided by the domestic packaging and test factories is not complete yet in Mainland China. Some of the products are rather high-end and some are too low-end. The processing service for products with regular specifications and with the greatest demand like TSOP/SOP of 50 and 54 tubes are seldom provided by domestic factories. Compared with Taiwan, the IC industrial chain in Mainland China needs to be improved." Nevertheless, early signs are encouraging. We have seen more and more local IC design companies choosing to produce, encapsulate, and test their products in Mainland China because they have such demand. There are already six OEMs available with 0.25 μm line-width technology in the Shanghai area. Once they mature, more and more companies will approach them for production. In addition, the offshore OEMs set a higher ordering quantity of chip production that is difficult for the newly established IC companies to accept. Therefore, they will turn their concentration more towards domestic IC OEMs.

8. Analysis of Mainland China's IT Industry

In recent years, the IT industry has advanced rapidly, powerfully pushing the development of social productivity. Firstly, the high-speed development of micro electronic technology causes the calculating ability and the performance/price ratio of a chip to increase by geometric progression. Such an increase creates an environment for large-scale and multiple digitalized information processing, transmission, and communications. Secondly, the highly developed software technology gives chips and computer hardware intelligence. Thus, the functions and applications of computer science are enlarged. Thirdly, being promoted by micro electronics, software and laser technologies, changes in telecommunications are accelerated from analog to digital, from low speed to high speed, and from single language media to multimedia. Fourthly, the interactions and combinations of computer, communications and media technologies are pushing the development of communications and information technology into a new phase.

The new generation of public information networks featured with IP technology will develop rapidly and become the uniform platform for the next generation of information networks. New Internet applications arising from e-commerce, distance learning, distant medical treatment, e-government, mobile office and home-based platforms will continuously emerge and develop. Information technology is widely rooted in social productions and in different fields of our daily lives. This will facilitate the progression of the IT industry at an even faster rate in the waves of digitalization revolution.

8.1. Current situation of the IT industry in Mainland China

Beginning from the 1990s, the digital technology that centered on computers and software began showing drastic development. The world became overwhelmed by waves of information technology. Now, computer science has entered a computing age — otherwise known as the post-PC era — with the Internet as a widespread entity. As the Internet rapidly develops and becomes used more widely, the traditionally technical patterns of communications have become altered. More and more people have accepted the developing trend of the comprehensive IP net. Mobile computing and common computing help to combine digital and wireless communications and make them an important direction for the development of the IT industry. On the one hand, micro engineering keeps moving towards high performance, multimedia, lower price and the Internet; on the other hand, portable computers, various mobile multimedia information terminal products and a variety of personal palm information terminal products have already combined mobile computing and digital communications. The products that combine computer science and communications have been widely promoted and have great potential in the market. The functions of the Internet will be further strengthened and the information service providers which support the Internet will also be developed with the establishment of broadband, Internet and wireless communications. A new market space is thus explored especially for a variety of application solutions in the domestic software industry. This combination will directly influence the development of 3G mobile services and the progress of comprehensive IP Internet.

Currently, the mobile network and regional wireless network have become more important. The products, facilities, and versatile information terminals connected with high-end Internet servers and routers will become an important route of development for the computer science industry in combining with other industries. The combination of Internet and digitalization is also applied to consumer fields which continuously bring out various 3C digital products integrating computer, communications, and consumer electronics. In Mainland China, there is a huge potential market. This represents an important potential development for the electronic information industry and electronic products. The market in Mainland China also makes the IT industry an important investment in the Tenth Five-Year Plan. Furthermore, new types of input and output facilities such as digital cameras and webcams make the computer peripherals more intelligent and user-friendly. Special products and application systems for computers are being developed in order to handle requirements from different industries. Technologies applied to different fields are combined together, creating new applications. Business opportunities and energy grow endlessly. In domestic and international environments like this, the IT industry maintains its fast growth and brings global attention to China's achievements.

At this present stage, the IT industry in Mainland China includes mainly the production of electronic information products, the development of software, and the application of IT services. After reform, opening up and rapid development, the IT industry in Mainland China has formed a better industrial production system. The gross output value of the national electronic industry is ranked fourth in the world. Major electronic products have been put into mass production. The production scope includes answering machines, telephones, color TV sets, color tubes, acoustic facilities, VCD players and some basic elements, and is ranked first in the world.

According to the report by the Statistics Bureau in Mainland China, the revenue of finished products in the IT industry totaled 2,655 billion USD in 2004, which is an increase of 41.7% compared with the same period in the previous year. The sales revenue of finished products from the manufacturing industry accounted for 2,412.6 billion USD, at an increase of 40.3%. The sales revenue of the software industry totaled 242.4 billion USD, at an increase of 48.4%, with a realized profit of 112 billion USD, an increase of 49.3%. The tax paid was 38 billion USD, an increase of 31.9%. The industrial finished import-export gross value totaled 388.4 billion USD, at an increase of 41.6%, in which export gross value accounted for 207.5 billion USD, at an increase of 46.0%. The electronic IT industry remains the main driving force for the development of the second industry in Mainland China. In 2004, the finished industrial increment of the IT industry totaled 565 billion USD, at an increase of 41%. The growth rate of the IT industry is still 24.3% higher than that of the national industry, and the IT industry has a contribution rate of 20.9% to the growth of national industry. In other words, the IT industry is responsible for 3.5% of the 16.7% growth rate of national industry.

In terms of economic classifications, the scale of overall investment in the IT industry made by suppliers from Hong Kong, Macau, Taiwan and other countries is higher than that of the average industrial level. The accumulated finished investment made by tri-capital enterprises in the IT industry in Mainland China totaled 82.31 billion dollars in 2004. The projects that began in 2004 totaled 577, and the number of projects specializing in production was 326. The annual finished industrial increment totaled 386.58 billion USD, at an increase of 58.7%, which was higher than 12.2% of the overall IT industry and accounted for 74.4% of the total increment of the IT industry.

In terms of the industrial structure, the pre-tax profit made by the communication facilities manufacturing industry reached 38.32 billion USD, at an increase of 60.3%, in which the finished profit accounted for 31.14 billion USD. The pre-tax profit made by electronic element manufacturing industry totaled 22.59 billion USD, at an increase of 37.6%. There was an increase of 40% made in the radar manufacturing industry. The gross profit made by the electronic elements manufacturing industry totaled 19.19 billion USD, at an increase of 64.6%. The growth rates of these four industries are higher than the average rates shown nationally.

The competitiveness of the software industry in the market is clearly increasing after passing through a gradual adjustment of industrial structure. In 2004, the scale of software industry had obviously expanded as the policies relating to the software industry were promoted and underwent constant improvements in the market environment in Mainland China. Software and information technology were widely used in national economies and different social fields, which further facilitated the adjustment of the industrial structure and pushed the progress of establishing information infrastructure in Mainland China. From January to December 2004, the final sales value of software in the electronic information industry totaled 242.4 billion USD, an increase of 48.4%, in which the sales revenue of software products accounted for 149 billion USD, or 61.5% of total sales revenue of the software industry. The sales revenue of system integration totaled 93.4 billion USD, at an increase of 75.73%.

In terms of exports, electronic information products remained 10.6% higher than the average export growth rate in Mainland China. Starting from the fourth quarter of 2004, the export of electronic information products grew rapidly. The accumulated import-export value from January to December reached 388.43 billion USD, increasing by 41.6%, in which the finished export value accounted for 207.5 billion USD at an increase of 46.0%. This was 10.6% higher than the 35.4% average growth rate of national export value. The contribution rate of electronic information products to national export growth was 42.2%, or 14.9% of the 35.4% national export growth rate. In short, the IT industry, as the core industry in modern technologies, will become the driving force behind long-term economic growth and social development in Mainland China.

8.2. Prospects of the IT industry in Mainland China

The market in Mainland China is incredibly huge, and is currently undergoing a transformation and adjustment of its industrial structure. It is inevitable that China must expand investment, change operational patterns, approve new technology, reduce national macro-management cost, enhance governmental efficiency, and reduce the percentage of administrative expenses to GDP, etc. Enterprises now generally pay attention to enterprise efficiency, operational risk control, and cost control. Economic globalization brings the enterprises in Mainland China certain opportunities and challenges, and further provides the IT industry in Mainland China with the chance to serve enterprises and the society.

According to the latest data estimated by IDC, Asia-Pacific Region's growth rate in 2004 increased by 50% compared with the previous year, and this growth comes mainly from Mainland China. Mainland China is the main base of expansion for the IT market in the Asia-Pacific Region, and therefore plays an important role in the IT market, both in the Asia-Pacific Region and worldwide.

The whole Mainland China market will be developed based on the following six major directions of digital TV, industrial storage management software, products with security, application of IP technology, new generation Internet, and the digital home.

8.2.1. Digital TV will bring industrial structural changes

In 2005, the digital TV market driven by the Central TV station came into effect in Mainland China. The current analog network will gradually be replaced by the digital network by 2015. As for IT hardware, software, and service providers, a great number of opportunities will arise in such a process. The rapidly growing applications of digital media are one of the most important structural changes in the developing IT industry.

8.2.2. Industrial storage management software market will open up

The application of disk storage is expanding quickly, triggering a highly growing demand for storage management solutions. In 2004, the overall disk storage capacity in the market increased by 54% compared with that in 2003 in the whole of the Asia-Pacific Region except Japan. In time, customers will make more investments in storage management software and related services. IDC anticipated that there would be an increase of 26% in revenue derived from the storage management software market in the Asia-Pacific Region by 2005. In Mainland China, the financial, communication and manufacturing industries will become big buyers of the storage management system. The demand by the above industries will facilitate the growth of the storage management software market.

8.2.3. Products with security — low price and high efficiency

Looking at the global trend, the integrated solutions with low cost and high efficiency will become popular among SMEs. However, this is a huge challenge for the service providers. Domestic key suppliers and transnational enterprises are positively performing their market strategies and putting their attractive solutions into practice.

The market for products with security systems in the IT industry will increase by 25% in 2005 compared with 2004. Security application software and tools, security solutions, security operations, and management platforms will be of enormous demand. The biggest market lies in the governmental units, financial industry, and telecommunications industry. This is a market which will definitely become larger than before.

8.2.4. Full development of the application of IP technology

A greater demand for the IP telephone will appear in Mainland China. The number of IP centers will increase. IP-VPN service demand will also emerge in 2005. Viewed globally, IP communication is expected to find its final profitable niche in enterprises. Single reserved exchange, long distance calling storage, convenient remote access, cost saving, etc., are the biggest attractions for enterprises. IDC estimates that traditional communication will begin to promote solutions based on IP technology. Major market opportunities still lie in small and medium scale markets. There is one more thing worth paying attention to — VoIP application will become popular in 2005. VoIP has already been around for four or five years. The VoIP products are now in their fourth or fifth generation. In the past, VoIP technology was considered a rim technology. In the consumer field, it was regarded as something that would merely make technophobes curious. However, the phenomenon caused dramatic changes in 2005. IDC suggests that VoIP has obtained recognition from many CIOs. It is an important trend for enterprises to apply VoIP. As for the consumer market, traditional suppliers will accelerate the promotion of mainstream VoIP products based on the foundation established in 2004.

8.2.5. The rise of wireless applications and the new generation internet

Wireless application in business, trade, and consumer markets are very popular. There is great potential for wireless LAN, especially in the business field. It is an inevitable trend to issue more 3G licenses. This issuance will have huge impacts on the competitive environment. Meanwhile, the use of broadband will increase rapidly. Great potential is shown in connection services and value-added services. For example, services like IPTV, online games and streaming media will become standard facilities. Suppliers not only provide facilities to end-users, but also provide net maintenance, net planning or net optimization, and consulting services. In the following years, the new generation Internet will find its market position.

8.2.6. Digital homes will experience partial market breakthrough

In 2004, more digital products quickly entered consumers' families and home offices. HP, Microsoft, Dell and Gateway continue to expand their product ranges, which are taking over the territory of traditional consumer electronic leading suppliers like Sony and Samsung. In 2005, digital homes showed a partial and rapid growth in the market. However, the profits have not grown as fast as their expansion. The products related to digital homes such as digital cameras, digital video cameras, MP3 players and LCD TVs will increase by 30%. The profit rate will obviously decrease due to competition.

From the perspective of the overall market in Mainland China, the growth of software and services will rise faster than that of hardware. Software and service markets will increase by over 20% in the following years. Consumers will make investments in comprehensive solutions instead of single products. The price competition among hardware suppliers will intensify. A great number of high-end international suppliers will further take up the market share of middle and low-end domestic suppliers. The market of SMEs will become the main battleground for IT suppliers. In recent years, the strict economic situation led by economic globalization has been pushing

SMEs towards optimized operations in order to become more internationally competitive. Therefore, global IT enterprises will concentrate on unique solutions demanded by these SMEs.

In the following years, more suppliers will compete to promote solutions based on the best technological support and services with a smaller budget. Moreover, suppliers will provide a series of preferential services and support so as to cover the network of the whole area. Based on such a strategy, the suppliers expect to create a variety of integration, combination, and adjustment for different business structures. The global IT market will keep on growing amidst this positive and vibrant situation, but should move cautiously in the process of various integrations, combinations, and adjustments. The IT market in Mainland China will obtain great opportunities and challenges in such a situation.

9. Analysis of Mainland China's Electronic Components Distributors

According to a survey conducted on buyers by ESM-Mainland China, 52% of electronic components were purchased through distributors in Mainland China in 2003. The average purchasing amount of electronic components reached 15.67 million USD. In 2004, the above mentioned rate rose 58.2%, and the average purchasing amount reached over 31.33 billion USD. These figures indicate that distributors play a role in the supply chain of the electronic industry in Mainland China that is as important as that of the components manufacturers. For components buyers, many enterprises take advantage of distributors to help stock their supplies, so as to solve problems otherwise encountered while receiving small lot orders or obtaining financial support, thus increasing flexibility. Looking at marketing strategies of components manufacturers, distributors' close relationships with customers drive manufacturers to cooperate closely with distributors. In addition, Mainland China is a huge country and its industries are widely spread across the land, so the customs procedures are complicated. All the above factors induce manufacturers to choose a distribution strategy in order to take advantage of the flexibility of distributors, which in turn will help them satisfy the unique requirements of domestic customers in Mainland China.

According to the survey by ESM-Mainland China, distributors from different areas specializing in different fields adopt different methods to serve their customers based on buyers' needs and conditions. More and more distributors have started targeting small and medium electronic manufacturers instead of large-scale manufacturers. In the face of the challenges of "no competitive orders in distribution market" and customer purchasing behavior, distributors will provide more value-added services for their customers and establish partnerships with suppliers to raise their competitiveness. Logistics, technical support, credit capability, and IT resources provision are four different business models. Distributors have to decide whether they should choose a single business model or combine the business models in the future.

9.1. Foreign and domestic distributors in the Mainland Chinese market

In terms of quantity, distributors are the main channel for sales of electronic components in Mainland China. Among the distributors surveyed, over 60% of distributors are domestic ones. The distributors from Taiwan, Hong Kong, and other foreign countries account for 9%, 15%, and 10% respectively. Looking at the sales derived from Mainland China/Hong Kong areas, the market is dominated by both foreign and domestic distributors. The market share owned by domestic distributors has decreased to 42%.

Because Mainland China is so large, customers are not concentrated in any one location. Distributors have to set up their service centers in more cities in order to expand their market coverage. Apart from Shanghai, Shenzhen and Beijing, cities located in the mid-west area like Chengdu, Wuhan and Xian are popular cities for distributors expanding their services.

Over half of the distributors choose to set up their offices in Shanghai and Shenzhen so that they can set up trade companies or logistic centers in the tax-protected zones of these two cities.

Distributors from Hong Kong, Taiwan, North America and Europe are aware that in the process of making arrangements in Mainland China, they have to achieve localized logistic services and accept RMB services if they plan to explore the local market and provide services to local customers. Setting up offices in or near tax-protected zones makes business easier for them.

Chen Ta Wei, Asia-Pacific Senior Vice-CEO and CEO of China Region of Arrow Electronics says,

> "More and more North American and European OEMs, contract manufacturers, and other transnational companies are moving their production facilities to Mainland China. We have seen that the electronic manufacturers from Taiwan and Japan that enter the market in Mainland China earlier are gaining profits. We will establish two extra marketing offices in Mainland China. In addition, we are going to set up an inside-sales team to provide services for small and medium-scale customers in Mainland China. Furthermore, we will expand and improve the offices in Shanghai and Beijing and increase our investment to improve logistic services and warehouse facilities in Shanghai. Preparations must be made in advance for the potentially increasing services in Mainland China."

The local distributors in Mainland China started very late in marketing arrangements compared with those performed by foreign distributors. Compared with Arrow Electronics, Anfuli Electronic, or WPI Group, which have offices in over ten cities, the average number of offices set up by domestic distributors is less than two. Most distributors are small ones active in a particular city. However, many domestic distributors expect to achieve more. Driven by the growth of market demand and increasing competition, the domestic distributors have stepped out for expansion.

"The distribution industry is similar to 'enclosure'. Whoever obtains more customers can have more opportunities to develop," says Hsiao Ching, general manager of SECOM Tel Shenzhen, who thinks that some component manufacturers may take advantage of foreign distributors in the early period to explore the market in Mainland China. As they get to know the market better, they will seek domestic distributors that are good at forming contacts with small and medium OEM customers in Mainland China.

9.2. Many competitors share the 3C market — different distributors specialize in different aspects

A large number of distributors are attracted to communication systems, consumer electronics and computers (3C). Few distributors are involved in fields such as industrial application, medical care, automobile, and aero electronics. Among the distributors, over 50% make communication systems or consumer electronics their core business item. Over 30% of distributors are attracted to computer and peripherals due to the higher standardized components, higher degree of modules, and a more transparent supply chain (see Figure 4.5).

Distributors from Mainland China, Taiwan, Hong Kong and other areas have different developing foundations; so their business models are different. Therefore, they choose distribution markets differently.

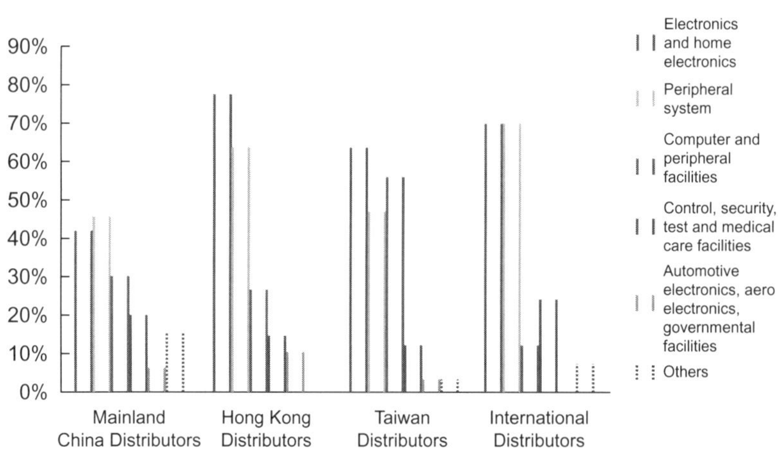

Figure 4.5. Percentage of Distributors in Different Areas Engaged in Major Product Fields

Source: ESM-Mainland China.

Domestic distributors are developing their supply channel as local electronics manufacturers. The core business depends on their primitive customers. Therefore, they lack independence while choosing markets. Compared with foreign distributors, most domestic distributors are not discriminate; they do not care to whom they sell or what they sell as long as they can make money. As the focus in the market changes and transforms, some distributors will become leading supply channels with their customers in a certain market while some may disappear because there are more distributors facing the pains of transformation.

The distributors from Hong Kong have been developing their companies in Mainland China's market for years. They started by going into businesses related to consumer electronics and home electronics. Moving with the trend, they then invested new resources into communication systems which show the fastest growth in the market. According to a survey, nearly 80% of Hong Kong distributors are involved in consumer electronics and home electronics markets. Communication systems and applications also attract 65% of Hong Kong distributors.

The overall advantages of the Taiwanese computer industrial chain led to a clear preference among Taiwanese distributors for the computers and peripherals market. However, they have not given up their investment in consumer electronics and communication fields. "We try to maintain current businesses related to PC products and peripherals. The business here is an extension of the services provided to Taiwanese manufacturers," says Chang Jung Kang, Vice Chairman of WPI Group and General Manager in the China Region. He suggests that "we will take the communication systems, which is just taking off, as a priority to develop. As for the consumer electronics market, we will develop the applications related to digital TV." International distributors from North America and Europe also place their focus on communication systems and consumer electronics. As a matter of fact, 70% of international distributors indicate their main business products lie in these two fields. As for computers, the applied components are mostly standardized. Therefore, manufacturers are inclined to make direct sales. Only 13% of international distributors regard this field as their main service.

9.3. Provision of technical support is the long-term development direction for distributors

There are two ways to increase distributors' income: expand their agent lines or obtain more income from a certain line. The former requires that distributors have strong capital support and market coverage and good relationships with original manufacturers. The latter involves providing technical support to their customers.

For most of the electronic components distributors in Mainland China, it is difficult for them to obtain the chance to act as an agent for the product lines with rich content due to a lack of sufficient capital and customers. In a 2003 survey, 48% of distributors claimed to be authorized distributors (see Table 4.8). Only 16% were independent distributors without representing any authorized brands. The final 36% of distributors were of a mixed type which operates authorized and non-authorized brands. Observing the number of authorized brands, about 50% of distributors had less than five authorized brands. Narrower product lines and excessive competition lead many distributors to sacrifice the bottom line, as some distributors have little choice but to accept a gross profit lower than 3%. Consequently, more and more distributors are trying to explore a space where they can survive by providing customers with more technical services and other value-added services. The smaller distributors who cannot provide more services will quickly disappear. According to the

Table 4.8. Percentage of Distributors and Numbers of Authorized Brands

Percentage of Distributors	Authorized Brands	
	Number	Percentage[a]
	1	3%
Mixed distributors 36%	2–5	48%
Authorized distributors 48%	5–10	17%
Independent distributors 16%	10–20	8%
	Above 20	8%

[a]16% unauthorized brands.
Source: ESM-Mainland China.

statistics derived from the market survey by ESM-Mainland China, the number of electronic components distributors in China decreased from 1,500 in 2002 to 1,350 in 2003. 71% of distributors claim that they can provide their customers with a certain type of technical support.

"The competition in the distribution market is endless. In order to face the challenges, we need to create more value for our suppliers and customers. We have to maintain them and keep on creating such values," says Cheng Min Chiang, General Manager of Hong Kong/Mainland China Region, Memec United. "Our aim is to provide a whole set of solutions for our customers, help them accelerate listing the products in the market, and aid original manufacturers in exploring market demand."

E-Smart CEO Yen Bin also suggests the same idea: "Large-scale distributors like Arrow Electronics and Anfuli take advantage of their scale. They provide complete services and multiple products. However, the problem many customers constantly face is the short life span of electronic products. Distributors need continued devotion to research and have to develop new products so as to gain and maintain market share." Yen Bin adds, "Good distributors must have a concentrated field. Furthermore, they have to be concentrated and specialized. We do not want to be like Arrow and act as an agent for a complete range of products. We plan to put our sources in a limited field in order to provide specialized solutions including software, PCB wiring, and final turnkey."

Generally speaking, more provision of technical support not only helps distributors and original manufacturers fulfill customer needs, but also increases overall revenue. For the electronics manufacturers purchasing components from distributors, they not only lower their development cost and accelerate the products entering the market, but also simplify their purchase flow. With these advantages, it is believed that there will be more distributors providing technical support and positioning themselves as professional and technical service providers, much like applied wireless technical distributors and portable digital technical distributors. Buyers may benefit from their technical support and "one-stop-shop" purchases (see Figure 4.6).

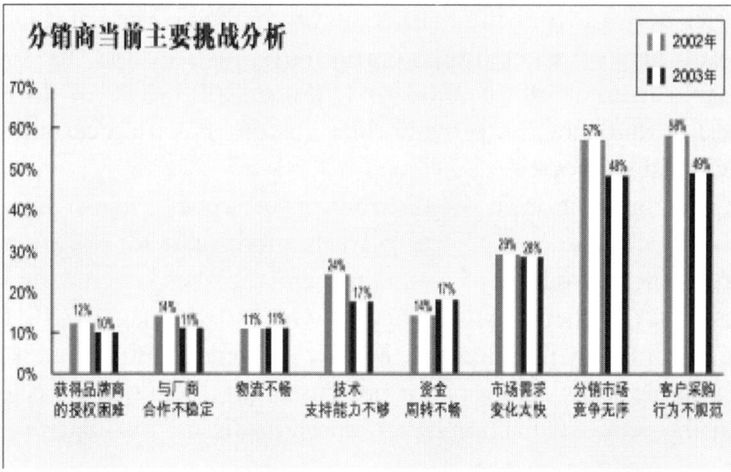

Figure 4.6. Current Challenges Faced by Distributors

X (from left to right): Hard to obtain authority from brand suppliers, Unstable cooperation with suppliers, Poor logistics, Insufficient technical support, Poor capital turnover, Fast shifting of market needs, Disordered competition of distribution market, Customer purchase behavior without norm; Y: Percentage

Source: ESM-Mainland China.

9.4. Lack of trust is the major challenge faced by distributors at the moment

Although there is evidence indicating that the distribution market of electronic components in Mainland China is moving from a disordered to an ordered condition, factors like "disordered competition of distribution market" and "customer purchase behavior without norm" are still the two major challenges faced by distributors (see Figure 4.6). Foreign distributors have the most complaints.

The disordered competition and price competition among distributors are one of the main factors hindering the development of the industry. Too many enterprises want to be the leader in Mainland China and too many small companies create interference in the market.

Compared with the disordered competition among distributors, the harm caused by customers' purchase behavior without norm is even more serious. According to Chien Shih Tsai, General Manager of SAC (Shenzhen), the problems arising from credit are the most complicated:

> "Currently, the most credit we grant is to Taiwanese manufacturers. The average credit term can last for 120 days. And Taiwanese manufacturers are honest and will pay the money back by the expiry date. We dare not grant credit to manufacturers in Mainland China even for a period of 30 to 60 days because Chinese manufacturers lack credit. Some manufacturers indeed have money to pay back but they want to revolve distributors' money for doing their own business."

A Taiwanese distributor unhappily says, "The development of the electronics industry in Mainland China is excellent, but it is a buyer's market. Therefore buyers do not consider the values and importance of distributors, nor do they treat distributors as partners. It is believed that cooperation will be established on a better foundation after going through one or two market cycles. Actually, we got to realize the meaning of cooperation after several development cycles earlier in Taiwan."

Not only do foreign distributors criticize the buyers in Mainland China; even domestic distributors feel great pressure too. "The position of distributors in Mainland China's market is sort of distorted. Suppliers request distributors to adopt standardized operation methods, while customers ask distributors to provide flexible services," says Chou Chi Kuo of CEASC. "Even though China has joined the WTO, it will take a long time for China to connect with the international society in the overall environment, financial system and trade regulations."

The above opinions reflect the immaturity of buyers in Mainland China. In fact, if they knew how to use and cherish the value of distributors, the distribution partners might eventually become strong backups. Most distributors are willing to explore the market with customers and share the risks. However, the premise is that both parties must have an equal foundation for cooperation. The distributors do not want to be a "free warehouse and cheap bank" for buyers.

9.5. The construction of the IT system is emphasized

Logistic services, technical support, creditability and provision of IT resources are four different business models. Distributors have to decide whether to choose one model or make a combination out of the four. If logistic services and technical support are the two legs supporting distributors to develop their business, then creditability and IT system applications are the hands pulling up customers at the upper and lower ends of the supply chain. The problems arising from creditability are honesty and the bottleneck caused by the financial system. Therefore, distributors are starting to place the construction of the IT system at a strategic position.

Drastic competition leads to a continuous decreased profit which brings the distribution industry into a micro-profit age. Distributors have to do something to improve and seek more space for their development. Chang Jung Kang of WPG says, "First of all, we have to make investments in IT. We have to put the high efficiency management including information, inventory, orders, and finance into practice." WPG has invested in IT system construction for over ten years and is one of the distributors that introduced Roset taNet standards at an early stage. Chang Jung Kang thinks that a good distributor should be able to provide basic services such as spare products, logistics, and technical support. In addition, a good distributor is supposed to help manufacturers establish the IT system. A distributor should be able to precisely reflect the supply and demand conditions at upper and lower streams through its and customers' information systems. In terms of industrial development, simply following the direction of IT, logistics, or technical support does not seem right. A balanced development combining these three elements is required in order to meet practical needs.

9.6. The bottleneck of finding suitable talent

Positively developing the marketing network, establishing partnerships with suppliers, and providing more value-added services for customers are the three major strategies taken by distributors in order to survive in the new competitive environment. However, the problem of finding suitable talent has become more serious in the process of adopting the above expansion strategies.

Chen Ta Wei of Arrow Electronics says, "Arrow will continue to make investments in Mainland China. We plan to increase the number of marketing engineers and product engineers. Meanwhile, we are going to hire more site application engineers to serve our customers in Mainland China. We have already chosen Shanghai as the headquarters in Mainland China. We will also transfer the management personnel from the Hong Kong headquarters of the Asia-Pacific Region to Shanghai to aid our management team. The biggest challenge now is how to hire more talent to fulfill the needs generated from such rapid expansion, especially in Beijing and Shanghai." His opinion is representative of the majority of distributors. Many distributors strengthen the market coverage rate and personnel while competing with other distributors. Now, finding talent has become the biggest challenge for them.

Huang Yeh Ching of Longteng Electronics has made comparisons between the talent in Taiwan and Mainland China. He says, "There is a lot of talent in the electronic industry in Mainland China. However, they lack experience. The graduates of the electronics departments in Taiwan usually stay in the same industry for a long time. They join manufacturers such as master boards, communication products, sound cards, etc. They accumulate related experience and learn how to design. The technicians graduating from schools in Mainland China often engage themselves in businesses not related to electronics. Therefore, they do not have electronics related experience." Huang Yeh Ching also shows his concern about the stability of talent. He says, "The development of the market in Mainland China is brilliant, resulting in many good opportunities. Many people can easily look for new jobs without keeping their previous jobs."

Generally speaking, the electronic information industry started later in Mainland China, and the foundation of the industry is still rather weak. Nonetheless, it has huge potential and has developed very quickly thus far. There is still tremendous room to develop and expand.

Chapter 5

Analysis of the Information Technology and Electronics Industry Market in Taiwan

1. Current Status of Information Technology and Electronics Industries in Taiwan

According to a survey conducted in July 2004, among the world's top 100 companies in the global information technology industry, there are fifteen enterprises from Taiwan which are ranked right after the top 46 enterprises from the US. The 2003 domestic and foreign output value of hardware produced by the IT industry in Taiwan totaled 57.1 billion USD at a growth rate of 17.9%. It was estimated that the 2004 domestic and foreign output value of hardware produced by the IT industry in Taiwan would total 67.6 billion USD at a growth rate of 19.9%.

According to the forecast by Nomura Research Institute, the 2005 total output value of the IT and electronics industries would reach 2,756.7 billion New Taiwan Dollars (see Table 5.1).

The information technology industry in Taiwan is clustered in science parks, and has reached economic scale. Table 5.2 shows the distribution of industries in Hsin Chu Science Park.

2. Development Background of Industries

2.1. Taiwan's semiconductor manufacturing industry

IC products are applied to various electronic products such as PC, servers, computer peripherals, Internet applications, wire and wireless communication products, automotive and industrial facilities, and a

Table 5.1. 2005 Output Value Forecast of IT and Electronics Industries

	Output Value (100 million NTD)	Structural Ratio (%)
Communications	2,923	10.60
Information Machinery	12,894	46.77
Semiconductor	11,750	42.63
Total	27,567	100.00

Source: Nomura Research Institute.

wide range of consumer electronics including DVD players, game boys and digital cameras. As innovative terminal electronic products develop, customers will widely use the advanced manufacturing process and production offered by professional chip OEMs.

The revenue of semiconductor manufacturers was very good in 1999 and 2000 because of the strong market demand in communications and information products. Over-capacity and decreased average selling price of global semiconductors occurred in 2001 and 2002 due to a great amount of capital expenditures and deterred results from production expansion. According to the IEK research carried out by ITRI in February 2004, the global semiconductor market declined by 32% in 2001, maintained its level in 2002 and grew by 18.1% in 2003 (see Table 5.3). The semiconductor industry in Taiwan recovered in 2003. The output value of chip OEM industry in 2003 grew by 25.3% compared with that in 2002. The general growth rate of product output value reached 25.7% in 2003. According to the IEK research in 2004, both price and demand of the semiconductor industry was steadily increasing.

The functions of electronic products have become increasingly innovative. The technology applied to semiconductors is also progressing, with the size of chips enlarged accordingly. According to a survey conducted by Dataquest, OEMs first produced 12-inch chips in 2002, which accounted for 3% of chip OEM production. As the economy boomed, the demand by Fabless for chips increased. Tight production has gradually emerged, which facilitates manufacturers accelerating their expansion of factories for producing 12-inch chips. In 2003, the OEM production of 12-inch chips rose to 6%. It was

Table 5.2. Distribution of Industries Located in Hsin Chu Science Park in June 2004

Industry	No. of Approved Companies/%	No. of Companies Stationed in HCSP/%	Registered Capital Amount (million NTD/%)	Paid-in Capital Amount (million NTD/%)	No. of Employees/%
IC	175/39.3	156/41.3	998,337/74.0	751,486/74.7	64,431/60.1
Computer and peripherals	60/13.5	57/15.1	100,450/7.4	65,146/6.5	12,676/11.8
Communications	62/13.9	56/14.8	44,886/3.3	32,143/3.2	6,660/6.2
Electro-optical	74/16.6	60/15.9	190,114/14.1	146,015/14.5	21,344/19.9
Precision Machinery	34/7.6	21/5.6	8,825/0.7	5,890/0.6	1,065/1.0
Biotechnology	40/9.0	28/7.4	6,936/0.5	4,836/0.5	998/0.9
Total	445/100	378/100	1,349,547/100	1,005,516/100	107,174/100
Others	2	1	1,990	1,594	69
Grand total	447	379	1,351,537	1,007,110	107,243

Source: Website of Hsin Chu Science Park (http://www.sipa.gov.tw).

Table 5.3. Performance of the Semiconductor Industry in Taiwan (100 million NTD)

	2000	2001	2002	2003	2003/2002
Industrial output value					
IC design	7,144	5,269	6,529	8,188	25.4%
IC manufacturing	1,152	1,220	1,478	1,902	28.7%
OEM	2,966	2,048	2,467	3,090	25.3%
IC packaging	978	771	948	1,176	24.1%
Domestic IC packaging	838	660	788	976	23.9%
IC test	328	253	318	409	28.6%
Product output value	2,872	2,179	2,796	3,514	
Growth rate of product output value	44.5%	−23.5%	27.3%	25.7%	
Growth rate of global semiconductor	36.8%	−32.0%	1.3%	18.1%	

Source: IEK-ITRI (2004/02).

estimated that the production would reach 13% in 2004. A percentage of 20% would further be reached in 2005. In contrast, the production of 6-inch chips decreased from 30% in 2001 to 15% in 2005. The production of 8-inch chips also dropped slightly. Therefore, 12-inch chips will become the mainstream product.

2.2. Taiwan's IC packaging and test industries

The IC packaging in Taiwan is generally divided into foreign suppliers and Taiwanese suppliers. Dating back to the 1970s when Taiwan had low-cost manpower to attract foreign investment, Microchip was the first company to set up a transistor packaging factory, followed by Philips, TI and Motorola. These companies introduced IC packaging technology and tests which established the foundation for the development of the semiconductor industry in Taiwan. After the 1980s, the Taiwanese government intended to further develop the upstream and downstream industries. ASE and SPIL were then established. In recent years, a few upstream semiconductor companies have recapitalized downstream packaging factories to vertically integrate upstream and downstream businesses, as the IC industry developed prosperously. For example, Nanya Technology recapitalized FATC, and Winbond recapitalized Walsin Advanced. Such recapitalization further triggered the cooperative pattern of strategic alliance among IC design companies, chip manufacturers, and downstream packaging and test factories.

As Taiwanese manufacturers are engaged in packaging and tests, they receive orders from the IC design and manufacturing industries. Regarding sources of customers, manufacturers in Taiwan and North America are two major groups with orders that account for 90% of the total orders. This is a result of TSMC and UMC cooperating with the IC design industry in North America through strategic alliances or joint investments. In the past few years, Taiwan has had the advantages of a complete information and electronics manufacturing system and professional division of work. Therefore, the IC industry in Taiwan is in its golden period.

The growth of packaging and tests comes from the growth of services in integrated packaging and tests, and the orders released by IDM. The packaging and tests have lower added values in the IDM value chain. In addition, the industry underwent a slump in 2001. The capital expenditure for packaging and tests have become higher. As a result, IDMs now stop making investments in establishing the production capacity of packaging and tests by taking the economic effectiveness into consideration. However, new semiconductor products are still introduced to the market. New packaging technology arises under the requirements of product characteristics and performance. The packaging facilities which IDMs owned before gradually failed to support the requirements of new products. Consequently, the packaging and test procedures are commissioned to contractors.

2.2.1. IC packaging

The so-called IC packaging means to seal the chips with plastic or ceramic materials for protective purposes after completing the production of chips. Plastic or ceramic materials are also made as the interface transmitting the signals between chips and systems. By observing the history of global packaging technology, SSOP/TSOP and TQFP/FQFP, we can see that these smaller and thinner products have become mainstream and are driven by the strong requirements of consumer products and portable electronics. When high-end products such as Chipset and Graphic have a greater number of gates and complicated design, the requirements on the number of packaging pins and performance become stricter. Therefore, BGA (Ball Grid Array), with a higher number of pins and better performance, has become the major packaging type. The introduction of DRAM also means the era of DDR II will be arriving. The test service suppliers that aim at traditional packaging have to develop CSP (Chop Size Package) packaging in order to handle future needs. It is anticipated that flip chip will replace wire bonding. Flip chip can meet the requirements of more pins and the high performance of high-end products. Driven by smaller consumer products and lower cost in packaging, it is expected that the wafer level package will become mainstream in the next generation. Besides, memory with large capacity for communication and consumer products also emphasizes smallness and lightness. The packaging types of stacked die, multi-chip package or system in package will play important roles in the 21st century.

As for the development trend of the packaging industry in Taiwan, large-scale enterprises will continue to drive the trend of integration. Furthermore, it will become a trend for the Taiwanese packaging industry to move to Mainland China and make investments by establishing factories there. Factors such as lowering cost, customer requirements, and global logistics will be highly considered. Foreign packaging factories like Korean IC Packaging AMKOR will remain in Taiwan to run the high-end packaging market. Such a phenomenon will stimulate Taiwanese packaging factories to head towards advanced packaging. In the future, the packaging market will consist of a few large-scale factories aimed at high-end packaging and the test market, and several minor-scale factories aimed at the niche market through mergers and integration. As for the technical aspect, large-scale packaging and test factories will concentrate on CSP, flip chip, wafer level packaging, and system in package (see Table 5.4).

In 2002, BGA, with a higher level of technology, was the major source of revenue for packaging, accounting for 35.5%. QFP applied to the optical disk driver, mobile, etc., was the second major source of packaging revenue, accounting for 25.2%. Influenced by high-end products that request higher number of pins and better performance,

Table 5.4. Packaging Types

	Packaging Type	Products Applied	Variable Types
Insert Pin	DIP	Consumer products, logic IC	PDIP, SK-DIP
Surface Stick	SO	Memory, logic IC	SOP, TSOP, SSOP, SOJ
	LCC	Programmable logic IC	LCC
	QFP	Logic IC	TQFP, LQFP
	Others (BGA, TAB)	Chipset, LCD	BGA, TAB, F/C

Source: ITIS.

and driven by domestic and foreign chipses and high-end graphic chip suppliers, the ratio of flip chips will become higher. Traditional packaging types such as DIP, SO and PLCC will gradually become less applied as IC functions get more complicated, given that traditional packaging types cannot provide sufficient pins and the prices are low. According to the data in *2003 Review and Prospects of Taiwan IC Industry* published by IEK-ITIS, the overall economic environment has begun to show positive signs. As the expenditures in IT made by enterprises increase, demand is activated by the PC and semiconductor market. Therefore, demand is recovering. As for supply, the capital expenditures were reduced by semiconductor manufacturers in 2001 and 2002, which effectively aided the control of global IC chip production capacity. Under the premise of demand recovery and effective control over supply, the global semiconductor market was expected to increase by 16.6% in 2003 compared with that in 2002. The expected output value in 2003 was 164 billion USD. The growth rates of IC packaging and IT test industries in 2003 increased by 21.9% and 28.6% respectively.

According to the *2003 Analysis on Domestic IC Packaging Types* presented by IEK-ITIS, since 1999 Taiwanese-invested packaging factories had maintained the trend of the output value from BGA which is applied to graphic chips with more pins and chipset. BGA is still the major source of IC packaging revenue due to market demand recovery and tight global high-end packaging production capacity in 2003. The said output value accounts for over 30% of IC packaging revenue, as shown in

Figure 5.1. High-end chips for graphics show obvious demand in more pins. Some products have been applied with flip chip packaging since 2002. As the flip chip technology matured and product demand became more obvious, flip chip packaging accounted for a higher percentage at 2.9% of the total revenue in 2003 compared with that in 2002.

QFP (quad flat package) applied to products related to optical storage components and mobiles still remained the second major source of 2003 revenue gained by Taiwanese-invested packaging. In the application of mobiles, a substrate packaging type is being developed as the demand for more pins and smaller chips is increasing. Chip scale package for more pins and smaller chips is expected to become the major contribution to mobiles in the next wave. Regarding the performance of the 2003 Taiwanese-invested packaging, CSP increased by 4.6% compared with that in 2002. In 2003, the delivery of Driver IC was good in Taiwan due to TFT-LCD's popularity. TFT-LCD gave better TCP (tape carrier package) packaging revenue in 2003 compared with 2002. In the trend driven by Driver IC packaging, the pixels of monitors kept increasing and there was an urgent need for more pins and fine pitch. Traditional TCP packaging type cannot break through the bottleneck of

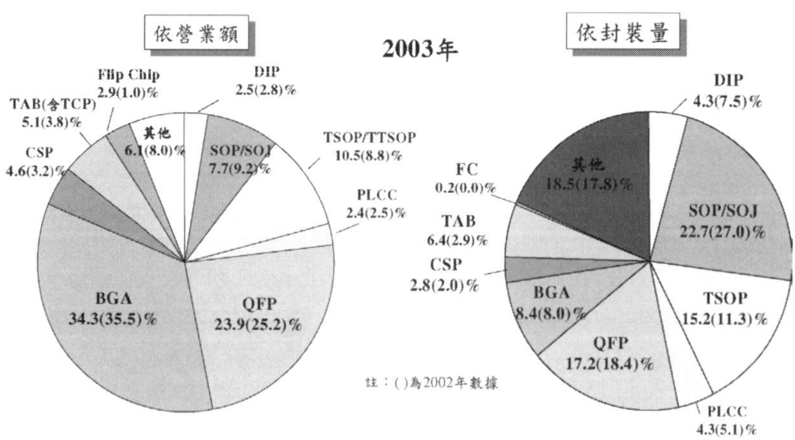

Figure 5.1. 2003 Packaging Types Applied by Taiwanese IC Packaging Manufacturers

Left circle: By sales amount; Right circle: By packaging quantity

Source: IEK-ITIS (2004/05).

dealing with pitch less than 45 μm. In the future, chip on film will replace TCP by being applied to larger panels.

The sales quantity dealt by Taiwanese packaging factories in 2003 totaled 11.79 billion USD, an increase of nearly 30% compared with 9.07 billion USD in 2002. SO type applied to memory related products still accounted for the biggest share at 40% of total sales quantity. TSOP with thinner and smaller encapsulated volume remains the mainstream technology for memory packaging. As the mainstream product of DRAM moves from DDR 333 to DDR II, CSP will become the standard packaging by considering the requirements in speed, volume, and electric characteristics. In addition, the chips applied to mobiles gradually moved away from QFP to meet the demand for more pins and smallness, and the quantity of CSP packaging increased to 2.8% in 2003 in Taiwan.

2.2.2. IC test

IC test uses a test program to control a set of precise and sophisticated electric circuits. The test program is able to simulate many possible environments and methods in use, for example, higher temperature, lower temperature, unstable voltage, low voltage, or high voltage. When IC is placed under a simulated environment, the test program will inspect if the chips are working within the specified range so as to ensure the stability of product quality.

The test on IC products is divided into the circuit probe (also called the wafer sort) and the final test (also called the package test). The chip test is performed under the form of a wafer aimed at distinguishing if dies are in a good condition before packaging in order to avoid unnecessary waste. While testing memory-related products, laser repair can be applied to dies, which are considered repairable after a chip test so as to increase the yield rate. A final test aims to ensure that chips meet the specifications after packaging. Regarding the test service in Taiwan, the final test acts as the major business. In 2002, the final test accounted for 80% of total test sales amount, while the circuit probe only accounted for 20%. Regarding the structure of test products, memory acts as a dominant product, accounting for

Table 5.5. Analysis of Product Structure of the IC Test Industry in Taiwan

	1999	2000	2001	2002	02/01 Growth Rate
Memory IC	55.7%	52.2%	51.0%	52.4%	29.1%
Logic IC	31.4%	29.8%	27.9%	22.8%	2.7%
Mix-signal IC	10.4%	12.2%	14.9%	22.1%	86.4%
Linear IC	0.6%	1.4%	1.5%	1.1%	–7.8%
RF IC	—	1.3%	0.7%	0.8%	43.6%
Others	1.9%	3.3%	4.0%	0.9%	–71.7%

Source: IEK-ITIS (2003/03).

52.4% (see Table 5.5). In addition, the development of chips is moving towards higher frequency, with more pins and multi-functions. Therefore, the demand for the mix-signal test has grown to 22.1% which is at an increase of 86.4% compared with that in 2001. The figures reflect an adjustment of product structure as the test industry faces the future development of SoC (System on a chip) industry.

Further classifying the memory test, the test service in Taiwan is targeted mainly at DRAM products. The DRAM test accounted for 71.9% of memory tests in 2002 (see Table 5.6). The popularity of consumer electronics (such as digital cameras) and computer peripherals (such as portable disks) activated the application of flash memory among non-volatile (NV) memory, which resulted in the dramatic increase of flash test revenue ratio.

The output value of the test industry dropped to 25.3 billion NTD in 2001 due to the regression of global semiconductors which is a decrease of 22.9% compared with the value in 2000 (see Table 5.7). However, the ratio of the test industry to the overall IC industry increased slightly. In 2002, the output value of the Taiwanese-invested test industry increased dramatically compared with the value in 2001 because the price of DRAM stabilized and IDMs placed more orders. In addition, many suppliers tightly controlled their capital expenditures and reduced their investment in advanced facilities in order to fight against the industrial recession in 2001. Suppliers generally hold a conservative attitude towards the expansion of production capacity.

Table 5.6. Memory Test Product Structure

Product Structure		2000	2001	2002
DRAM	Below 4MB	1.0%	0.0%	0.4%
	4MB	5.3%	0.1%	1.4%
	16 MB	10.1%	7.9%	3.4%
	64 MB	55.4%	44.0%	4.2%
	128 MB	10.3%	23.5%	21.8%
	256 MB	1.9%	9.5%	40.7%
	Total	84.0%	85.0%	71.9%
SRAM	Low speed	2.6%	4.6%	7.3%
	High speed	2.5%	2.5%	2.6%
	Total	5.1%	7.1%	9.9%
NV Memory	ROM	1.7%	1.3%	1.6%
	Flash	6.4%	6.1%	13.2%
	Others	2.8%	0.5%	3.4%
	Total	10.9%	7.9%	18.2%

Table 5.7. Essential Indexes of the IC Test Industry in Taiwan

	1998	1999	2000	2001	2002
No. of test suppliers	26	33	37	36	35
Output value of test industry (100 million NTD)	131	185	328	253	318
Growth rate of output value (%)	23.6	41.2	77.3	–22.9	25.7
Ratio (%) of test industry to overall IC industry	4.6	4.3	4.6	4.8	4.9
No. of employees in test industry	5,800	7,389	12,800	11,500	12,850
Ratio (%) of R&D fees to output value	5.3	7.4	2.6	2.7	3.0
Ratio (%) of capital expenditures to output value	104.3	117.4	7.5	28.3	33.4
Export ratio (%)	39.9	38.9	39.6	41.1	42.0

Source: IEK-ITIS (2003/03).

As the economic development recovers, the test capacity is still insufficient. The amount of capital expenditure in the test industry in 2002 grew by 48.3% compared with that in 2001.

2.3. Taiwan's electric industry

The electric industry can be classified into the heavy electric equipment industry, wire and cable industry, household electrical appliances manufacturing industry, and lighting equipment manufacturing industry (see Table 5.8).

2.3.1. Heavy electric equipment industry

Electric engines are commonly called motors, which are the source of power for machinery. Motors are widely applied in important components for industrial use. Motors are mainly used in industrial machinery, transportation tools and vehicles, tool machinery, etc., apart from the carbon brush motors used in hair dryers and juicers, the sensor motors used in electric fans, and the water pump motors for pumps. Recently, the global industry suffered from an economic recession that led to major constructions being cut back or delayed in Taiwan. The related downstream manufacturers kept moving out of Taiwan, which reduced the demand for motors there. Faced with such market impact, the heavy electric equipment manufacturers in Taiwan tried their best to obtain high gross profit orders from large motors overseas and explored the market in Mainland

Table 5.8. Changes of Output Value in the Electric Industry in Taiwan (million NTD)

	1998	1999	2000	2001	2002
Heavy electric equipment	39,354	48,310	48,585	33,493	33,208
Wire and cable	88,603	100,017	95,617	82,951	90,115
Household electrical appliances	72,274	72,601	59,191	51,939	48,239
Lighting equipment	18,461	18,609	18,157	14,560	12,689

China, in addition to developing smaller inverters so as to lower production cost and maintain the growth of revenue and profits. The motor industry belongs to the high-tech collective and capital-collective industry, and early development in Taiwan was slow. Now, the production range is complete, with products varied from 1/8 horse power to 60,000 horse power. Even motors belonging to heavy electric equipment with large voltage and large current for large capacity can be produced by the manufacturers in Taiwan. This means Taiwanese manufacturers have solid manufacturing capability for heavy electric equipment.

2.3.2. Household electrical appliances industry

The household electrical appliances industry in Taiwan produces products like air conditioners, refrigerators, color TV sets, washing machines, etc. As the economy develops in Taiwan, the living standard and quality are enhanced. A variety of household electrical appliances have become daily necessities in Taiwan. The market penetration rate is quite high, which implies a mature market. The demand for overall household electrical appliances has increased slightly due to the introduction of various technological household electrical appliances and the replacement of products. The drastic competition among the whole household electrical appliances market in Taiwan is either in market competition or in production and marketing technologies. Washing machines, refrigerators and air conditioners with mature manufacturing technology will develop with high added values.

Though the penetration degree of household electrical appliances is high enough, the demand quantity is still increasing due to the fact that each household may hold more electrical appliances. Therefore, a steady growth will be shown in the market. In addition, the deregulation of the overseas market in Mainland China and Eastern Europe will aid in the expansion of the export market for household electrical appliances. Taiwanese household electrical appliances manufacturers will increase the self-made ratio and levels of products by strengthening research and development. Moreover, the upstream and downstream production and marketing systems should be integrated.

After-sales service will be improved in speed, quality, and technology. Then, a strong growth may be expected. In recent years, information household electrical appliances (IA products) keep being developed. The market has been gradually accepted and recognized by the public. To obtain market share, IA manufacturers will endeavor to adapt themselves to the changing environment and provide various hardware and software solutions.

2.3.3. Wire and cable industry

The wire and cable industry in Taiwan has become mature and stable. The market structure consists of dominant domestic sales supplemented by export. The development of the wire and cable industry is closely associated with the domestic demand market in Taiwan. Copper is the upstream material used in the wire and cable industry. The copper material is imported, and therefore, the price of copper fluctuates based on the international copper price. The profits of midstream and downstream products offered by the wire and cable industry are thus influenced by the price of copper.

2.4. Taiwan's information industry

2.4.1. PC industry

The PC that Compaq first introduced in 1997 sold for less than 1,000 USD, and was soon popular in the US. This success stimulated other PC system suppliers to follow Compaq's direction. Further aided by the Internet, the market pattern began to change. System suppliers came to seek partnerships in Taiwan while facing the trend of low-priced computers. These system suppliers enlarged the purchase scale in Taiwan. The international system suppliers came up with an ideal approach — global logistics management — to reduce the pressure brought on by inventory depreciation and to increase the gross profits of products. The so-called global logistics management requires that OEMs bear part of the risk arising from depreciation of inventory. In the past, OEMs only had to be responsible for production. The ownership of products after delivery belonged to customers.

Now, the international suppliers ask OEMs to establish delivery warehouses or assembly factories near their target markets. Whenever the international suppliers place orders, the delivery is made from the warehouse. The result of such an approach eliminates the risk in depreciation of inventory and cost for warehouse management. This way, the competitiveness in price and product gross profit can be increased. However, OEMs have to absorb the expenses arising from such warehouse management. The profit earned by OEMs becomes even thinner.

To increase the profit of OEMs, they have to expand production capacity to lower the cost. Limited by these conditions, only large-scale OEMs are capable of taking orders. Small and medium-scale OEMs are gradually eliminated through such competition. The international suppliers are in favor of the unique management capability and flexibility in manufacturing offered by Taiwanese manufacturers. They enlarge the purchase quantity in Taiwan, which directly facilitates the development of the information industry in Taiwan. The information industry in Taiwan has an important position in the world. The industries that produce notebooks, master boards, monitors, etc., are ranked first in the world. As the information industry matures, however, development will slow down. The PC age will gradually be replaced by the post-PC age, which is centered on information household electrical appliances. The information industry will move out of Taiwan to seek cheaper costs elsewhere.

Observing the history of world computer development, the following stages may be marked. In 1960, the beginning of the computer industry centered on large computers. Manufacturers like IBM produced very large computers at that time. Users also had to spend a lot of money on them. Therefore, the users of computers were limited to governmental units or large-scale industries. There was a breakthrough in the industry in 1970, with the arrival of the super mini computer era. The major manufacturers were Prime and DEC. The characteristics of computers during this period were decreased investment, smaller scale computers, easy operation, suitable for medium enterprises, etc. In 1980, the industry entered the PC age with major manufacturers like Apple and IBM. Workstations were also popular during that period with major manufacturers such as SUNMICRO and APOLLO. The features during this period were mature and stable technology, establishment of standardized software and hardware, cheaper prices, and suitability for the public. Therefore, manufacturers could adopt mass production in order to lower the costs. In 1990, world information technology entered a brand new professional age. Super computers, mini computers, micro computers and special-purpose computers came to an "integration" state. Integration in the Internet system helped to enhance the added values of software and hardware. During this stage, PC and workstations still dominated the market. After the millennium, professional processing systems were formed. The compound Internet system became widely applied to management. Household electrical appliances provided comprehensive solutions through the global logistics management system. The development of computerized technology was facilitated. So far, the growth of PCs in the commercial market has slowed down due to the rise in the concept of the digital home. The consumer PC will dominate the market in the future. Besides general computer functions, household computers will allow users to enjoy convenient personalized audio and video entertainment. In the meantime, notebooks will be developed to provide more portability and home entertainment.

(1) Portability

The high-tech core system and highly efficient power-saving feature are outstanding functions for wireless technology. The advantages formed by an extended battery and mature wireless environment will gradually replace the thick and heavy PC.

(2) Home entertainment

Bright TFT-LCDs will be developed at a larger size so as to provide a more authentic video effect. In sound support, the 5.1 channel sound system will be upgraded to a 6.1 channel sound system. The TFT-LCD is equipped with a brilliant graphic display capability and high-quality 3D sound effects. The mobile multimedia will play an important role in future digital AV homes through home servers and the functions of home entertainment.

2.4.2. Notebooks

The increasing popularity of notebooks has affected the desktop computer market. According to an investigation made by MIC, the supply of notebooks accounts for 25% of the global PC market. As the differentiation of the prices and functions between desktop computers and notebooks becomes smaller, the supply of notebooks is expected to increase to 33% in 2006. It is foreseen that the prospects of notebooks are good. The position of the Taiwanese market which is already ranked first in the world will become even more important when the demand for notebooks becomes higher. Currently, the price for notebooks is falling. Brand suppliers will strictly demand that notebook manufacturers keep the production cost low so they can lower the selling price. Compared to Japanese and Korean manufacturers, Taiwanese manufacturers are famous for their fast and low-cost production. The competitive advantage of Taiwanese manufacturers becomes even more obvious when low-price computers become popular. The suppliers in America and Europe choose Taiwanese manufacturers as their biggest OEM partners. Japanese brand suppliers also place orders with Taiwanese manufacturers after considering the cost. The notebooks produced by Taiwanese manufacturers now account for 60% of the market. According to MIC, the American and European regional markets remain the biggest in the growing trend of notebooks, accounting for over 60% of the global market. In contrast, the growth of the market in the Asia-Pacific region is more limited. Among the products in the growing market, low-price products are the most favored by consumers. The market price of notebooks keeps on decreasing, while the ratio of desktops with LCDs is rising. Therefore, the price of a desktop is stable and its price difference with notebooks is smaller. In low-priced notebooks, the products at 1,000 USD are growing fastest. The products below 1,500 USD reach over 13% of PC sales, which accounts for 50% of total notebook sales. Notebooks below 1,500 USD have become the mainstream products. The price of commercial notebooks has decreased more than those of notebooks for other uses. However, notebooks enjoy a greater growth in the consumer market under the situation of the rising consumer market and lower unit price of consumer notebooks.

3. Industrial Structure and Characteristics

3.1. Taiwan's semiconductor industry

3.1.1. Industrial structure

The semiconductor industry in Taiwan can be divided into three parts: upstream — IC design, midstream — IC manufacturing and foundry, and downstream — IC packaging and testing (see Figure 5.2). In the overall IC industry, packaging and testing are the latter part of the whole process of manufacturing. IC packaging refers to the connection of chips with lead frame by gold wire and isolation encapsulated with a plastic or ceramic case. It functions as the IC bearing seat for connecting the IC and external wiring.

The IC manufacturing process starts with the IC design. The establishment of specifications, circuit design, partial consequent testing, and confirmation of specifications are part of the IC design. After the IC design is completed, it is handed to the optical mask companies to transfer the IC circuit distribution to a surface of a plane glass according to pre-set chip manufacturing steps, resulting in an optical mask. This process is like producing an IC module. After finishing the optical mask, the IC manufacturers apply lithography imaging technology to copy the fine wire diagram from the optical mask to the silicon wafer by using a material like photoresist. Nitric acid is then used for cleaning and etching. Then, the production of wafers is complete. Afterwards, the wafers are cut into dies. Gold wire is used to connect the dies and circuits of the lead frame. Then, the plastic or ceramic case is used to package the wafers for isolation so as to protect the IC chips and achieve IC electric connection. The packaging is aimed at supporting, protecting, and enduring and radiating heat. The last step is to test if the IC functions well. The test is classified into: digital logic IC, memory IC, analog IC and mix IC.

3.1.2. Industrial characteristics

(1) Capital and technical collective industry

As the level of IC products is upgraded, the IC industry is in need of research and development for new technologies. A huge amount of

Facilities and Equipment Capital and HR

Computer Design TSMC VIS

Auxiliary Design UMC ProMos

SIS Winbond LiPin

ALI SIS Nanya

Realtek photomask

Mediatek Taiwan photomask ASE

Minsheng SPIL

VIA Huatai

ITE Tachung

Taipin Manufacturing Packaging Packaging Substrate

ASE

PPT

Huatung

Yaowen

Unimicron

Dayshine

Material Chip

MEMC EMerk Chemicals Testing VATE

ShinETsu Furei

FPG ChipMos

KYEC

Figure 5.2. Structural Figure of the Semiconductor Industry in Taiwan

Source: IEK-ITIS Plan.

capital and great human effort and energy are required for the research and development of new technology, manufacturing processes, and new products. Therefore, the IC industry is a capital and technical collective industry.

(2) Professional division of work

The amount of investment made in IC chip factories and packaging and testing factories is growing. The requirements of facilities, factories, manpower and space keep increasing. The management systems, manpower cost, and HR quality of IC manufacturing and downstream packaging and testing are quite different. Therefore, professional division of work in the IC industry becomes necessary so as to avoid complicated systems and inconvenient management. So far, the trend of IDMs' commission packaging and testing to contractors has been formed. This trend brings great business opportunities to Taiwanese manufacturers.

(3) Strategic alliance between upstream and downstream manufacturers

IC packaging and testing belongs to the latter process of IC manufacturing. Therefore, packaging and testing factories are closely associated with IC design and chip manufacturers. Taiwanese IC design companies and chip manufacturers are recapitalizing downstream packaging and testing factories. The cooperation between upstream and downstream factories leads to the provision of complementary services and increase of customer satisfaction. Packaging and testing factories also actively search for opportunities to cooperate with upstream semiconductor chip manufacturers or IC design companies through strategic alliances so as to ensure that no idle capacity occurs after the expansion. Therefore, strategic alliances have become the best model for packaging and testing suppliers to enhance their competitiveness.

3.2. Taiwan's IT industry

Semiconductor Industry	Electric Components I	Other Upstream Industries
IC design	Passive components	Light emitting diode
IC manufacturing	Resistors	CD/Disk
Chip OEM	Capacitors	PCB
Packaging and testing	Inductors	Connector
Photomask lead frame	Rectifier diode	

Electro-optical Industry	Electric Components II	PC Peripherals	Internet & Communication
Monitor	Master board	PC frame	Modem
TFT-LCD	Power supply	Mouse	Network card
Scanner	Transformer	Keyboard	Communication module
CD player	Interfaces		
	Chipset		

	Computer		
	Desktop computer		
	Notebook		
	Industrial computer		
	Server		
	PDA		
	Mobile		
	Digital camera		

Figure 5.3. Association of Industries

4. Business Opportunities

4.1. Taiwan's semiconductor industry

The overall output value of the semiconductor industry in Taiwan first exceeded 1,000 billion USD in 2004. It is expected that IC related industries such as design, manufacturing, chip OEM, packaging, testing, etc., can achieve equally impressive results. Though the latter half of 2004 was influenced by the economic environment, 2004 still remained the best performing year since 2000.

According to the estimation made by IEK-ITIS, the total IC industrial output value (including design, manufacturing, packaging and testing) in 2004 in Taiwan reached 1138.4 billion NTD, which is an increase of 39% compared to the output value in 2003. From the total industrial output value, the output value of the design industry accounted for 254.9 billion NTD at an increase of 34%; the output value of the manufacturing industry accounted for 665.7 billion NTD at an increase of 41.6%; the packaging industry accounted for 159.6 billion NTD at an increase of 35.7%; and the testing industry accounted for 58.2 billion NTD at an increase of 42.36%.

In the IC design industry, the reason for the growth of revenue lies in the consumer chips used in cell phones, which have become popular in the market. The market share of consumer chips has been rapidly increasing. Furthermore, the low-power SRAM used for in-hand devices (such as cell phones, PDA, etc.) enjoyed an increase in delivery in the first half of 2004. In addition, the demand for high-end graphic chips rose, which activated the demand for niche memory. However, when the supply capacity of memory in the latter half of 2004 was achieved, the pressure of supply over demand was generated and caused niche memory price to drop. The trend of LCD replacing CRT remained unchanged. The drives and control chips used for LCDs kept growing, and the revenue of related industries increased as well. As the LCD TV market did not perform as well as expected, the LCD panel price rapidly fell in the third season which depressed the performance of LCD driven chips. Still, 2004 was an abundant year for the IC design industry.

The production of chip OEM benefits from the growth in communication and consumer electronic products. The production capacity is very tight, and the use rate is almost full. Manufacturers seek contractors when there is no new chip OEM being planned in the near future. The OEM preferential price is thus canceled. In addition, the demand for the high-pressure manufacturing process driven by LCDs contributes to a great growth in revenue by the second-line manufacturers. TSMC and UMC broke the records all the way during 2004. The 12-inch production capacity and yield rate of DRAM have been improved a lot as well, and the time spent on the manufacturing process has also been shortened. Therefore, the quantity of DRAM keeps on increasing. The price for the 256MB is stable at 4.5 USD, which has contributed to record levels of revenue in the memory industry in Taiwan.

According to the estimation by Dataquest, a high growth rate was in store for the chip OEM market in 2005. The estimated growth rate reached 36%, or 25 billion USD. However, with the positive expansion in production capacity of the semiconductor industry in the following years, a negative growth rate of the chip OEM market would most likely appear in 2006. It was estimated that the growth rate of the 2006 chip OEM market would decrease slightly by 3.8% to 24.1 billion USD. In 2007, there will be a rebound in the chip OEM market. The growth rate will increase by 14% to 27.4 billion USD.

The foreign large-scale suppliers have released a great amount of orders for packaging and testing, hence the growing revenues gained by packaging and testing manufacturers in Taiwan. The orders related to high-end packaging such as BGA, CSP, FC, etc., and testing types such as mix-signal and high frequency are increasing. The memory packaging and testing market is facilitated by the increasing quantity of DRAM and its stable high price; therefore, the revenue is impressive.

The demand for packaging relies on the demand for semiconductors. It was widely forecasted that there would be great growth created by the semiconductor market in 2004. However, Dataquest forecasted that the growth for the market relating to communication products would slow down in 2006 because of the great recession of the digital cellular products market. Therefore, the growth rate of the semiconductor market in 2006 would decrease at a rate of 4.6% (see Table 5.9). According to the IC-Insights' report, the estimated decreased GDP growth rate of 2% in 2005 would activate a decreased growth rate of the semiconductor market at 8.4% in 2005. However, the practical market value remains higher than that in 2001 when the industrial development was at a low point.

ETP estimated that in 2004, the global IC packaging output value would reach 16.752 billion USD at an increase of 12.4% compared with that in 2003 (see Table 5.10). The CAGR (compound annual growth rate) from 2002 to 2007 is 7.91%. Driven by the semiconductor output value, the capital expenditure of packaging and

Table 5.9. Estimation of the Scale of the Global Semiconductor Market (billion USD)

Unit	Date of Publication	Item	2001	2002	2003(e)	2004(f)	2005(f)	2006(f)	2007(f)
Dataquest	2003.08	Semi market scale	152.2	155.1	173.2	209.8	248.7	237.2	257.1
		Growth rate	–32.8%	1.9%	11.7%	21.1%	18.5%	–4.6%	8.4%
In-Stat	2003.04	Semi market scale	139.0	140.7	164.2	206.8	234.7	209.8	236.9
		Growth rate	–32.0%	1.2%	16.7%	25.9%	13.5%	–10.6%	12.9%
IC-Insights	2003.08	Semi market scale	139.0	140.7	160.1	196.3	179.9	192.7	224.8
		Growth rate	–32.0%	1.2%	13.8%	22.6%	–8.4%	7.1%	16.7%
WSTS	2003.05	Semi market scale	139.0	140.7	156.9	185.7	200.4	207.9	**
		Growth rate	–32.0%	1.2%	11.5%	18.4%	7.9%	3.7%	**
SIA	2003.06	Semi market scale	139.0	140.7	154.9	180.9	191.5	204.9	**
		Growth rate	**	1.2%	10.1%	16.8%	5.9%	7.0%	**

Source: Research Institutes, IEK-ITIS Plan (2003/09).

Table 5.10. Global Packaging Market — Revenue Distribution (2002–2007) (million USD)

Packaging Type	2002	2003(e)	2004(f)	2005(f)	2006(f)	2007(f)	CAGR(%)
DIP	360	354	339	312	316	334	−1.49
SO	3,365	3,562	3,851	3,671	3,866	3,939	3.20
CC	265	281	310	297	313	332	4.61
QFP	2,794	2,882	3,092	2,980	3,252	3,527	4.77
PGA	2,465	2,756	3,015	2,895	3,126	3,299	6.00
BGA	3,022	3,670	4,415	4,287	5,021	5,665	13.39
CSP	1,095	1,394	1,730	1,858	2,118	2,461	17.58
Total	13,366	14,899	16,752	16,300	18,012	19,557	7.91

Source: Electronic Trend Publications (2003); IEK-ITIS plan (2003/06).

testing manufacturers keeps on increasing. Although the result in 2004 was estimated to be better than that of 2003, the actual growth rate should have been a little higher.

In terms of the amount of packaging revenue, high-end packaging types at higher unit prices are the major source contributing to the packaging revenue. PGA and BGA applied to CPUs, graphics and chipsets are estimated to account for 46% of the total packaging revenue in 2007. The demand for thinner, smaller and multifunctional IC products is growing. The number of I/Os required by packaging is increasing accordingly. Traditional lead frame no longer satisfies the urgent needs arising from BGA and CSP. According to the estimation by ETP, the CAGR of the BGA and CSP packaging market scale may reach 13.39% and 17.58% respectively. BGA and CSP are the items, with most potential in the variety of packaging types worldwide. Among all packaging types, the revenue created by BGA accounts for the most. BGA revenue was estimated to account for 26.36% of total packaging revenue in 2004.

In the global packaging market, SO type applied to memory products still remains the major packaging type, accounting for about 60% of the total packaging quantity. Driven by the demand for thinner and smaller products, and a DRAM market gradually entering the

Table 5.11. Global Packaging Market — Quantity Distribution (2002–2007) (million USD)

Packaging Type	2002	2003(e)	2004(f)	2005(f)	2006(f)	2007(f)	CAGR(%)
DIP	6,740	6,906	7,027	6,745	7,114	7,836	3.06
SO	48,132	53,740	62,017	63,237	69,298	75,205	9.34
CC	1,705	1,854	2,050	2,023	2,186	2,372	6.83
QFP	7,958	8,642	9,758	9,999	11,263	12,551	9.54
PGA	255	284	302	295	320	354	6.78
BGA	2,793	3,338	3,971	4,149	4,798	5,512	14.56
CSP	4,731	6,465	8,802	10,274	12,702	15,302	26.46
DCA	6,243	7,247	8,368	8,733	9,834	11,051	12.10
Total	78,557	88,476	102,295	105,455	117,515	130,183	10.63

Source: Electronic Trend Publications (2003); IEK-ITIS plan (2003/06).

DDR II age, CSP, which is more suitable for product requirements, will show dramatic growth in the following years. The CAGR from 2002 to 2007 of CSP packaging quantity will reach 26.46% (see Table 5.11). The growth rate of CSP will be the greatest among various packaging types.

Under the premise of economic recovery in the semiconductor industry, the orders placed by IDM will become a driving force for OEM packaging. According to the estimation by Dataquest, the sales amount of the contracted OEM packaging and testing market reached 14.15 billion USD in 2004 (see Table 5.12). The market will grow to 20.64 billion USD in 2007. One of the major causes contributing to such growth is the IDM's increasing and fast-growing contracted orders.

According to the estimation made by ETP, the size of the OEM packaging market will keep rising. The CAGR of OEM packaging market scales from 2002 to 2007 is estimated to be 16.76% (see Table 5.13). The ratio of OEM packaging to total global packaging will reach 62.6% in 2007.

In 2005, communication and plane video products became the major force triggering future growth of the IC industry. Meanwhile, the integration of chip functions is inevitable when considering cost and effectiveness. DDR II became the mainstream product in the

Table 5.12. Growth of Semiconductor Packaging and Testing Market from 2001–2007 (million USD)

	Overall Packaging and Testing Market Scale						
	2001	2002	2003	2004	2005	2006	2007
Overall packaging and testing market scale	24,893	26,751	29,560	35,005	41,770	40,953	44,617
Overall packaging market scale	19,954	21,476	23,605	27,968	33,194	32,610	34,344
Overall testing market scale	4,939	5,275	5,955	7,037	8,576	8,343	9,273
	IDM Semiconductor Packaging and Testing Market Scale						
	2001	2002	2003	2004	2005	2006	2007
IDM internal packaging and testing market scale	17,750	18,383	19,355	20,855	23,625	22,654	23,982
IDM internal packaging market scale	14,200	14,691	15,550	16,735	18,969	18,166	19,187
IDM internal testing market scale	3,550	3,692	3,805	4,120	4,656	4,488	4,795

(*Continued*)

Table 5.12. (***Continued***)

	Professional OEM Packaging and Testing Market Scale (SATS)						
	2001	2002	2003	2004	2005	2006	2007
Professional OEM packaging and testing market scale	7,143	8,368	10,205	14,150	18,145	18,299	20,635
Professional OEM packaging market scale	5,754	6,785	8,055	11,233	14,225	14,444	16,157
Professional OEM testing market scale	1,389	1,583	2,150	2,917	3,920	3,855	4,478
	Ratio of Packaging and Testing OEM						
	2001	2002	2003	2004	2005	2006	2007
OEM ratio (packaging and testing)	28.7%	31.3%	34.5%	40.4%	43.4%	44.7%	46.2%
OEM ratio (packaging)	28.8%	31.6%	34.1%	40.2%	42.9%	44.3%	45.7%
OEM ratio (testing)	28.1%	30.0%	36.1%	41.5%	45.7%	46.2%	48.2%

Source: Dataquest.

Table 5.13. Global OEM Packaging Market — Revenue Distribution (2002–2007) (million USD)

Packaging Type	2002	2003(e)	2004(f)	2005(f)	2006(f)	2007(f)	CAGR(%)
DIP	136	133	131	127	123	122	–2.15
SO	855	918	1,027	1,152	1,286	1,390	10.21
CC	130	144	164	143	152	161	4.37
QFP	1,267	1,404	1,563	1,469	1,635	1,817	7.48
PGA	552	841	1,299	1,792	2,000	2,245	32.39
BGA	1,989	2,436	2,968	3,100	4,032	4,700	18.77
CSP	714	938	1,166	1,379	1,582	1,810	20.45
Total	5,643	6,814	8,318	9,162	10,810	12,245	16.76

Source: Electronic Trend Publications (2003); IEK-ITIS plan (2003/06).

memory sector in 2005. The total output value of the semiconductor industry in 2005 was forecasted to increase by 15% by extending the power driven in 2004, with an output value of 1,309.1 billion USD.

4.2. Taiwan's PC industry

According to the statistics provided by Dataquest, in 2003 the delivery growth rate of PCs increased by 10.8%. As notebooks gradually replaced PCs, the global growth rate of mobile PCs reached 26.2%. In 2003, the PC market recovered comprehensively. In addition to the boost of the consumer market, desktop PCs and mobile PCs both grew in 2003 due to the increase of IT expenditures by enterprises, which accounted for 70% of the commercial market. It was estimated that desktop PCs would grow by 10% in 2004, with an increase of 22.2% for mobile PCs. The total PC output value will show a slow growth due to the increasing quantity and decreasing price of PCs in the future.

4.3. Taiwan's communication industry

For wireless communication products, the overall output value in Taiwan in 2003 restored its growth by 87%, according to IEK-ITIS. The total output value reached 200 billion NTD. The dominant products include mobile phones and WLAN which contributed about 80% of this output value. However, GPS and Bluetooth show great potential too. The output value of these four products grew several times over.

Taiwanese manufacturers started actively making arrangements just a few years ago. The results of their efforts showed in 2003. The overall output value of wireless communication facilities totaled 198.27 billion NTD. From the above total output value, mobile phones contributed 107.87 billion NTD, which accounted for 54%. The production capacity of mobile phones reached 44.62 million. High-end models of mobile phones were delivered continuously in the latter half of 2003, which made the average delivery unit price increase from 60 USD to 70 USD. The output value of the communication facilities industry in Taiwan reached 398.4 billion NTD in 2004, at an increase of 27.9% compared with 2003.

Mobile phones have already become an irreplaceable communication tool in modern society. Many functions are integrated on cell phones. Cell phones have become a consumer commodity equipped with video and audio effects. As the transmission speed, quality, and service contents continue to improve, it is believed that cell phones will retain an importance in the daily lives of human beings.

According to the data provided by MIC, the global delivery amount of cell phones in 2003 totaled 0.45 billion USD at an increase of 6%. Under the great demand to keep replacing cell phones, 2.5G (GPRS, CDMA20001X) will become a mainstream product in the market. European, American, Japanese and Korean manufacturers are currently devoting themselves to 3G cell phones (WCDMA, CDMA20001X-EV).

As cell phones play multiple roles in our daily lives, the appearance and ringtones that differentiate them are considered to be important features in attracting consumers' attention. Furthermore, the screen display, MMS function and commercial PDA functions are becoming more important. The opportunities and challenges for cell phone suppliers lie in anticipating the multiple needs of consumers.

The average unit price of cell phones has increased due to the increasing delivery of mid and high-end products such as color camera phones and intelligent cell phones. The output value reached

133.3 billion USD in 2004, with an annual growth rate increase of 23.5% compared with that of 2003. The performance of WLAN was not good in the first half of 2004. However, the price dropped and a greater quantity of delivery occurred because of the beginning of a new school semester in the North American region and the birth of dual-band with low-price chips. Furthermore, the delivery quantity of VoWLAN products decreased in the fourth quarter.

It is now the time for enterprises to replace their old network facilities. The price of products at Gigabit level has dropped. Therefore, the market demand for net exchange among LAN products is increasing. European and American suppliers come to Taiwan to seek OEMs. In addition, Japan is working hard to promote FTTH broadband service, which also stimulates the demand of LAN net exchange.

The VoIP equipment (including voice gateway, video phone, IP phone) is facilitated by the promotion of Internet telephone services by many countries and the purchase of international large-scale suppliers in Taiwan. The performance of VoIP Gateway, IP phone, and Video phone has so far been good and is full of potential.

Japan, Europe, and Mainland China continue to promote broadband access to the Internet. Therefore, xDSL CPE keeps on growing. The delivery quantity of cable modems shows a dramatic increase due to the decreased Internet access fees and rapidly increasing number of subscribers.

As for the prospects of the development of the communication industry in Taiwan in 2005, the major products such as cell phones, WLAN, DSL CPE, SOHO router, GPS, ethernet chips, etc., would keep growing. Bluetooth, VoIP, Home gateway, FTTx, etc., are also products we should pay attention to. According to the forecast made by IEK-ITIS, the output value of communication facilities in 2005 would reach about 454.8 billion USD with a growth rate of 20.8%.

4.4. Taiwan's electro-optical industry

The performance of the overall electro-optical industry was excellent in 2004. The output value totaled 987.4 billion USD at an increase of 33.9% compared with that of 2003. The year 2004 had the greatest growth in recent years. The ratio of monitors to the electro-optical industry grew higher and higher. The ratio was 44% in 2003 and over 50% in 2004. The growth of optical storage used to have the best performance but eventually slowed down. Therefore, its contribution to the output value has decreased.

In regards to monitors, the driving force comes from large-scale of TFT-LCD. The production lines for fifth generation TFT-LCD expanded positively in 2003 and 2004 and started mass production. The production capacity is thus obviously increased. The price of panels has dropped due to greater demand. However, the overall output value continues to grow. Meanwhile, medium and small panels have grown rapidly due to the demand of color portable devices.

There has been great progress in projection technology in recent years. DLP and LCD groups promote the projection components with better performance and lower cost. Under such benign competition, projectors have become affordable products which consumers and many organizations can purchase. Therefore, the market for projectors has grow rapidly. The rapid growth and keener competition reflect the importance of channel management and overall operational efficiency. After the promotion of commercial projectors, consumer projectors are the next potential product in the market.

According to the data provided by Display Search, the global delivery quantity of TFT-LCD in 2003 totaled 4.5 million USD at an annual growth rate of 194%. The suppliers of information household electrical appliances in Europe, America, Japan, Korea, and Taiwan endeavor to run the market which will inevitably cause the replacement of TFT-LCD in the consumer market. It is estimated that the delivery quantity of TFT-LCD will account for 20% of total TV delivery quantity in 2007.

As for the optical storage industry, the driving power for growth comes from a large delivery quantity of DVD products with high unit prices. The focused products include DVD recorders and DVD-R/+R for the industry. The above products have entered the stage of price competition and enhanced recording speed. However, the price competition will depress the total output value of the optical storage industry.

In terms of optical input, manufacturers of digital cameras expanded too fast in their production capacity in 2003. Thus, they faced a harsh year in 2004. In addition, the Japanese suppliers did not release their orders as expected and low-end cameras were squeezed out by camera cell phones. Consequently, the performance of the optical input industry was not as good as expected.

Optical communication has stepped out of recession in the last few years. In early 2004, the specifications for the components related to FTTH construction in Japan were altered. The product quality and preferential prices which Taiwanese manufactures can provide are still considered the priority for Japanese suppliers. After being accredited on new specifications for half a year, the products with new specifications have started to be delivered to Japan in the third season, contributing to the growth of output value accordingly.

Among the electro-optical components, the ratio of brighter LEDs has increased to 52% due to the continuous growth of mobile phones and the substitute effect arising between the products with high brightness and common brightness. However, the average price has dropped to about 25% to 40% due to the slower demand of the LED market in Mainland China, price competition from Korean manufacturers in the Korean market, and the expansion of production capacity by 50% by Taiwanese manufacturers. As a result, the LED market scale in Taiwan reached 40.5 billion USD in 2004, an increase of 23% compared with 2003.

The future development of the electro-optical industry will focus on consumer electronics, especially the increasing demand for digital video products such as digital TV and optical fiber communications required in a digital home. This demand will make a great contribution to the development of the electro-optical industry. In 2005, the output value of the electro-optical industry was expected to reach 0.1263 trillion dollars at an increase of 27.9% compared with that in 2004.

4.5. Taiwan's wire and cable industry

The development of the wire and cable industry is influenced by the continuous release of the sixth power supply plan made by

TaiPower, public construction projects promoted by the government, and the gradual recovery of the economy. The wire and cable in relation to the power supply will be beneficial due to the above factors. A growth of 10% due to increasing market demand is expected. The demand for hook-up wires and enameled magnet wire will decrease by 15% because of the industry's move to Mainland China. The communication cable market faces the greatest impact after joining the World Trade Organization (WTO). Though the government continuously creates business opportunities of network construction to upgrade the national living quality and strengthen national infrastructure, the growth will decrease with an imbalanced demand and supply and keen price competition.

Chapter 6

The Business and Investment Environment of the Information Technology and Electronics Industries in Mainland China

Since 1978, Mainland China has been executing a series of reforms and "open door" policies. The economy in Mainland China has become a part of the international economic cycle by stepping out of its enclosed status. Absorbing foreign capital and technology, and encouraging foreigners to make investments in Mainland China are examples of several important policies. The Chinese government and the public have shown a positive attitude towards foreign investments. Mainland China is gradually abandoning its previous socialist ideology and accepting the requirements of the process of economic development. Particularly after Deng Xiao Ping's trip to southern China, the reform in the country entered a new era. In order to meet the needs of the market economy, the government made new criteria and adjusted foreign investment policies. Mainland China improved the environment for investment based on international regulations. As development of policies and social economies became stable, the infrastructure was gradually completed and much progress was made in a variety of soft environments, thus Mainland China began to attract more foreign investors. Generally speaking, a comprehensive and multi-level opening system and pattern have been formed. Currently, the series of terrorist and counter-terrorist threats following the events of 9/11 and the Iraq War have great influence on the world economy. Many people are worried that America will be attacked by the terrorists again. Therefore, many transnational investors consider national security and stability as top priorities when they evaluate potential investments. In Mainland China, the political situation is stable and its economic development is fast.

Such competitive advantages will make Mainland China attractive as an environment for investment. Furthermore, Mainland China's participation in the WTO will facilitate further system reforms and structural adjustment in the country. This, too, will encourage foreign investors to put their capital in Mainland China.

According to a 2004 survey by Commercial Affairs Department on the top 1,000 global enterprises listed in *Business Weekly*, the key factors for foreign investors placing their investment in Mainland China lie in Mainland China's market scale and high growth rates. This is especially obvious in the IT and automotive industries. The IT market in Mainland China reaches 242.4 billion USD. In 2003, the overall IT market accounted for 30.8% of the Asia-Pacific market, excluding Japan. This figure will increase to 38% in 2008, which is far above the average global growth rate. The number of automobiles owned in Mainland China will reach 8.5 million, an increase of 30%. Policies, agglomeration effect, labor force, and the cost factor are other important elements that encourage foreign companies to invest in Mainland China or expand their existing investments there.

The Chinese Government kept its word about opening its markets and further removed a variety of limitations and restrictions, all of which provide foreign investors good opportunities and an environment in which to make investments. Many foreign companies have indicated that they will become involved in retail markets after Mainland China opens its marketing channels. This will thus encourage the logistics industry to invest in Mainland China.

Some product suppliers, gather in a certain location to trigger a positive influx of enterprises. Such a gathering attracts other suppliers and finished product manufacturers to move into this area. When upstream and downstream industries cluster in the same area, it is convenient for manufacturers to integrate related businesses and operations. As a result, the market in Mainland China becomes even more attractive to foreign investors.

The labor cost in Mainland China is low. It accounts for only 1/3 of that in Mexico and Brazil, and equals the cost in India. This is very attractive to foreign investors who are seeking efficiency and low cost in labor. Other conditions that foreign investors take into prior consideration include the various preferential policies made by local governments and the continuity and degree of realization of those policies. Next, they would consider the efficiency and transparency of the local government.

Since the reforms made in Mainland China, local-level governments have been authorized to take charge of their own administration, and are entitled to manage their economies. To attract foreign investors, local governments need to strengthen a variety of preferential policies. However, the succession of terms of local governments, the effect of central macro policies on local policies, and excessively preferential policies made by local governments will influence foreign investors' decisions. Foreign investors will focus on the substantial realization of preferential policies made by local governments. Generally, foreign companies hope to obtain more efficient services provided by local governments with which they might connect. Foreign companies generally hope to proceed with normal and fair business operations and market competition.

Of course, many problems still exist in the environment in Mainland China, especially in the soft environment. For example, the current economic system still carries some habits of the traditional system. A series of negative problems and contradictions of depth have not been solved yet. In addition, the laws, regulations, rules and policies of Mainland China are quite different from internationally accepted regulations. For many foreign investors, it is important to have a thriving soft environment. Foreign investors care about the security of capital when they make investments in other countries. There is no doubt that foreign investors choose countries and regions with low risks and better soft environments. A soft environment also refers to the sum of invisible elements in an investment environment, including politics, law, social and economic environment, cultural and human concepts, and customs of related governments.

In the following sections, we will analyze and evaluate the business and investment environment in Mainland China from four aspects: index of economic environment; systematic factors in the investment environment; policy and laws; and infrastructure and natural factors of the investment environment.

1. Index of the Economic Environment

1.1. Analysis of economic factors in the investment environment

The factors relating to the economic environment of an invested country directly influence and dominate the flow of investments, operations, and revenues made by foreign investors.

1.1.1. Market scale

Market scale greatly influences the inflow of investment made by foreigners. After all, the sales profits earned by enterprises can only be realized by the market. If the invested country has a market of larger scale and sufficient consumer ability to purchase the products manufactured by the investors, then the country (or region) possesses great advantages for foreign investors no matter what motivations they hold. When making investments in a developing country which has large economic scales and market capacity, investors can obtain market shares step by step even though they may not receive instant feedback. The investors can receive huge benefits when the invested country gradually shows its market potential at a later stage. Therefore, market scale is the most important factor in an investment environment. Enormous market scale is the most prominent advantage in Mainland China.

Generally speaking, market scale is regarded as an aggregate index. It is the performance of the total purchase capacity. In the evaluation of an investment environment, the overall economic scale reflects market capacity or market scale.

(1) The realization of the potential market scale lies in the total population of a country. It is considered a big country or region if a country has a population of over twenty million people. According to the fifth population census in 2002, the population in Mainland China in 2001 was 1.276 billion. The population increases at a rate of 6.5%.[4] Therefore, there is no doubt that Mainland China is a huge potential market.

---

[4] *2001 National Economics and Social Development Statistical Report.* Available at http://www.stats.gov.cn

(2) The substantial market scale is evaluated by the average GDP, the international accepted index. The GDP in Mainland China in 1999 reached 4112.2 billion USD.[5] Mainland China has reached the growth rate of 8.8%, 7.8%, 7.1%, 8.0% and 7.3% respectively in the five years since 1997.[6] The economic performance is ranked first in the world. Such a huge market scale and excellent potential have become unique conditions that attract foreign investors.

1.1.2. Globalization

An excellent macro-economic environment allows Mainland China to further its entry into the world economy. In the meantime, the integration of Mainland China's economy with the world economy improves the macro-economic environment as well. In fact, Mainland China's participation in the WTO symbolizes its further step towards globalization. People have already recognized the important functions activated by trade in facilitating productive investment and economic growth. There are different methods to evaluate the "openness" of trade, including relative trade scale (share of import and export) and distorted degree (average customs and dispersion degree). Studies show that a greater degree of openness is associated with higher economic growth, both in advanced and developing countries. Sachs and Warner (1995) found that the degree of economic openness is a very important factor in determining economic growth. The proper combination of openness and property rights can become a sufficient condition to realize economic growth in a poor country. Kang and Sawada (2000) found a similar influence of economic opening on economic growth. They believe that the economic growth rate of a developing country can be enhanced by decreasing the labor force cost in the process of combining openness and financial development.

Direct foreign investments and trade act as good indices for globalization. In the past ten years, there has been a great increase in net FDI in Mainland China from 3.487 billion USD in 1990 to 60.630 billion

---

[5] *World Development Report 2000/2001.*

[6] *1997–2001 National Economics and Social Development Statistical Report.* Available at http://www.stats.gov.cn

USD in 2004. This growth is amazing. Viewed from the ratio of GDP, the net FDI flowing into Mainland China is much higher than those in other Asian countries, though a little less than those in Latin American countries. Mainland China still remains the developing country absorbing the greatest amount of FDI in terms of absolute amount.

At the same time, tariffs in Mainland China have been decreased to about one third of the rate twenty years ago, from 49.5% in 1982 to 16.8% in 1998 (World Bank, 2002) and then to about 11% in 2004. As a result, the ratio of trade to GDP increased from 15% in 1980 to 50% in 2000. (Mainland China's GDP at this time is much larger than it used to be.) As for imports, the output value increased from 36 billion USD in 1980 to 412.8 billion USD in 2003. The import amount in relation to foreign investments accounted for 56% of the total import value at 231.9 billion USD. Similarly, the export value increased from 27 billion USD in 1980 to 438.3 billion USD in 2003, in which the export value contributed by foreign invested enterprises accounted for 55% of the total export value at 240.3 billion USD.

1.1.3. Industrial structure

Industrial structural factors refer to the composition of the industry. The influence of the industrial structure in a country or region on foreign investors is shown through two aspects. Firstly, an industry under the stage of maturity can often bring in the synergy generated by the aggregated industries. Such synergy is shown in external economies, technology, knowledge, gathering of talents, flow effects, and the development of satellite industries (e.g., Chung Kuan Village in Beijing, Soochow High-tech Science Park, etc.). In particular, SMIC and ACE located at Chang Chiang Science Park in Shanghai smoothly realized their capital increment plans in early 2005, followed by the chip design projects signed with world leading companies, National Semiconductor and Infineon, and Chien Teng's commencement on the factories manufacturing related package products. On 11 March 2005, the world's largest TFT-LCD facilities manufacturer, Tokyo Electron Ltd., and the leader of the domestic IC industry, Huahong Group, signed an agreement to authorize the use of land with Chang Chiang Group. They agreed to move their headquarters into Chang Chiang Science Park and begin the construction of new factories and R&D centers. Another big manufacturer Weiyu, engaged in packaging and testing, is formally devoted to production. It is reported that Grace Semiconductor, Belling, ACE, Chien Teng, etc., will sooner or later devote themselves to producing chips. The IC industrial output value in Shanghai has already accounted for half of the market in Mainland China. In 2002, the export value reached 2.26 billion USD, which made Shanghai the most attractive region for foreign investors. It is estimated that the annual growth rate of the IC industry in Shanghai will be 40% to 50% in 2005.

Secondly, the level of industrial competition can affect the method of investment for foreign investors. When a domestic industry shows its maturity in a developing country, foreign investors often adopt methods like acquisition or mergers to enter the market. However, domestic industries usually operate under unfavorable conditions in a developing country, and foreign investors often have to adapt their methods of investing given these constrictions. Therefore, we can estimate that the direct investments in mergers and acquisitions made by foreigners will increase as domestic economic development and the maturity of related industries are improved.

One way to evaluate the maturity of the industrial structure is to examine the ratio of the third industry to GDP. The industrial structure in Mainland China is shown in Table 6.1.

According to the data, the ratio of the third industry to GDP in Mainland China has remained at 30% for a long time, which is far lower than the 60% to 70% shown in the countries or regions with mature market economies. Meanwhile, the figure is also lower than some of those in other developing Asian countries. As a result, the capability in providing comprehensive package services in the investment environment in Mainland China is not competitive.

1.1.4. Factors for foreign investors' participation and withdrawal

Free participation and withdrawal of enterprises are very important in developing healthy economies. Barriers on enterprises' participation and withdrawal interfere with the emergence of enterprises with innovation and higher productivity. Djankov *et al.* (2002), Emery *et al.* (2000) and Friedman *et al.* (2000) record such barriers faced by enterprises in

Table 6.1. Ratio of Industrial Structure to GDP in Mainland China from 1990 to 2000

	1990	1991	1992	1993	1994	1995	1996	1997	1998	1999	2000
First Industry	27.1	25.4	21.8	19.9	20.2	20.5	20.4	19.1	18.6	17.6	15.9
Second Industry	41.6	42.1	43.9	47.4	47.9	48.8	49.5	50	49.3	49.4	50.9
Third Industry	31.3	33.4	34.3	32.7	31.9	30.7	30.1	30.9	32.1	33	33.2

Source: White Paper on Foreign Trade and Economic Cooperation in 2001.

initiating their business in developing countries. For example, Djankov *et al.* (2002) organizes the information in connection with the number of regulations for enterprises participating in the economic activities in 85 countries. Great differences are found in the number of rules and regulations in different countries — these varied from two in Canada to 21 in the Dominican Republic (20 in Bolivia and Russia). Similarly, the application period to set up an enterprise varies from two to 152 working days (see Figure 6.1). These procedures for setting up a new business (excluding bribes) create great cost to economies. The study further finds that stricter participation regulations are usually associated with more corrupt and larger scale of underground economies.

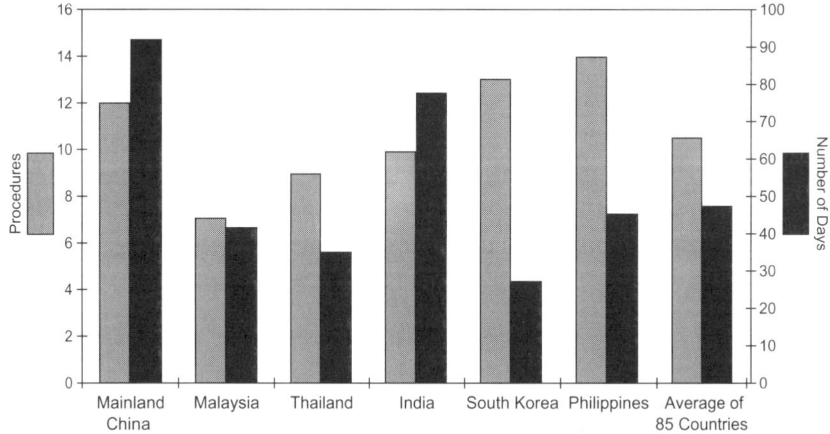

Figure 6.1. Procedures and Number of Working Days Required to Set Up a New Enterprise

Source: Djankov *et al.* (2002).

If we use Djankov's evaluation standards on market participation, there are twelve procedures to be completed before an enterprise can enter the Mainland China market, which is higher in number than the average 10 procedures. The time spent applying to set up an enterprise is 92 days, higher than the average 47 days. The information shows that it takes a longer time to set up an enterprise in Mainland China than in India, Thailand or Malaysia. However, according to the IMD *2001 World Competitiveness Report* for 75 countries, the survey contradicts with the above evaluation standards. Although in 2001 Mainland China ran with an estimation of six, which was worse than the average set up permission among 75 nations, the study also indicates the average time needed to set up an enterprise is thirty days in Mainland China, which is much better than that shown in India.

1.1.5. HR analysis

Human resource (HR) is one of the essential factors influencing an investment environment. This is shown in two aspects — labor force cost and labor force quality. The HR environment directly influences the performance of foreign investment effectiveness and the choices of investment sectors.

(i) Labor force cost: the foreign investments in Mainland China are mainly outward FDI (i.e., export-oriented FDI). Foreign companies basically seek advantages in the cheap and abundant labor force of Mainland China, which helps reduce production costs. Meanwhile, foreign investors are entitled to ownership. They can use the labor force more effectively than domestic enterprises. The advantage in ownership of production rate combined with cheap labor force facilitates the direct investments in collective laboring.

When a country wishes to upgrade its industrial structure, the priority for an enterprise making transnational investments is the cost of labor force. In the 1970s, Japan made investments in the four dragon nations (including South Korea, Singapore, Taiwan and Hong Kong) and ASEAN countries because it wanted to lower production cost by exploiting the differences of labor cost in different countries so as to obtain the competitive advantage in price. In the 1980s, the labor cost was surely one of the priorities for Japan when it made investments in foreign countries after the appreciation of the Japanese Yen. As the labor cost in the four dragon countries and ASEAN countries rose, enterprises transferred their investment target to Mainland China to establish labor-collective industries.

In 1997, Chunlai Chen examined the substantial industrial distribution in Mainland China based on amount of FDI in the manufacturing industry. He found that the capital/labor ratio of a certain industry and amount of FDI has a negative relationship.[7] That is, the industry with a highly collective labor force would be more attractive to foreign investors. According to the Hecker-Ohlin (H-O) principle, when a country has a unique advantage, it will use this advantage intensively for professional production in the international division of work and export-related products. This is one of the important factors that explain why the highly collective labor in Mainland China attracts export-oriented foreign investors. Therefore, Mainland China's impressive development of FDI results from its competitive advantage of labor cost under the trend of international industrial transfer. Table 6.2 shows Mainland China's tremendous and incomparable advantage in labor cost compared with other Asian countries.

(ii) Labor force quality: the analysis of labor force cost cannot be separated from the analysis of labor force quality. The population size in

Table 6.2. Wage Rate for the Labor Cost in Asian Countries (USD/hour)

	1997	1998	1999	2000
Mainland China	0.34	0.41	0.42	0.44
Taiwan	5.90	5.27	5.62	5.63
Hong Kong	5.42	5.47	5.19	5.17
Philippines	0.84	0.66	0.73	0.71
Thailand	2.38	1.86	2.08	1.95
South Korea	7.86	5.39	6.71	7.62
Singapore	8.24	7.77	7.18	7.76

Data Source: Economist Intelligence Unit (2002).

[7] Chen, Chunlai (1997), "The Location Determinants of Foreign Direct Investment in Developing Countries," The University of Adelaide, Working Paper No. 97/12.

Mainland China is huge, so the labor resources are relatively varied, but the overall level of the quality of labor force is relatively low. This is a serious disadvantage for the HR environment in Mainland China, and is one of the reasons for the limited quantity of FDI in technology-intensive industries. According to the data from the fifth population survey in Mainland China, the number of people who have university degrees per 100,000 inhabitants is only 3,611, those with senior high school degrees only 11,146, and the total national illiterate population (the illiterate person or person with less reading capability than a 15-year-old) is 85.07 million people. This is equivalent to a high illiteracy rate of 6.72%. However, educational development in Mainland China is occurring very fast. In 2004, the number of Mainland China's annual recruitment of graduate students was 326,000, the number of progressing students in masters programs 820,000, and the number of students that graduated from graduate school 151,000. The number of annual recruits for general higher education (university and college) was 4.473 million, the number of progressing students in schools (university and college) 13.335 million, and the number of graduates from schools (university and college) 2.391 million. The number of annual recruits for each type of middle and vocational education (middle and vocational schools) was 5.481 million, the number of progressing students in schools (middle and vocational schools) 13.679 million, and the number of graduates from schools (middle and vocational schools) 3.510 million. The number of annual recruits for entire general senior high schools was 8.215 million, the number of progressing students in senior high schools 22.204 million, and the number of graduates from senior high schools 5.469 million. The number of annual recruits for all general junior high schools was 20.782 million, the number of progressing students in junior high schools, 64.750 million, and the number of graduates from junior high schools 20.704 million. The number of annual recruits for all elementary schools was 17.470 million, the number of progressing students in elementary schools 112.462 million, and the number of graduates from elementary schools 21.352 million. The number of recruits in special education was 51,000, and the number of progressing students in special education 372,000. The number of kindergarten students was 20.844 million.[8]

We shall now focus on the notable binary structure of Mainland China's human capital. Despite the enormous number of low-education level and non-skilled labor force, Mainland China still possesses a relatively large high-quality and low-cost skilled labor force. Among some technology-intensive industries, there are R&D talents who have the same or similar research capabilities as those in developed countries, but with the lowest wage level globally for the same type of labor. However, the regional allocation of skilled human capital is centralized in the developed cities of Beijing, Shanghai, Guangzhou and Suzhou; this is the reason the direct investment of R&D type by many multinational companies has increased rapidly in Mainland China recently. During 1994 to 1999, the amount of R&D investment that multinational companies invested and guaranteed in Mainland China was 350 million USD, with the Beijing region alone attracting an investment of 200 million USD.

1.1.6. Channel of finance capital

The general economics theory states that enterprises will invest in projects when the expected earning is more than the investment cost. In addition, investment is more apt to occur when enterprises do not have to face the tightness of credit loans which are not related to its beneficial result. In countries with well-developed financial systems, this situation does not arise. In fact, many researchers have proved the importance of well-operated financial systems to growth (World Bank, 2001). Generally speaking, countries that possess "deeper" financial systems (banks, stocks and securities market) will develop more rapidly than countries with "shallow" systems. A stable, well-operated financial system is able to promote growth if it allocates financial resources to the investment projects with higher production efficiency, as this is more effective than simply increasing the total investment. According to a survey conducted by the World Bank, the number of regular

[8] National Bureau of Statistics. *2004 National Economics and Social Development Statistical Report*. Available at www.stats.gov.cn

finance opportunities that Mainland China's enterprises acquired was less than the enterprises of other Asian countries. About 29% of operating capital for the major enterprises in Mainland China was acquired from bank loans, which was less than the enterprises in Indonesia, Malaysia, Philippines, Thailand and South Korea (see Figure 6.2).

For small- and medium-sized enterprises, the situation seems even worse (see Figure 6.3). SMEs in Mainland China only acquired 12% operating capital from bank loans, compared to 21% in Malaysia, 24% in Indonesia, 28% in the Philippines, and 26% in Thailand and South Korea. In addition, if the scale of enterprises is smaller, they will find it even more difficult to acquire regular financing loans (see Figure 6.4). Mainland Chinese enterprises with more than 100 employees had about 27% of capital coming from bank loans, and Indian enterprises had 39%; Mainland Chinese enterprises with 20 to 100 employees had only 13% capital from bank loans, and Indian enterprises had 38%; while

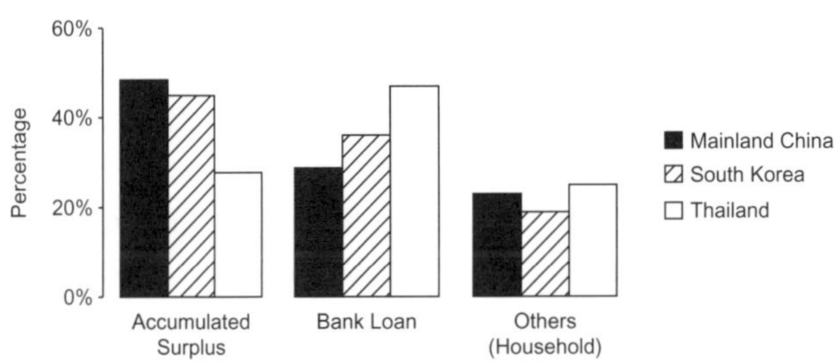

Figure 6.2. Financing Sources for Major Enterprises in 2000

Source: www.worldbank.org

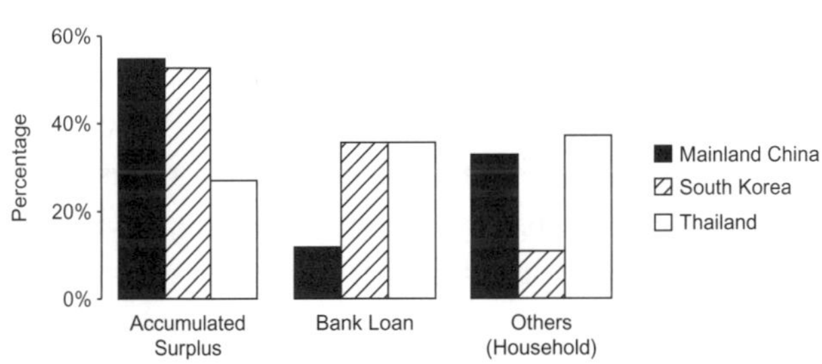

Figure 6.3. Financing Sources for SMEs in 2000

Source: www.worldbank.org

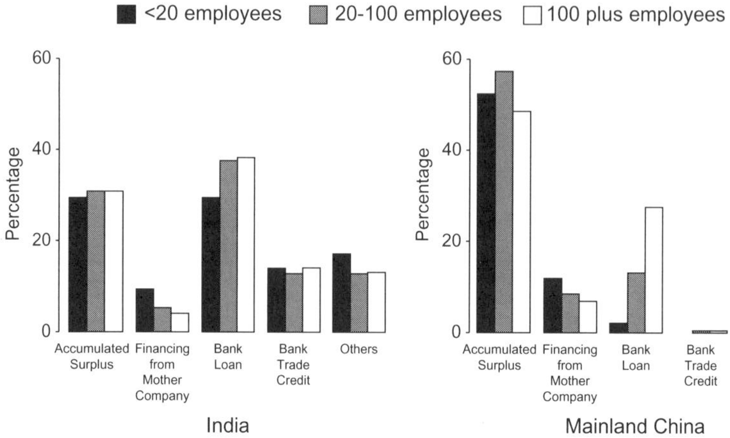

Figure 6.4. Finance Sources in 2000

Source: www.worldbank.org

Mainland Chinese enterprises with less than 20 employees had only 2.3% of capital on average from bank loans compared with 29% in India.

It is usually thought that an enterprise of smaller scale will acquire less regular finance. Because small enterprises have a shorter harvest period and because their prospects are full of uncertainty, regular banks will be more hesitant to grant loans to them. Moreover, these risk factors will cost more to acquire the loan. On the other hand, the ratio of major enterprises in Mainland China that depend on financing to resolve their operating capital is also much lower than other countries, and the difference of such ratios between Mainland China and other Asian countries is greater when the scale of the enterprises is smaller. As a result, the data strongly expresses the tightened monetary policy that Mainland Chinese enterprises may face in the future.

The result of this tightened monetary policy is very serious. The major enterprises may find it difficult to expand or upgrade, and the entrepreneur may not be able to enter the market. Such obstruction to entering the market, especially with increased economic costs, will be regarded as a type of entrepreneurship (for those who take the risk of providing innovative products and services to the market), which is a key composition of economic growth (McMillan and Woodruff,

2002). Allowing an entrepreneur to acquire financing, and then enabling them to implement their new ideas and develop the new business, is a global issue. However, if regular financing sectors continue to refuse to provide loans to enterprises because of their smaller scale, this problem will be even worse for individual entrepreneurs.

1.1.7. Inflation rate

The inflation rate is the major index that evaluates a country's economic stability. A higher inflation level is linked to lower unemployment and higher economic growth, but, generally speaking, the inflation rate that investors can accept is within 10%. On the other hand, the inflation rate also expresses the stability of the currency value. If the host country has a higher inflation rate, then the degree of depreciation of its currency is bigger. Serious inflation (above two digits of inflation) will make investors face the risk of depreciated investment. Table 6.3 is a survey of the inflation environment in Mainland China.

1.2. Analysis of an economic environment's potential

The increasing potential of a country's economy can be evaluated by inspecting its GDP growth rate, savings rate and domestic investment rate. GDP growth rate is expressed as the increasing speed of the total economic output and total consumption for a country, and also the measurement of the increasing market scale. For developing

Table 6.3. Mainland China's Inflation Rate, 1984–2004 (%) (GDP Price Index)

Year	Inflation Rate	Year	Inflation Rate	Year	Inflation Rate
1984	4.9	1991	6.7	1998	−1.1
1985	10.1	1992	7.9	1999	−1.4
1986	4.6	1993	14.6	2000	0.4
1987	5.1	1994	19.9	2001	0.7
1988	12.1	1995	13.2	2002	0.8
1989	8.8	1996	5.9	2003	1.2
1990	5.7	1997	0.8	2004	3.9

Source: *Mainland China Statistical Yearbook*, various issues.

countries, more FDI will be attracted when the GDP growth rate is higher and the prospect of profit for FDI is better. We will first take a look at the comparative data in Tables 6.4 and 6.5.

Mainland China's economic growth trend has a pattern similar to that of the emerging industrialized Asian countries. Currently, the economic growth speed of the East Asian Tigers, also known as Asia's Four Little Dragons, has entered a post period of high increase. The economic growth potential is still powerful in Mainland China, which will maintain the relative advantages for a longer time.

The savings rate and domestic investment rate reflects the host country's capability of using foreign capital in a way that reduces the risk and burden of FDI and increases the usage effectiveness. Thus, the effect of the savings rate and domestic investment rate on FDI is long-term. To further understand this, please refer to Table 6.6.

Table 6.4. GDP Growth Rate in Asian Countries (%)

Country	Period	GDP Growth Rate	Period	GDP Growth Rate	Period	GDP Growth Rate
Mainland China	1952–1978	6.15	1978–2000	9.52	—	—
Taiwan	1951–1962	7.92	1962–1987	9.48	1987–2000	6.59
Hong Kong	1966–1968	2.61	1968–1988	8.69	1988–2000	4.14
South Korea	1953–1962	3.84	1962–1991	8.48	1991–2000	5.76
Singapore	1960–1965	5.74	1965–1984	9.86	1984–2000	7.18

Table 6.5. GDP Growth Rate in Developed Countries (%)

	US	UK	France	Germany	Japan
1950–2000	3.46	2.5	3.64	3.94	—
1960–2000	3.45	2.45	3.28	3.01	4.9
1970–2000	3.21	2.3	2.49	2.54	3.21
1980–2000	3.21	2.48	2.08	2.45	2.61
1990–2000	3.24	2.65	1.79	2.65	1.23

Source: "Mainland China's Future Economic Growth and Its International Position Prospect", *Research Reference Data*, 2002, 2nd Issue.

Table 6.6. Comparison of Recent Savings Rate in Some Countries (%)

	1995	1996	1997	1998
Mainland China	43.1	41.5	42.7	42.6
Hong Kong	30.5	30.7	31.8	30.5
South Korea	35.5	33.5	32.8	33.4
Malaysia	33.7	34.7	32.8	33.4
Global Average	22.1	22.0	22.3	—

Source: *2000 International Statistical Yearbook.*

The table above shows that the savings rate in Mainland China, which is over 40%, is much higher than the average global level. Recently, the average domestic total investment rate in Mainland China has also been higher than the level in Hong Kong, South Korea, and Malaysia within the same period. This data expresses that the continued high savings level in Mainland China and stronger domestic investment rate form the motivation for long-term economic increase, and provide a sufficient guarantee for the sources of the follow-up capital of FDI to become an advantageous factor in the investment environment.

1.3. Analysis of foreign-related factors in the investment environment

The foreign-related economic factors in a country's (region's) investment environment reflects the depth and scope for its participation in the international disintegration and global trade. The related indices of the international trade, international balance, and international financial affairs not only express its capability of international competitiveness, but also reflect the country's FDI risk.

1.3.1. Analysis of foreign debt in Mainland China

The size of foreign debt of a country and its international credibility are very important for attracting FDI, and the size of foreign debt should be within the limit of the national economics bearing capacity. The influence of foreign debt on attracting FDI has two aspects: in general, countries or regions that are in debt will be actively attracting foreign capital, but high liability may incur a debt-paying burden, and exchange and default risk that will cause foreign investors to hesitate. In other words, the host countries have the willingness to attract direct investment, but the investing countries will be worried by the high liability that may result in the uncertainty of the expected income; both aspects form two opposite trends. When the host country's liability level is within the security level, then its attraction is greater than its repulsion and the FDI tends to enter the host country. When the host country's liability level is above the warning limit of the security level, however, its attraction is smaller than its repulsion, and then most FDI will tend to cut down or even withdraw completely.

The following two tables measure the size of foreign debt of Mainland China. The economic liability ratio indicates the ratio of foreign debt to the GNP; it reflects a country's liability level and repayment capability. According to international experience, generally, it should be maintained at around 10% to 20%. The debt service ratio is the ratio of repayment of capital with interest to the total export amount in the same year. It reflects a country's actual debt-paying capability in a certain year. This index should be maintained at around 20% or below 25%, and it might appear as the default risk if it exceeds the security level. From Table 6.7, we can see that the money supply of international debt in Mainland China is within the security level. The debt crisis or the country's default risk is small, and therefore Mainland China possesses a higher country credibility and good investment environment compared to the Latin American countries.

1.3.2. Analysis of Mainland China's international reserves

As shown in Table 6.8, the international reserves of Mainland China have maintained an increase over the years, and currently ranks at second place globally. This also reflects a good country credibility,

Table 6.7. Status of Mainland China's Foreign Debt, 1995–2004

Year	Foreign Debt (100 million USD)	Foreign Exchange Income (100 million USD)	Economic Liability Ratio (%)	Debt Ratio (%)	Debt Service Ratio (%)
1995	1,065.9	1,472.4	15.2	72.4	7.6
1996	1,162.8	1,716.8	14.2	67.7	6.0
1997	1,309.6	2,072.5	14.5	63.2	7.3
1998	1,460.4	2,075.9	15.2	70.4	10.9
1999	1,518.3	2,185.0	15.3	69.5	11.3
2000	1,457.3	2,795.6	13.5	52.1	9.2
2001	1,701.1	2,994.1	14.7	56.8	7.5
2002	1,685.4	3,654.0	13.8	46.9	7.9
2003	1,936.3	4,850.0	13.7	39.9	6.9
2004	2,286.0	6,550.0	13.8	34.9	3.2

Source: State Administration of Foreign Exchange of the PRC, www.safe.gov.cn.

Table 6.8. Status of Mainland China's International Reserves, 1995–2005 (100 million USD)

Year	Reserve	Year	Reserve	Year	Reserve
1995	735.97	1999	1,546.75	2003	4,032.51
1996	1,050.29	2000	1,655.74	2004	6,099.32
1997	1,398.9	2001	2,121.65	March 2005	6,591.44
1998	1,449.59	2002	2,864.07		

Source: State Administration of Foreign Exchange of the PRC, www.safe.gov.cn.

and forecasts a lower country contract-violating risk. Further to the advantageous conditions of attracting FDI, however, we need to consider the potential for the appreciation of local currency exchange rates that result from high foreign exchange reserves. The perspective on product exports brings a certain pressure, thus this is also found in the advantages of export costs.

2. Analysis of Systematic Factors in the Investment Environment

2.1. Sociocultural factors

Sociocultural factors are the basic aspects of a country's soft environment. These include language, race, religion, customs, traditional values, moral codes, and population quality. Investors will generally place emphasis on the host countries' socioculture due to the fact that socioculture greatly influences the international investment behavior, production and operation management, as well as its wide range of influencing aspects that involve the people's lifestyle, consumption tendency, working pattern and management behavior. In addition, in order to conduct investment activities and follow-up production and operation in the host countries, international investors have to interact extensively with the local government, employees, suppliers, sales agencies and customers. Thus, it is also important to evaluate each sociocultural factor in the host countries. However, the influence of sociocultural factors is often speculative and difficult to quantify. Generally speaking, if the cultural difference between the two countries is large, then the difficulty will also be large in terms of mutual communication and business collaboration, which will be unfavorable to FDI; on the other hand, common cultural origins and conventional values make it easier to advance the progress of direct investment.

Among the cultural commonality advantages within the investment environment in Mainland China, language is the most significant, stable and typical characteristic factor in the socioculture. The Chinese language is the only pictographic language in the world, while all other languages are phonetic languages, and its influence should not be underestimated. It is due to this characteristic that the 50 million overseas Chinese people who reside in every corner of the world, even if they speak different dialects (such differences can be even greater than the difference between Spanish and Italian), are still able to interact and communicate with each other using the Chinese characters and jointly use the thousands of years of historic heritage of Chinese culture. Another factor is the Chinese traditional cultural values, which have deep roots. After overseas Chinese establish a foothold in their residing country, they usually tend to return to Mainland China to conduct investment and trade. These overseas Chinese hold two unique advantages when investing in Mainland China, due to the characteristic sociocultural factors. The first is in reducing information access costs and mutuality of information, through official and nonofficial contact. The network of Chinese people advances the flow and mutuality of business information, and such information dispersal is very important in Mainland China which has a low policy transparency. It is beneficial for reducing the investment risk and default risk that results from asymmetric information, thus the contractual cost can be cut down. Secondly, the commonality of cultural values results in a convergence of behavioral manners. This decreases the disputes and loss during business operations, and also reduces the costs of implementing collaboration.

The disadvantages of the cultural differences within Mainland China's investment environment include the influential power of the Chinese language, which is still limited globally, and the ratio of business personnel in major Western developed countries who are familiar with the Chinese language, which is also very low. This restricts the channel of acquiring information about Mainland China's market for Western multinational companies, and makes it difficult for them to sufficiently understand the investment environment in Mainland China. At the same time, there also exists a bigger barrier for Chinese people when they try to use English to engage in international commercial exchange. In addition, the sociocultural traditions of Chinese people are very different from those of Western countries that inherited the ancient Greek and Roman cultures. This results in unavoidable disputes and contradictions between the personnel that come from these two different cultural backgrounds during commercial cooperation, and will be a negative factor in restricting the investment of Western countries to Mainland China to a certain extent.

2.2. Political environment

A survey conducted by the Political and Economic Risk Consulting Co. in 1997[9] showed that the participating companies thought the ratio of political risk counted for more than 30% in their evaluation of Asian countries' investment environment, with the average ratio of Mainland China's political risk at around 68.55%. The results of the survey were influenced by the political events surrounding the return of Hong Kong to Mainland China at the time, thus it unavoidably has a certain deviation. However, these results do reflect the differences between economic systems which greatly influence the evaluation of the investment environment. Since Mainland China became a member of the WTO, though, the developed Western countries have uniformly lowered their expectation of Mainland China's political risk.

Mainland China's political environment can be analyzed from three perspectives. Firstly, the political system, which, in light of recent economic reforms, can be categorized as an authoritative centralized socialist transforming economic country. Within the Western countries' evaluation of its systematic environment, the nature of "socialism" and "transformation" will objectively become an unstable factor influencing the inflow of FDI, especially when considering the opposition to the ideology and the change in international political environment. Secondly, is stability of the political situation. Mainland China's enormous national economic size and sound economic development has reduced the risk of an unstable political situation. In addition, the risk of major unrest or unexpected events that could influence the stability of the regime is very low in Mainland China compared to other developing countries. Thirdly, there is continuity and prediction of policies. As a centralized country, the integrated continuity of Mainland China's macro-management policy is very high. The prediction of policies related to FDI will be mainly expressed in the planning and implementation of internal regulations and rules, and the case management of FDI projects. The ratification of some investment projects may be the result of negotiation between the government and multinational companies. In addition,

[9] Available at www.perc.com

disputes over benefits may originate from the inconsistency between the government's industrial protection policy and the profits of multinational companies. Such case management usually has no significant guidance for the ratification of follow-up projects. Therefore, the existence of such policies results in the lower evaluation of the prediction for Mainland China's policy toward foreign organizations.

2.3. Governmental control over the environment

Factors such as the suitability of policies and regulations, administrative efficiency of governmental behavior, and the existence of corruption are closely associated with daily operations of enterprises. These factors have impact on the operational costs of enterprises. Foreign investors are especially sensitive to the system environment, due to their limited understanding of the national conditions of the invested country and their inferior position for getting relevant information.

2.3.1. Governmental administrative efficiency

We follow the Heritage Foundation's evaluation system of the governmental administrative environment. The concrete evaluation standards are shown in Table 6.9.

In the report of economic freedom presented by the Heritage Foundation in 2002, the management environment in Mainland China is ranked at level 4, which means higher interference and ineffective management. The reason lies in Mainland China's lack of transparent systems in policies and regulations, and inconsistency of execution according to international standards. A review published by *The Wall Street Journal* suggests that "the authorities and officials understand the changes of regulations and laws. But the investors and other governments don't." As a result, the Chinese government generates operational obstacles in administrative management to a certain extent. Such institutional disadvantages constrict the direct investments made by Western countries such as America and West European countries in Mainland China.

Table 6.9. Evaluation Standards of Governmental Management Factors

Level	Governmental Interference Degree	Evaluation
1	Very low	Current regulations are transparent and simplified. All enterprises are applied with uniform standards. Governmental management does not form barriers for the operations of businesses. There is almost no corrupt behavior.
2	Low	Review procedures are simple. The existing management is concise and transparent. Most of the cases are applied with uniform standards, but sometimes governmental management creates burdens for businesses. Very little corrupt behavior happens.
3	Middle	Review procedures are complicated. Governmental management creates great burdens for businesses. The existing regulations and policies are randomly applied to certain circumstances. Corrupt behavior exists and creates burdens for enterprises.
4	High	Government places restrictions on production. There is a national plan to a certain degree. Greater barriers exist for newly established enterprises. Review procedures are complicated, and handling charges are high. Sometimes, bribes are necessary. Corrupt behavior exists and creates greater burdens for the operation of enterprises.
5	Very high	Government places restrictions on the establishment of new enterprises. Corrupt behavior is common. Policies and regulations are applied randomly.

Source: The Heritage Foundation (2002), "Index of Economic Freedom."

2.3.2. Governmental incorruptibility

The degree of governmental incorruptibility has great influence on direct international investment. In terms of economics, the cost of commercial contracts between enterprises and the government, the local business license obtained, or permission may rise due to the corruption of the government which will cause a waste of economic resources. The corruption of governmental officials obviously carries negative influence on direct foreign investments (Wei, 2000). Direct international investments are inclined to flow to governments, countries, or regions with integrity and high transparency (Lu, 1997). By observing concrete situations, Andy Xie of Morgan Stanley estimates that the corruption cost in Mainland China accounts for approximately 2–3% of GDP, while Doctor Hu An's estimation of the corruption cost in Mainland China reaches 13–16% of GDP. In 2000, a judicial investigation was conducted to uncover governmental officials associated with corruption. 45,000 cases were discovered, an increase of 15% compared to that in 1994. According to the 2001 corruption index of worldwide countries presented by Transparency International, the corruption degree in Mainland China was rated 3.5 (10 for the least corrupt). This shows that Mainland China is more corrupt than average, and the incorruptibility is ranked 57 in the world.

The existence of corruption in Mainland China negatively affects international direct investment. The corruption has a particularly large influence on American enterprises because the Foreign Corrupt Practice Act in the US discourages American enterprises from making investments in a country with serious corruption. Wei's study (2001) indicates that Mainland China can attract more direct investments, especially investments from America, if the Chinese government can improve its corruption situation, without changing other conditions.

2.3.3. Legal protection

The legal environment in Mainland China is quite different from that in western countries. By observing the legal system, it is clear that the judicial mechanism in Mainland China faces strong interference from administrative concepts and power. The judicial mechanism is not an independent process. The law enforcement in Mainland China is more like a completion of an administrative mission instead of an independent process. Conflict of interest exists everywhere, especially in the cases involving governmental interests (Index of Economic Freedom, 2002).

The major international capital export countries are market economic countries. The market economy is based on private ownership. Therefore, the enterprises from these countries value the property rights under the protection of the laws stipulated by the invested countries when making transnational investments. The rating made by the

Heritage Foundation on Mainland China's legal system environment is 4 (1 for the best protection). This means that the protection of property rights offered by the laws in Mainland China is weak. The ownership of private property is poorly protected. Furthermore, the legal system is extremely inefficient, given that the judiciary is influenced by other governmental units. The factors described above are disadvantages in attracting direct foreign investments from advanced western countries. The Chinese government is trying its best to improve the above situation. For instance, the government has adopted practical measures to offer legal protection for the enterprises invested by foreigners. The PRC Constitution, Article 13, has specified that the nation shall protect the ownership of legal property and the right of succession. The most updated Property Rights Act will initiate a new start for protecting civil rights. Furthermore, related laws and regulations in connection with foreigners explicitly indicate the legal protection for various rights involving foreigners. Substantially speaking, the Chinese government protects foreign partnership, the property ownership of foreign investors, and disposal rights in accordance with laws. In Article 2 of the Chinese and Foreign Joint Ventures Act, the Chinese government commits to protecting foreigners who invest in Mainland China in accordance with the accords, contracts, agreements and articles approved by the Chinese government, regarding the aspects of investments, profit shares and other legal rights. The Chinese government also signs bilateral investment protection agreements with 71 countries, which is an important measure to protect each legal right to which foreigners are entitled.

3. Economic Policies and Laws

Among the factors contributing to FDI, the economic policies and laws involving foreigners act as the functions for guiding and adjusting investment relationships, protecting the benefits and safety of investors, and adjusting the investing behavior. The attitude towards international capital export and import held by a country and the relative incentives, protections and restrictions are explicated through laws and regulations. These policies and laws form an important part of the environment for direct international investment.

3.1. Trade and customs policies

In recent years, when people make evaluations on the environment for international direct investment, they have become increasingly aware of the importance of trade and customs policies in the invested countries. The global production, management and marketing of a transnational company closely combine international trade with the international direct investments. A transnational company has great import needs when dealing with its machinery, facilities, raw material, components and parts. Therefore, the trade and customs policies of an invested country determine the investment policies made by the transnational company.

Since the reforms began in Mainland China, the customs policies have been changed from self-sufficiency to outward-opening policies. The tagline for customs policies has become "obey the nation's opening policies, facilitate import and export, ensure the supply of daily necessities, push economic development, and guarantee the financial income of government". Since joining the WTO in November 2001, Mainland China has kept its promise to lower import customs for over 5,000 items. On 1 January 2002, the average tariff was lowered from 15.3% to 12%; the average tariff rate for industrial products was reduced from 14.7% to 11.3%; and the average tariff rate for agricultural products (exclusive of aquatic products) was reduced from 18.8% to 15.8%.[10] As for non-customs barriers, Mainland China has been steadily eliminating or modifying the restrictive trade measures against related regulations. The regulations and laws in connection with foreign trade and foreign investment have been undergoing great modifications. By December 2001, the number of abolished department regulations and documents reached 356. This situation indicates that the Chinese government is gradually promoting free trade.

3.2. Tax collection

Mainland China is a country with heavier tax — the highest margin tax rate of personal income reaches 45%; the average tax rate is

[10] www.moftec.gov.cn

20%; and the highest margin tax rate of business income reaches 33%. However, there are certain exemptions for enterprises involving foreign investments. As for business income tax, there are a variety of preferential business income tax rates such as preferences by business, region and industry. For example, enterprises with foreign investments engaged in production can enjoy the tax collection policy of "two exemptions and three reductions". Most foreign-invested enterprises are production-oriented. Therefore, they enjoy preferential tax rates that vary from 15% to 24% based on the nature of business or location. The foreign-invested enterprises that are located at Hai Nan Special Economic Zone or Shanghai Pu Tong Special Zone and are engaged in airports, ports, railway, freeway or power stations for a period of 15 years can further obtain the preference, "five exemptions and five reductions".[11]

3.3. Intellectual property rights

From the perspective of the legal environment of intellectual property rights, trademark law, patent law, copyright law and illegal competition law constitute the basis of intellectual property rights protection in Mainland China. There are another 30 administrative regulations and rules relating to intellectual property rights. In addition, the 1997 revised criminal law specifies the criminal contents violating intellectual property rights. The above-mentioned regulations and laws form the legal system protecting intellectual property rights. To adapt itself to the international society, Mainland China has joined major covenants relating to intellectual property rights.

Since becoming a member of the WTO, Mainland China has come under the jurisdiction of TRIPS (Trade-Related Aspects of Intellectual Property Rights). Generally speaking, TRIPS reflects the benefits of advanced countries led by America. It specifies certain regulations which are of high standards and high requirements, and are therefore difficult to perform. TRIPS is forcing Mainland China to revise many regulations specified in its related laws of intellectual property rights. From 2000 to 2001, Mainland China finished the modifications

[11] Ministry of Finance (1999), *Taxation Law*, People's Republic of China.

regarding patent law, patent law enforcement, trademark law, copyright law, and computer software protection regulations, etc. Mainland China also stipulated IC design protection regulations and made modifications on trademark law enforcement, copyright law enforcement, and pharmaceutical management law enforcement. After the completion of these regulations and laws, the intellectual property rights protection is expected to meet the requirements specified in TRIPS.

3.4. Policies and regulations relating to foreign investments

The main feature of policies and regulations in the investment environment in Mainland China is the coexistence of preferential investment measures and investment restrictions in its export-oriented policies. These policies and regulations aim at encouraging enterprises that can sufficiently use the domestic competitive advantages (cheap labor cost, low land price and sufficient natural resources), supplement the disadvantages (lagging technology and insufficient infrastructure), and facilitate regional development (in central and western areas). The restrictions aim at reducing the potential impact on weak domestic industries. Controls are conducted over the foreign investments made in the industries such as media, telecommunication and finance, which are sensitive to national economies.

Substantially, since Mainland China promulgated the Chinese and Foreign Joint Venture Law, over 500 laws and regulations stipulated have influenced direct foreign investments. The department regulations and rules promulgated by the Internal Affairs Ministry and local governments have exceeded 1,000.

Being a member of WTO has brought great impact on Mainland China's policies and laws. The domestic laws and regulations that contradicted the agreements with WTO have been modified. In 2002, the government promulgated the Industrial Guidance for Foreign Investments, which increased incentive projects from 186 to 262 and reduced restrictive projects from 112 to 75. In terms of industrial restrictions, the public businesses including natural gas, heat power, water supply, etc., are now listed open to foreign investments. Mainland China also followed up on its agreement to deregulate the restrictions in region, quantity, business range and ratio of shareholdings relating to the banking, insurance, professional services (accounting, auditing and laws), and telecommunications sectors. For example, the Chinese government is now encouraging more foreign investments in the western areas, such as infrastructure of energy resources. The ratio of foreign shareholding has also been raised. The newly promulgated and revised laws and regulations reflect the current orientation of policies in Mainland Chinese industries, i.e., to further take advantage of the influx of foreign investments to improve the infrastructure of energy resources and transportation, facilitate the reform and upgrading of industries, and strengthen the investments in the fields of electronic information and bio-engineering. Moreover, Mainland China endeavors to provide a more open and fair investment environment for foreign investors. This trend is becoming increasingly beneficial to transnational investors.

4. Infrastructure and Natural Factors in the Investment Environment

Basic constructions are also called infrastructure. Its basic function is to provide services for the complex of production areas, industrial hubs and cities, industrial zones, and other special economic zones. These are the material requirements which these areas rely on for their development. Infrastructure acts as the hardware factor of an investment environment, and is an important consideration when foreign investors make investment decisions. Foreign investors engaged in production activities in an invested country will need transportation services, mailing and telecommunication services for external contacts, and power and water supplies.

4.1. Energy resources infrastructure

A stable and sufficient energy supply acts as the basic protection for normal industrial production activities and is a top priority for foreign direct investors. Lu Ming Hung (1997) suggests that international foreign direct investment tends to flow into countries with better energy resources infrastructure. The per capita energy consumption reflects the significant influence of infrastructure on the inflow of FDI. Whenever the per capita energy consumption increases by 1%, the current FDI inflow increases by 0.292% accordingly. The aggregated FDI in the next period increases by 0.3014%.

According to the data in the *World Development Report 2000–2001* presented by the World Bank, per capita petroleum parity energy consumption is one of the indices for national energy infrastructure. Table 6.10 selects some countries for comparative data.

From the above data, it can be seen that the energy infrastructure in Mainland China is not in accordance with per capita energy consumption. There is still a great gap compared with that in Southeast Asian countries, Hong Kong and Taiwan. More effort should be devoted to achieve the levels shown in advanced countries.

Looking at the energy production flexible coefficient, energy growth is slower than economic growth in Mainland China (see Table 6.11). Therefore, problems of energy shortage will occur in Mainland China in the future.

Receiving reliable power supply at a reasonable price is considered the first priority for many manufacturers. Power generation is obviously increasing in Mainland China. However, the newly added power generation capacity is still lower than the population growth. In this aspect, Mainland China is falling behind other countries. Thailand, Uruguay and Malaysia have a higher per capita power generation. Now, Mainland

Table 6.10. Per Capita Petroleum Parity Energy Consumption by Country

Country	Per Capita Energy Consumption by Petroleum (kg)	Country	Per Capita Energy Consumption by Petroleum (kg)	Country	Per Capita Energy Consumption by Petroleum (kg)
Bengal	197	Uruguay	883	Taiwan	3,405
India	479	Mainland China	907	Italy	2,839
Philippines	520	Thailand	1,319	Japan	4,084
Vietnam	521	Chile	1,574	Canada	7,930
Peru	621	Hong Kong	2,172	Germany	4,231
Indonesia	693	Malaysia	2,237	US	8,067

Source: *World Development Report 2000–2001.*

Table 6.11. Energy Production Flexible Coefficient for Mainland China

Year	Growth Rate Last Year (%)			Flexible Coefficient	
	Energy Production Ratio	Power Production Ratio	GDP Ratio	Energy Production	Power Production
1985	9.9	8.9	13.5	0.73	0.66
1989	6.1	7.3	4.1	1.48	1.78
1990	2.2	6.2	3.8	0.58	1.63
1991	0.9	9.1	9.2	0.10	1.00
1992	2.3	11.3	14.2	0.16	0.80
1993	3.6	15.3	13.5	0.31	1.13
1994	6.9	10.7	12.6	0.55	0.85
1995	8.7	8.6	10.5	0.83	0.82
1996	2.8	7.2	9.6	0.29	0.75
1997	−0.2	5.0	8.8		0.57
1998	−6.2	2.9	7.8		0.37
1999	−12.2	6.3	7.1		0.89
2000	−2.0	9.4	8.0		1.18
2001	13.0	8.6	7.5	1.73	1.15
2002	14.4	11.5	8.3	1.73	1.34
2003	15.8	16.5	9.3	1.70	1.77

Source: 2004 Mainland China Annual Statistics.

China is working hard on a series of large-scale wiring constructions which will dramatically enhance the power generation. Power generation capacity is an important index to evaluate power, as a reliable power supply means the transmission of power with few problems. If the power fails constantly, it will not be helpful to enterprises even if the power is provided at a very low price. According to a survey by the World Bank, only 30% of Chinese enterprises are equipped with emergent power generators. In India, the power reliability is not good, but 69% of enterprises are equipped with emergent power generators. Chinese enterprises claim that the average loss arising from a problematic power supply accounts for 2%. In Pakistan, it accounts for 6%. Tables 6.12 and 6.13 present the situation of power supply for the investment environment in Mainland China.

Table 6.12. Per Capita Power Consumption by Country in 2000

Country	Per Capita Power Consumption (kWh)	Country	Per Capita Power Consumption (kWh)	Country	Per Capita Power Consumption (kWh)
Bengal	76	Uruguay	1,710	Taiwan	—
India	363	Mainland China	714	Italy	4,315
Philippines	432	Thailand	1,360	Japan	7,241
Vietnam	203	Chile	2,011	Canada	15,829
Peru	607	Hong Kong	4,959	Germany	5,626
Indonesia	329	Malaysia	2,352	US	11,822

Source: *World Development Report 2000–2001.*

The above tables show that the overall development of power infrastructure is not sufficient in Mainland China. There is a substantial gap in power infrastructure between Mainland China and advanced countries and new industrialized countries. Mainland China's world ranking of the above index is at No. 75. On the other hand, the energy infrastructure is under rapid construction and development. The prospects brought by such improvement on power supply are generally good. In 2001, Mainland China increased the production generation capacity with newly added energy facilities. The supply of raw coal increased by 10.81 million tons; the supply of crude oil increased by 18.98 million tons; the natural gas increased by 2.1 billion cubic meters; the capacity of middle and large-scale power generators increased by 14.98 million kW; and the capacity of substation facilities over 110,000 volts increased by 53.47 million kV.[12]

4.2. *Transportation infrastructure*

Transportation infrastructure also plays an important role in the investment environment. A well-developed transportation system serves as an essential premise for a smooth and highly efficient flow of human power, property and materials in the process of realizing investment.

[12] *2001 National Economics and Social Development Statistical Report.* Available at www.stats.gov.cn

Table 6.13. Power Balance Sheet in Recent Years in Mainland China (0.1 billion kWh)

	1990	1995	2000	2001	2002
Availability	6,230.4	10,023.4	13,472.7	14,632.6	16,330.7
Power generation	6,212.0	10,077.3	13,556.0	14,716.6	16,404.7
Hydro power	1,267.2	1,905.8	2,224.1	2,774.3	2,879.7
Thermo power	4,944.8	8,043.2	11,164.5	11,767.5	13,273.8
Nuclear power		128.3	167.4	174.7	251.2
Import	19.3	6.4	15.5	18.0	23.0
Export (-)	0.9	60.3	98.8	101.9	97.0
Consumption	6,230.4	10,023.4	13,471.4	14,633.5	16,331.5
Among consumption:					
1. Agriculture, forestry, livestock farming, fishery, water conservancy	426.8	582.4	673.0	762.4	776.2
2. Industry	4,873.3	7,659.8	9,653.6	10,444.7	11,793.2
3. Architecture	65.0	159.6	154.8	144.9	164.1
4. Transportation and delivery, storage and mailing, telecommunications	105.9	182.3	281.2	309.3	338.0
5. Wholesale and retail, trade and catering	76.2	199.5	393.6	444.9	500.0
6. Others	202.4	234.2	643.2	688.1	758.6
7. Daily consumption	480.8	1,005.6	1,672.0	1,839.2	2,001.4
Among consumption:					
1. End consumption	5,795.8	9,278.9	12,534.7	13,600.0	15,162.8
Industry consumption	4,438.7	6,915.3	8,716.9	9,411.2	10,624.5
2. Loss of power transmission and distribution	434.6	744.5	936.7	1,033.5	1,168.7

Source: *2004 Mainland China Annual Statistics.*

(i) Railway transportation: the railway system has been playing a very important role in the industrial production in Mainland China. Even though rail construction has undergone rapid development in past decades, it still cannot fulfill the needs of passenger and cargo transportation arising from national economic activities. By the end of 2004, the rail operating mileage in Mainland China totaled 73,000 thousand meters. In 2004, the passenger traffic by rail accounted for 6.12% of total passenger traffic in Mainland China, which is a decrease compared with that in the previous year. Meanwhile, the figure is far lower than the passenger traffic by road, which was 92.24% in 2004. The freight traffic by rail accounted for 14.16% of total freight traffic, which is also far lower than freight traffic by road at 74.28%. Insufficient railway capacity has become an important factor restraining economic development, especially in underdeveloped areas.

(ii) Road transportation: the total road mileage in Mainland China in 2004 reached 1,800,000 thousand meters, which increased by 33.33% compared with 1,350,000 thousand meters in 1999. The mid-level roads (including highways) accounted for over 90%. By observing the information from 1999 to 2004, the passenger traffic by road accounted for over 90% of total passenger traffic. Meanwhile, the freight traffic by road is also rising.

(iii) Water transportation: there are many advantages shown in water transportation. For example, water transportation can bear a larger capacity at lower cost. There are many water systems in Mainland China which are mainly gathered in Jiangsu, Zhejiang, and Chongqing. The development of water transportation has been fast. In recent years, the freight traffic by water, which accounts for 50%, has exceeded the traffic by rail, and keeps on rising.

(iv) Others: air transportation in Mainland China has not been well-developed so far. Air transportation does not account for much, either in passenger or freight traffic. However, rapid development in air transportation is occurring in Mainland China, especially in airports and air routes. By cooperating with American and European countries, the aero development is further facilitated. According to the civil aviation statistics, the average annual growth rate of domestic passenger traffic increased by 15.9% during 1991 to 2001. The average annual growth rate of domestic freight traffic increased by 17.3%. In the experience of the international civil aviation industry, the growth period for civil aviation transportation generally lasts thirty to forty years. The civil aviation industry in Mainland China will maintain a higher growing speed over the next twenty years. In addition, Mainland China has over four hundred oil and gas transportation systems with a mileage of over 15,000 thousand meters.

Generally speaking, Mainland China has already built up a transportation infrastructure consisting of a passenger and freight transportation network by rail, road, water, pipe and air. However, the extent of such infrastructure still falls behind the requirements for national economic development. This is a disadvantage for the investment environment.

4.3. IT infrastructure

IT infrastructure has also been a key factor in an investment environment. IT infrastructure has become increasingly important as current IT develops rapidly. Table 6.14 compares the IT infrastructure of Mainland China with that of advanced countries.

There is still a great gap in IT infrastructure between Mainland China and the advanced countries. However, one of the outstanding features shown in IT infrastructure in Mainland China is the amazing speed of growth. For example, the average annual penetration rate of telephones reached 30% from 1998 to 2003. By observing the overall developing trend, the volume of postal and telecommunications services completed in 2004 totaled 979.1 billion USD, which is an increase of 34.9% compared with that of the year before. The postal service value reached 56.6 billion USD at an increase of 4.4%, while the telecommunications service value totaled 922.5 billion USD at an increase of 37.4%. The annual addition of exchange reached the number of 70.2 million gates with a total capacity at 0.42 billion gates. The number of newly added landline subscribers totaled 49.7 million.

Table 6.14. IT Infrastructure in Different Countries

Country	Telephone Penetration Rate (Sets/Thousand persons)	Internet (PCS/10 Thousand Persons)	Fax Machine (Sets/Thousand Persons)	Newspapers (Copies/ Thousand Persons)
Mainland China	105	0.21	16	23
Hong Kong	558	108	543	105
Taiwan	—	—	—	—
UK	557	149.06	339	351
Germany	567	140.68	731	317
France	570	73.2	474	237
Japan	503	107	1268	576
US	661	975	784	228

Source: *2000 International Annual Statistics.*

By the end of 2004, the number of subscribers reached 312.44 million. From this number, urban subscribers accounted for 210.85 million and rural subscribers accounted for 101.59 million. The number of newly added mobile phone subscribers totaled 647.27 million and 334.83 million at the year's end. The total number of landline and mobile subscribers reached 647.27 million, which is an increase of 114.57 million.[13] The skyrocketing development of IT infrastructure is the most obvious feature in the current and future investment environment of Mainland China. This feature is extraordinarily attractive to FDI of service industries.

4.4. Analysis of natural factors in the investment environment

The natural factors in an investment environment can be divided into three types: population, resource advantages, and geographic factors. The major influence of population on FDI comes from labor supply, which is part of the HR environment discussed previously.

[13] *2004 National Economics and Social Development Statistical Report.* Available at www.stats.gov.cn

The total population has a significant impact on the flow of international investments. Also, a huge population represents tremendous potential purchasing power.

In terms of resource advantages, agricultural resources are abundant in Mainland China (see Table 6.15). The industries relating to agriculture like forestry, livestock farming and fishery are well-developed, with complete production systems. China's mineral resources have brought it a ranking as one of the world's top countries. The storage of natural resources is large and is an excellent advantage for Mainland China in attracting FDI. In the meantime, rich natural resources provide good support for the foreign services and manufacturing industries when they make their investments.

In China, the deposits of various natural resources and energy resources are vast. Among the variety of mineral energy resources, coal, uranium, petroleum and natural gas are ranked the third, seventh, eleventh and fifteenth respectively, which totals 13% of the grand deposits of mineral energy resources in the world. Hydropower resource accounts for 16.7% in the world, which ranks China as first globally. These percentages are higher than the percentage of land in Mainland China, which accounts for 7% in the world. Compared with advanced countries, the degree of geological exploration in Mainland China is still low. Therefore, there may be new major discoveries in the future. If per capita share is taken into consideration, the energy

Table 6.15. Rating of Natural Resources in Major Countries (full mark: 100)

	Former Soviet Union	America	Mainland China	Brazil	Canada
Area of cultivated land	15.7	12.9	8.8	5.1	3.2
Area of permanent grass land	11.8	7.6	9.0	5.2	1.0
Forest conservancy	25.0	7.7	2.7	19.5	8.0
Water resources	10.1	6.8	5.6	11.0	6.7
Available hydropower resources	11.9	8.3	16.7	4.0	4.2
Mineral resources	17.7	16.6	14.6	2.9	4.0
Accumulated percentage	92.2	59.4	55.4	47.7	27.1

Source: Cheng Hung, Editor-in-Chief, *Manual of Natural Resources in Mainland China*.

resources in Mainland China remain insufficient. Even though the hydropower resource in Mainland China is ranked first in the world, the per capita share only reaches 78% of world average, and coal, uranium, petroleum and natural gas reach 87%, 20%, 11% and 4% respectively. The statistics show that the per capita share of each energy resource in Mainland China is much lower than the world average.

5. Comparison of Regions within Mainland China

In the past fifteen years, the investment environment in Mainland China has improved a lot. However, the domestic investment environment is not consistent. National indices reveal the great differences existing in different regions. Mainland China is a country with varied cultures, foods, geographic characteristics and climates. Therefore, the numerous investment environments are different from each other. It is necessary to study the differences of the various investment environments in Mainland China. Beijing, Shanghai, Guangzhou and Chongqing are good examples, and their regional differences for making investments will be analyzed.

We adopt a series of indices to introduce and evaluate the suitability degree of an investment environment instead of one single index. Such a method reflects the complicated characteristics influencing enterprises in their investment decisions and also emphasizes the features or weaknesses of a region by comparing it with other regions.

The indices are classified into nine groups based on previous studies, and reflect the most important factors influencing enterprise performance. In the discussion of regional investment environments, the following nine items are taken into consideration:

- infrastructure;
- globalization;
- skills and technological advantages;
- barriers for participation in and withdrawal from the domestic market;
- flexibility of the labor force market;
- participation of the private sector;

- financial conditions;
- governmental efficiency; and
- tax.

5.1. Infrastructure

Infrastructure serves as one of the factors determining the production efficiency of an enterprise. Every enterprise depends on transportation to deliver its products and import raw materials. Meanwhile, a communication network is necessary for an enterprise to make external contacts. Gas, power and water supplies are essentials for the production of an enterprise. It is obvious that the quantity and quality of infrastructure form an important part of an investment environment. Newly established enterprises are likely to choose a location with complete infrastructure in which to place their investments. When an enterprise is not satisfied with the infrastructure at its location, it may move to another place. Therefore, infrastructure is an important determining factor for an enterprise when choosing its set up location and production efficiency. However, it is very expensive to build infrastructure, which is usually a part of public facilities. The ratio of investment in infrastructure to total social investment is 20%, and to public investments 35% to 55%.

According to the report by the World Bank, the following factors regarding infrastructure are taken into consideration — power transmission loss, loss of goods in transportation due to theft and damage (to total sales amount), quantity and quality of transportation by road and port, traffic jams, etc. The rating results are as follows: Shanghai and Chongqing are rated A; Guangzhou and Beijing are rated B–.[14]

5.2. Globalization

As discussed in the previous chapters, greater access to the international market is beneficial to a country's economic growth, on both the macro and mirco levels. The enterprises with foreign partnership,

[14] www.worldbank.org

participation in the international market, and pressure from import competition are more productive, especially in developing countries. The participation of foreign enterprises facilitates the transfer of technology and management experience and helps the domestic market integrate with the international market. The increase of market share by importers may bring greater pressure to domestic competitors. However, this will then force domestic competitors to enhance their production capacity. As a result, a friendly investment environment should invite foreign investments and welcome foreign products.

There are three standards adopted to evaluate the degree of globalization. Firstly, the degree of foreign ownership. This refers to the ratio of foreign investors with total ownership. The ratio also acts as an index of the existence of foreign investments. In general, a greater potential effect will be shown if the cooperation between foreign companies and local companies does not result in a wholly invested foreign subsidiary. Therefore, the second standard is to evaluate the ratio of enterprises with join venture partnership in each city. The aim of this standard is to evaluate other important cooperations besides ownership, such as research, development, training and marketing.

In terms of globalization, Shanghai and Guangzhou are at leading positions. In Shanghai, almost 40% of the enterprises cooperate with foreign companies. The ratio of foreign ownership to the asset value of the surveyed companies is nearly one third. 28% of the enterprises in Guangzhou cooperate with foreign investors, and foreign ownership accounts for 35%. Beijing is located somewhere in between. The inland city, Chongqing, is ranked last. Only 7% of the enterprises in Chongqing cooperate with foreign companies and foreign ownership accounts for 10%.

The third standard is the market share of import products to the products produced by domestic enterprises. The international competition faced by domestic enterprises becomes drastic whenever there is a higher import ratio. Based on such a standard, greater international competition is faced by Guangzhou, Shanghai and Beijing. The import market share accounts for 11.7%, 8.8%, and 8.2% respectively. The degree of market openness is low in Chongqing, with an import market share of 5.9%.

5.3. Skills and technological advantages

Generally speaking, investors are attracted to regions with a large mature labor force and advanced technology. In order to evaluate the status of skills and technology in each city, three indices are established by using the information on the investment environment in Mainland China. The first index is the ratio of workers with formal training. Training enhances the skills and production capability of workers. The second index is the quality of workers. This index is raised according to the proportion of technicians, management personnel and salespersons with educational degrees. The reason for emphasizing technology and management (and sales) personnel is that employers seek workers with better skills than other types of employees such as production workers or assistant workers. The last index is R&D. It includes per capita R&D expenditure, or the ratio of R&D personnel to total employees and the dependency on external research and development services.

Table 6.16 shows that no city is better than the others in every aspect. Overall, though, Shanghai, Beijing and Guangzhou perform better in skills and technological advantages. Among the surveyed enterprises in Shanghai, 57% of employees have received training (lower than Guangzhou). As for the quality of personnel and the degree of R&D, Shanghai is ranked first. The enterprises in Chongqing show a poorer result. Less than half of the employees there have received training, while the quality of personnel is ranked third and the degree of R&D last. 48% of employees working for enterprises in

Table 6.16. Skills and Technological Advantages among four cities

City	Percentage of Employees Receiving Training	Index of Personnel Quality	Degree of R&D
Beijing	0.48	0.61	0.58
Shanghai	0.57	0.66	0.63
Guangzhou	0.74	0.51	0.57
Chongqing	0.41	0.52	0.32

Source: www.worldbank.org

Beijing have received training, which is ranked third, the quality of personnel second, and the degree of R&D second. Though Guangzhou has the best trained workers, the quality of personnel and the degree of R&D are not good. This is because Guangzhou has a great number of migrant workers, and most enterprises in Guangzhou belong to low-technical fields.

5.4. Barriers for participation in and withdrawal from the domestic market

Many people have noticed the barriers for participating in and withdrawing from the market in Mainland China. The National Development Planning Committee established a team to study this problem and published a book in 2000 (SDPC, 2000). The Mainland China Team of the World Bank also conducted further studies on local protectionism. Their report aimed to evaluate the barriers for participating in and withdrawing from the four cities from as many angles as possible. The first index for evaluating the barriers is the border effect index recently established by PONCET. It directly analyzes commodity trade to assess the border effect, i.e., to make sense of the barriers of borders of each province on trade. By comparing the trade volume generated trans-province and in-province and taking the factors of transportation cost and geographical barriers into account, it calculates the essential customs for the trade of the same category by presuming all markets are uniform (Gilly, 2000). The customs level calculated is the border effect. Therefore, the trade barriers of a province are greater whenever the border effect is greater. According to the most recent available data from 2003, Chongqing has the highest border effect, followed by Beijing, Guangzhou and Shanghai. Other indices investigate the market share of major products produced by domestic enterprises. However, this method is not the best way to assess market competitiveness, since a higher share represents higher barriers for foreign investors entering the market. The 2003 World Bank Report indicates that the Guangzhou enterprises account for the lowest market share at 7.9%, followed by Chongqing at 10%. The next index is to evaluate the ratio of cost generated by contractors to the total cost generated by an enterprise. When the market is more flexible, it is not necessary for an enterprise to have all its activities completed internally. Contracting represents an easy participation for an enterprise in the market and a more concrete division of production work. Such a phenomenon makes the region more attractive. Shanghai is at the leading position, followed by Beijing, Guangzhou and Chongqing.

Finally, a good approach to evaluate the barriers is to examine the excess production capacity of enterprises. Investment, employment and unemployment cause cost. An enterprise usually runs at a certain excess production capacity. A larger excess production capacity means it is difficult to withdraw from the market. Based on this method, Chongqing and Beijing bear the greatest barriers to withdraw from the domestic market. The percentages are 32.1% and 30.5% respectively. Guangzhou and Shanghai are better at 27.5% and 24.9% respectively. Generally speaking, Guangzhou and Shanghai have lower barriers to participate in and withdraw from the market, followed by Beijing and Chongqing.

5.5. Flexibility of the labor force market

Besides product market, there is another important component for an excellent investment environment, namely, an effective labor force market. A sound labor force market should be flexible and include lower barriers for withdrawal from the market. In addition, a flexible labor force market effectively matches workers and employers. For example, if an employee cannot meet the technical requirements set by the employer or if an employee finds another job with a higher income, it would be better at this time to release both parties from prior agreements. Even though the flexibility of the labor force market can be categorized under barriers for participation or withdrawal, we still regard it as an independent factor because of its extremely important features. There are many barriers for withdrawal consisting of labor force factors, for example, restrictions on unemployment of labor force, governmental interference in the process of laying off redundant employees, strict regulations of seasonal employment or contracted employment, employee insurance, medical care protection, pension, etc. There are two ways to measure the withdrawal barriers in connection with labor

Table 6.17. Comparison of Labor Market Policies among Mainland Chinese Cities

	Shanghai	Beijing	Guangzhou	Chongqing
Excessive employees	0.075	0.175	0.061	0.072
Temporary workers	0.14	0.126	0.21	0.24

Source: www.worldbank.org

force. The first is the ratio of short-term employees. It is much easier to downsize temporary and seasonal workers. Therefore, one more short-term worker represents relatively lower withdrawal barriers. Furthermore, in a survey conducted on companies, they were asked one question: what is the tolerated ratio of redundant employees in a company if companies are not penalized by the government when they lay off the employees? Therefore, a more flexible labor force market can also be determined by the ratio of short-term employment and lower ratio of redundant employees. For a comparison of labor market policies in Shanghai, Beijing, Guangzhou and Chongqing, see Table 6.17.

5.6. Participation of the private sector

Consistent with the international competition, Mainland China is allowing the private sector more participation. With the expansion of the private sector, the growth and development of production capacity are facilitated. Therefore, one of the crucial policies of the Chinese government is to promote multiple ownership systems. The realization of such policies lies in the privatization of SMEs, participation of private enterprises, and the separation of business operations from politics. As private enterprises increase in Mainland China, economic growth and productivity are accelerated accordingly. Without budget limitation or guaranteed market, private enterprises encounter greater stimulation for innovation and response to market signals. In general, private enterprises have better production efficiency and investment rates compared with public enterprises. The competition among private enterprises tends to be fair, as the resources usually flow to the best enterprises. Public enterprises, on the other hand, can be allowed to obtain financial resources and governmental support, so they can still operate even though the efficiency is low. In doing so, the public enterprises occupy the resources and distort competition among enterprises. The cities good at separating businesses from politics are more energetic, and they can also provide a better investment environment.

The private sector can be categorized into three types: management ownership, individual ownership, and foreign ownership. The extent of these three types are shown in Table 6.18. Guangzhou has the leading position in private sector participation, followed by Chongqing, and the last is Shanghai. However, there are great differences in the distribution of the different types of private sectors. The main feature of Chongqing is that it has the highest individual ownership ratio, but the lowest ratio of foreign ownership. In fact, the great gap shown in foreign ownership offsets the advantages brought by the other two types. The example of Chongqing reflects potential problems faced by most inland enterprises. That is, it is very difficult for them to attract foreign investors. The main feature of Shanghai is that it has more foreign investment (only falling behind Guangzhou), but has the lowest ratio of individual ownership.

5.7. Financial conditions

A healthy financial service industry is extremely important to enterprises. The industry guides capital from the savings accounts of private investors to manufacturers, and also helps enterprises avoid financial restraints. The financial industry aids the enterprises in performing expansion when it exceeds its internal financial capability. Compared

Table 6.18. Degree of Participation of the Private Sector in Four Cities

	Beijing	Shanghai	Guangzhou	Chongqing
Ratio of management ownership	0.06	0.01	0.05	0.04
Ratio of individual ownership	0.16	0.14	0.20	0.35
Ratio of foreign ownership	0.24	0.26	0.34	0.11
Grand total	0.46	0.41	0.59	0.49

Source: www.worldbank.org

with situations in which enterprises have to obtain the capital to ensure their development — through informal loans, personal finance or financing from private investors — the financial industry can help to lower transaction cost. Obtaining capital is especially important for countries in transition. In fact, researchers often regard finance as the most important system to support the market.

Based on the importance of finance, we construct three indices to evaluate the functions of a financial market. The first is whether enterprises can receive formal bank loans for one year within a limited period of three years; the next is the ratio of enterprises using trade credit to obtain raw materials; and lastly, if it is necessary for enterprises to make informal payments (i.e., bribes) so as to obtain bank loans. It is important to get the approval for bank loans, and in many countries, the major financial broker is the banking system, with surplus and banking capital as the most important sources of capital. However, many people may not know that informal loans such as trade credits are also very important under many circumstances, especially for countries in transition like Mainland China (Cull and Xu, 2003). In fact, early studies show that trade credit positively influenced the performance of enterprises, especially private enterprises. It is difficult for private enterprises in Mainland China to receive formal loans from the banking system because most banks grant loans to public enterprises instead. It is essential for banks to distribute their loans ethically. If the officers responsible for granting loans in a bank are influenced by informal payment, then loans may not flow into the sector with the best production efficiency. In contrast, the loans will flow to the enterprises which offer bribes.

As shown in Table 6.19, the financial service industry in Shanghai ranks as the first, followed by Guangzhou, Chongqing and Beijing.

Table 6.19. Financial Service Industry among Four Cities

	Beijing	Shanghai	Guangzhou	Chongqing
Ratio of companies with loans	0.49	0.55	0.38	0.58
Ratio of companies with trade credit	0.15	0.22	0.19	0.19

Source: www.worldbank.org

5.8. Governmental efficiency

In the past decade, economists have become more concerned about the functions of government in developing healthy economies. In particular, if the government establishes a fair platform, officials then care more about the effective execution of rules instead of gaining benefits from the execution. Governments should provide sufficient resources to establish an environment offering and protecting public facilities (public security, infrastructure, etc.). This way, economic development will be improved (Shleifer and Vishny, 1999).

Four variables are selected to evaluate the efficiency shown in the government's provision of public facilities. The first index is informal payment, that is, the ratio of gifts or bribes given to the government and authorities relative to the sales amount. The second index is the time the top management of an enterprise spends on receiving governmental officials. The greater this variable, the higher the management cost brought by the government to an enterprise, since the government will be more inclined to interfere with the enterprise development. The third variable is the ratio of transportation loss caused by theft and damage. The government should take care of public security and port affairs, because the loss incurred during transportation reflects the inefficiency of governmental services and infrastructure. The last variable is the time consumed in solving commercial conflicts through the judicial system.

Table 6.20 shows the large gaps of the first three variables between each city. For example, Shanghai, Guangzhou and Beijing have less informal payment compared to Chongqing. However, if we take a

Table 6.20. Governmental Efficiency Index

	Beijing	Shanghai	Guangzhou	Chongqing
Informal payment	0.04	0.01	0.03	0.12
Reception time spent by government officials on management	0.15	0.13	0.12	0.07
Transportation loss	0.14	0.06	0.11	0.13

Source: World Bank Report.

look at the time that top management spends on the authorities, Chongqing is ranked least, followed by Guangzhou and Shanghai. When we further probe the transportation loss ratio due to theft or other factors, Shanghai is ranked the best, followed by Guangzhou, Chongqing and Beijing. Therefore, the efficiency of the government varies in different aspects.

Another important service provided by the government is the judicial service. As the economies grow and the manufacturers become more professional, relationship agreements (only signing contracts with various channels the enterprise is familiar with) becomes less important, while arms-length relationships become more important. The judicial system becomes more important when the industries are further professionalized and the demand on trade increases, as it helps ensure more effective planning and adjustment of production and execution of agreements. The relevant index here is the handling time spent by the judicial system on recent commercial disputes. The time spent on a case trial may be influenced by many uncertain factors, for example, sufficient judicial personnel, appropriate knowledge of commercial and business laws and awareness of other related regulations, and the degree of judicial division of work based on the related personnel's professional levels and capability. All these factors influence the judicial procedures and trial time. We should also take note that Mainland China is transforming into an economy ruled by law. It is a good sign when people go to court to solve economic disputes. However, people may be deterred from bringing disputes to the court if the judicial system takes too long to solve a case. The trial time is not the only approach to evaluate the effectiveness and performance of the judicial system. Other parameters such as judicial fairness and impartiality are also important. At present, we use only trial time to represent the importance of the judicial system in the overall investment environment (see Table 6.21).

5.9. Tax

Tax is an important factor in the investment environment, as enterprises are concerned about the revenue after tax. Investors have to

Table 6.21. Trial Time at Court (months)

	Beijing	Shanghai	Guangzhou	Chongqing
Average time for a case	NA	NA	NA	5.7

NA: Not available.

consider potential tax problems before they decide on the location to make investments. Usually, they encounter such problems in newly developed lands. High tax rates reduce the power of enterprises. In many countries, there are small gaps in taxation domestically. However, this is not the case in Mainland China. The local governments in Mainland China have been authorized to stipulate their own tax policies. Provinces and cities targeted for economic experiments are given a certain extent of decision-making power so as to attract foreign investors with liberal policies. Furthermore, the arrangement on taxation results from negotiation between the central and local governments. Such arrangements may bring different incentives for the local governments to create versatile economic effectiveness (Gordon and Li, 2002). Also, Mainland China's economies are dispersed. Local tax revenue forms an important part in taxation. Therefore, tax is an important factor influencing local investment environments in Mainland China.

6. Changes in the Investment Environment in Mainland China After its Entry into WTO

Mainland China's entry into the WTO signifies a more ideal investment environment for foreign investors. Since joining the WTO, the Chinese government has been honoring its agreements with the WTO — the investment fields are expanding; foreign investors can choose their investment targets with more freedom; the construction of infrastructure is reinforced; and the governmental administrative system is improving. Mainland China has also modified and improved its economic policies and legal system to create a stable, transparent and reputable environment that facilitates fair competition among foreign investors. These changes in the investment environment are extremely important to foreign invested enterprises. The following sections will introduce the various changes in the investment environment in Mainland China upon entry into the WTO, including changes in the domestic market, governmental administrative policies, legal policies, etc.

6.1. *Honoring its agreements*

For many years, Mainland China has been endeavoring to further open its market in accordance with international regulations so as to create a more stable, disciplined and attractive environment in terms of its policies, systems and laws. In the five years since Mainland China joined the WTO, it has been seriously and positively honoring its agreements.

6.1.1. *Opening up the regulations for entry into Mainland China's market*

The Chinese government has adjusted its policies relating to entry into Mainland China's market through the following ways:

(i) Mainland China further decreased the average tariff rate for trade of goods on 1 January 2002 and 2003 based on its agreement, from 15.3% to 11%. Mainland China will continue to lower tariffs according to the promised schedule. Since 1 January 2002, Mainland China has also canceled the quotas, license, and specific bidding management of items under 16 tax numbers including foods, wool, cotton, acrylic, polyester, P-chips, chemical fertilizers, tyre parts, motorcycle parts and related components, automobiles and related components, cameras, and watches in accordance with the agreement made with the WTO. According to the regulations of Import and Export Management Reinforcement on Commodities, the Chinese government stipulates a series of concrete regulations controlling imported commodities. This action tremendously increases the transparency of management policies on imported commodities. The import quotas, distribution principles, and application procedures for electrical engineering products, finished oil, automotive tyres, natural rubber and important industrial commodities have been announced to the public ahead of the promised time. Related quota distributions have been conducted in accordance with its promises.

(ii) Since 2002, Mainland China has also announced the following — Temporary Regulations for Customs Quotas of Import Agricultural Products, 2003 Reinforcement of Import Customs Quotas of Major Agricultural Products, and 2003 Reinforcement of Import Customs Quotas of Wool and Wool Tops. With the Temporary Regulations for Import Customs Quotas of Chemical Fertilizers and 2003 Total Import Quantity, Distribution Principle and Application of Customs Quotas of Chemical Fertilizers, the quotas of related products have been distributed to the end users by 1 January 2003.

(iii) Some important trade authorities providing related services have promulgated new rules and regulations in connection with reviewing and approving the foreign investments into Mainland China based on the Chinese government's promises. The laws and regulations announced so far include Reinforcement of Foreign Attorney At-Law in Mainland China and execution regulations, Regulations of Foreign Investment in Telecommunication Industry, Regulations of Foreign Invested Financial Institutes in PRC and related reinforcement, Regulations of Foreign Invested Insurance Companies, PRC International Marine Transportation, Regulations of Audio and Visual Products Marketing Industry of Foreign and Chinese Joint Ventures, State Council's Determinations on Regulations on Travel Service Agencies, Temporary Regulations on Foreign Investment in Commercial Industry, Review and Approval Principles on Foreign and Chinese Joint Investment in Securities Industry,

Review and Approval Principles on Foreign and Chinese Joint Investment in Fund Management Industry, etc. The stipulation of these laws and regulations provides a transparent environment for foreign investors to join the permitted industries. While formulating these regulations, the Chinese government created a great number of job opportunities. Since joining the WTO, three insurance companies have been approved to operate in China (including CIGNA from America). Meanwhile, the agreement to open regional restrictions has been advanced. The foreign invested banks in Mainland China have also experienced fast development. By the end of July 2002, the number of foreign banks permitted to provide RMB services reached 44 (13 more banks were approved after entry into the WTO). The total assets value generated by foreign invested banks reached 44.577 billion USD. The business and profits than grew rapidly.

So far, there has already been over 200 foreign invested marketing companies in Mainland China. The treatment extended to the foreign invested marketing industry even exceeds the range which the Chinese government promised, for example, the controlling stake offered to the foreign invested marketing industry. After China joined the WTO, the number of retail companies being approved reached eight, with a total of 28 shops. Carrefour and Walmart have already set up over 20 chain stores in Mainland China. The National Tourism Bureau also suggested that it would encourage foreign investors to enter Mainland China's tourism market through holdings or independent investment. Compared with the agreements that China made upon entry into the WTO, the above step occurred three years in advance. By the end of 2001, a total of 11 foreign and Chinese joint invested travel agencies were approved. Besides the above measures, Mainland China also approved the first audio and visual marketing company with Chinese and foreign joint investments. The aviation field was further expanded and open to foreign enterprises. Several academic institutes, hospitals and clinics with Chinese and foreign joint investments have also been approved.

6.1.2. Removal of operational restrictions

In 2002, the Chinese government promulgated the new Guidelines for Foreign Investment and Index of Industries for Foreign Investment. More operational restrictions on foreign investment were removed. Detailed definition on the industries which foreign investors can or cannot participate in are specified in these documents.

6.1.3. Maintenance of fair trade environment

After joining the WTO, the Chinese government began to use WTO regulations to protect the benefits of domestic industries. Mainland China established several institutes related to WTO and received many anti-dumping cases. Such measures have helped domestic industries and foreign invested industries avoid losses. Mainland China has also provided necessary guarantees on systems to uphold the rights and obligations of Mainland China after its entry into the WTO.

6.1.4. Transparent policies

To enhance transparency, the Chinese government has established the WTO Reporting and Information Bureau under the administration of the Ministry of Commerce. It constitutes Information Enquiry Regulations of WTO Reporting and Information Bureau in Mainland China. The Bureau provides consulting services for information relating to intellectual property rights, foreign exchange, etc. So far, the Bureau has received over 600 written enquiries from ambassadors, Chinese and foreign enterprises, and individuals. Thanks to the support of related governmental units, most enquiries have been replied. In the meantime, the establishment of the WTO consulting website further enhances the transparency of policies. The efficiency and quality of the WTO consultancy are further improved.

6.2. Changes in the market environment

After joining the WTO, the Chinese government further accelerated its reform of the economic system. The market environment experienced significant changes due to the adjustment of policies and modifications on laws and regulations.

6.2.1. Market economy orders

It is important to maintain good market orders in the development of healthy market economies. After joining the WTO, the Chinese government started a series of works to establish a unified, open and fair market.

(i) In order to break through local protectionism, the Chinese government established a management institute to break the blockage among provinces and regions. A variety of laws and documents have been announced, which includes State Council's Prohibition on Regional Blockade in Market Economic Activities, State Council's Determinations of Reorganizing Market Economy Orders, etc. These laws and regulations form the united market economy orders in Mainland China.

(ii) The Chinese Government conducts reforms on the traditional monopoly industries such as petroleum, power supply, telecommunications, civil aviation and rail through approaches like reorganization. The competitive mechanism is thus introduced to strengthen the vitality of enterprises. In the process of reorganization, the de-monopoly work in the petroleum chemistry industry was finished five years ago. Currently, CNPC and CINOPEC have been listed in the market. The management and operations are good in both companies. China Telecom has similarly divided into two groups, respectively located at northern and southern China at the end of 2001. Thus, a preliminary competitive condition in the telecommunications industry in Mainland China has been formed.

6.2.2. Acceleration of the adjustment of economic structure

The adjustment of the economic structure influences the changing directions and approaches in the market economic environment. After joining the WTO, the Chinese government adjusted its economic structure in three aspects:

(i) Adjustment of industrial structure. The Chinese government conducted comprehensive adjustments in poor agricultural production, big but weak industries, and service industries with backward development. Currently, the focus of the first industry is to adjust the commodity structure of agricultural products, accelerate livestock and aqua products, and enhance the level and efficiency of agricultural processed products. The focus of the second industry is to accelerate industrial reshuffle and reform and develop the hi-tech industry led by IT. The focus of the third industry is to accelerate the development of the service industry, enlarge the total amount, optimize the structure, expand the sector, and enhance the standards.

(ii) Adjustment of regional structure. In early 2000, the Chinese government proposed the strategy of developing western China. A preliminary achievement has been reached in recent years. The Chinese government is encouraging foreign investors to make investments in agriculture, forestry, water conservancy, transportation, energy, public facilities, environmental protection infrastructure, mineral development projects, and tourism development projects in western China. The restrictions on the shareholding of foreign investors in infrastructure and preferential industry projects in western China have also been removed, although this varies by industry.

(iii) Adjustment of ownership system. Mainland China's private sector, consisting of foreign and Chinese joint invested enterprises and independent foreign invested enterprises, has developed rapidly after joining the WTO. The benefits shown in economics, import and export trade, and increased employment brought by private ownership system has become even more obvious. The Chinese government explicitly indicates that it strengthens and develops the public ownership system while encouraging, supporting and guiding the private ownership system. Both systems will continue to undergo further development. Under such circumstances, greater space will be created for the private ownership system.

6.2.3. Strengthening infrastructure

A good market environment cannot be maintained without sound and high-quality infrastructure. In recent years, the Chinese government has provided a great amount of financing, materials and labor force for infrastructure construction.

(i) Increase the investment in infrastructure. Based on the arrangement specified in the Tenth Five-Year Plan of National Economic and Social Development by the Chinese government, the construction of infrastructure was the focus during 2001 to 2005. The mileage in newly constructed roads totaled 200,000 km. Of the total mileage, the highway accounted for 10,000 km. As for railways, 7,000 km new lines and 4,000 km overlapping lines were planned. Construction of new power supply stations would increased the supply to 77 million kW. 60,000 km of master wiring would be set up, and 130,000 km of low voltage power supply wiring would be built. 140 berths at deepwater ports would be constructed, and 40 secondary airports would be constructed and expanded.

(ii) Establish administrative institutes for infrastructure construction projects. To ensure the quality of infrastructure construction, the Chinese government set up auditing systems for major projects. The related institutes are under the administration of the National Development and Planning Committee. This system plays an important function in supervising the quality of infrastructure construction.

(iii) Encourage multiple enterprises to invest in infrastructure construction. The Chinese government removed restrictions on the investment in infrastructure and now encourages and offers incentives to the private sector and foreign organizations to participate in the construction of infrastructure.

6.3. Changes in the administrative system and legal environment

To adapt to the new environment caused by the participation in the WTO, the Chinese government changed its functions and works, strengthened the transparency of policies, and comprehensively adjusted and completed current laws, regulations and articles in connection with foreign investment. It initiated the needs required by market economies under the principles of socialism. It established a uniform, complete, and transparent foreign economic and trade legal system which fulfilled the requirements of WTO regulations and the national conditions of Mainland China as well.

6.3.1. Changes in the administrative system

Mainland China carried out self-reform, adjustment and optimization based on three aspects after joining the WTO.

(i) Reform of the administrative licensing system

Administrative licensing has long been the government's major approach to controlling economic activities in Mainland China. Complicated procedures and versatile items not only burden the government, but also restrain the development of enterprises. The vitality of the market is thus bound. In 2001, Mainland China was determined to reform the administrative licensing system in order to form relationships among enterprises, government and market. The core contents of the reform of the licensing system are: establishing scientific procedures in licensing, reducing unnecessary procedures, and strengthening the transparency of licensing. For example, reviews and seals from tens of divisions were required in the past to set up a company in Beijing, but now, only one division is needed to complete licensing. The other core content is that the reviewed items canceled will be replaced by market mechanisms or social agencies. According to statistics, Mainland China has already canceled 789 administrative licensing projects in which 560 projects involved economic management affairs.

(ii) Cultivation of social agencies

Modern agencies such as associations, societies, accountants, lawyers, etc., are important participants in market economic activities. It is meaningful when agencies can develop their functions of communications, coordination, information service, fair supervision and industrial self-discipline in order to create a thriving market environment. The Chinese government carried out much reorganization on the agencies to achieve these functions. The governmental units to which law firms and accountants are subordinate have been totally changed. The agencies have now established an image of independence and neutrality. Starting from 2001, the Chinese government has announced the cancellation of National Bureau-governing industries. Mainland China has also promoted relevant industrial associations. In the following years, Mainland China will continue to accelerate the development of agency services and endeavor to cultivate fair and independent agencies in order to provide important support for a good market environment.

(iii) E-administration

The execution of e-administration is beneficial to enhance the transparency of policies, facilitate the opening of governmental services, and avoid "under-the-table" operations. The establishment of e-administration helps to communicate governmental orders and enhance governmental efficiency. It is also beneficial to improve the legal and judicial systems and the relationship between the government and the public. The role of the government is transformed from a manager to a public service provider that is capable of distributing resources in the market to a greater extent. It is good for creating an environment in which everyone can use production elements fairly so as to facilitate liquidation of commodities and production elements.

In the 16th Session of National Representative Meeting convened by the Chinese Communist Party in November 2002, the policies regarding industrialization activated by IT and strengthening of e-administration were explicitly addressed. The reform of rooted administrative system was also raised. From this meeting, Mainland

China further changed its functions, promoted e-administration and lowered administrative costs. It is expected to form behavior criteria, operational coordination, fairness and transparency, and an efficient and honest administrative system. The Chinese Communist Party will use e-administration to activate a new governmental system that symbolizes a new development stage, or a revolution of governmental management. E-administration will build up a government which is more simplified, open, transparent, efficient, honest and diligent.

6.3.2. Changes in the legal environment

(i) Transparent foreign investment policies

The laws and regulations in connection with foreign investments have been modified to create an open and transparent legal environment. The contents that contradicted the regulations of WTO have been revised. The State Council of PRC has dealt with over 2,300 cases relating to foreign economic regulations, in which 830 cases have been abolished and 325 cases have been modified. Many of the regulations are associated with investments. A total of over 400 cases regarding foreign investment were modified and almost 100 cases were abolished. The local governments in Mainland China are following the requirements made by the State Council to proceed with related modifications so as to make local regulations and laws consistent with the criteria for national regulations and laws as well as with WTO regulations.

(ii) Modifications on three laws regarding foreign investment

Basic modifications have been made on three laws relating to FDI: PRC Chinese and Foreign Joint Ventures Law, PRC Chinese and Foreign Cooperative Business Law, and PRC Foreign Invested Business Law and related reinforcement. The most important agreement signed at WTO involving foreign investment is the Agreement on Investment Measures Regarding Trade, which specifies that the enterprises with foreign investment in Mainland China should not necessarily be with the share owned by local people and should also not necessarily balance foreign exchange holdings. These can be elaborated further.

(a) Clause regarding foreign exchange balance. The restriction of self-balance of foreign exchange by foreign invested enterprises has been lifted. There are two reasons for this. The first is that the self-balance of foreign exchange by the foreign invested enterprises was not compatible with the new trend under the foreign exchange management system and socialist market economy of Mainland China. In 1994, Mainland China started conditional foreign exchange under RMB current items. In 1996, foreign exchange arising from current projects became available. At this time, self-balance of foreign exchange was no longer necessary. The second reason is that such regulation is contradictory to WTO regulations. According to WTO regulations, the import of an enterprise should not be restricted by foreign exchange balance. The cancellation of self-balance of foreign exchange will not cause negative influence on the international balance of revenue and expenditure in Mainland China. The import and export from foreign invested enterprises have already accounted for nearly 50% of total import and export value in Mainland China. Since 1986, the obligation on foreign invested enterprises has been weakened. At the moment, foreign invested enterprises have already received foreign exchange services in the foreign exchange market and no sign of a worsened foreign exchange balance or greater trade deficit has appeared. The total self-balance of foreign exchange shows a positive figure every year. It is believed that the trend of foreign exchange surplus will remain as long as foreign investment in Mainland China increases.

(b) Conditions relating to local share. The regulations relating to purchase of shares by the Chinese government as a priority does not meet the principles of market economy of WTO regulations. Under the conditions of a socialist market economy, enterprises should be entitled to determine how they prefer to purchase based on market considerations, and the government should not intervene. Just like domestic companies of the same industries, foreign invested enterprises should also have the same freedom to purchase. The regulations of the Agreement on Investment Measures Regarding Trade specify that enterprises may not contain local shares and foreign invested enterprises should be treated like domestic enterprises.

(c) Requirements on export practices. The regulations stating that all or most products manufactured by foreign invested enterprises should be exported do not meet the conditions of a market economy of WTO requirements. Under the conditions of a socialist market economy, enterprises shauld have the right to decide whether their products will be exported or sold domestically, and the government should not intervene. Foreign invested enterprises should have the same rights as domestic enterprises when they sell their products. The regulations of the Agreement on Investment Measures Regarding Trade specify that all WTO members should not place restrictions on quantity, values or quotas of export products manufactured by enterprises. These regulations are part of the requirements on export practices prohibited by the WTO.

(d) Clauses on production plans of enterprises. The modifications of the clauses on production plans of enterprises came about because the government no longer requires enterprises to report their production plans. Under the conditions of a socialist market economy, enterprises are entitled to have decision-making rights. The Chinese government now plays the role of giving guidance from a macro perspective instead of intervening with the substantial activities performed by enterprises.

(iii) National treatment for foreign invested enterprises

National treatment principle is a very important principle valued by the WTO. The Chinese government has gradually canceled the existing non-national treatment given to foreign invested enterprises. For example, the enterprises are not allowed to set up branches in locations other than their registered addresses. However, the government will continue to execute preferential policies on foreign investments. The Director of the National Tax Bureau, Chin Jen Ching, once indicated that WTO indeed emphasizes the execution of "national treatment", but without objecting to "super national treatment". The preferential policies for foreign investments in Mainland China aim at domestic economic development. Therefore, foreign investments are encouraged in Mainland China and the government is expected to facilitate the development of foreign invested enterprises. The developing countries in the world are offering preferential policies to different extents. Mainland China will maintain its long-term preferential policies on foreign investment.

(iv) Newly modified Index of Industries for Foreign Investment

To fulfill the requirements arising from the strategic adjustment of national economy and the participation in the WTO, the State Council promulgated a newly modified Guidelines for Foreign Investment on 11 February 2002. On 11 March, the State Council further approved the Index of Industries for Foreign Investment promulgated by the National Statistics Committee, Economic and Trade Committee, and Foreign Economic and Trade Committee. The new industrial policies and indices continue to support the progress of foreign investments positively, reasonably and effectively. The new Index and Guidelines were put into effect on 1 April 2002.

The new Index is divided into four parts — incentives, permissions, restrictions and prohibitions — and is significantly open to foreign investment. The incentives increased from 186 to 262, while the restrictions decreased from 112 to 75. Second, the shareholding of foreign investors increased, for example, the cancellation of Chinese shareholding in common docks. Third, new fields for investment have been opened such as telecommunications, gas, power, drainage system, etc. Fourth, the service industry such as banks, insurance, commerce, foreign trade, tourism, telecommunication, transportation, accounting, auditing, law firms, etc., is opened up as well. Fifth, foreign investors are encouraged to make investments in western China. The percentage of shareholding in investments made in western China has increased, and industrial restrictions have been removed. Sixth, general industrial commodities are categorized into permissions so as to make the market competition more lively, and to upgrade industrial and product structure.

(v) Intellectual property rights protection

In accordance with the WTO requirement specified in Measures on Intellectual Property Right Regarding Trade, Mainland China has modified its regulations and laws in Copyright law, trademark law, patent law, etc. Bilateral cooperation and exchange regarding intellectual property rights have also been made with European countries.

(vi) A series of foreign investment policies and regulations

In 2001, the Foreign Economic and Trade Ministry, Technology Ministry, and National Industrial and Commercial Administration commonly promulgated the Temporary Regulations of Business Establishment by Foreign Investors. By referencing the laws and regulations regarding risky investment mechanisms and international principles of other countries, Mainland China established regulations for foreign investors to follow and provide services to unlisted high-tech enterprises in shareholding investment and establishment of new business. The Securities Supervision Committee and Foreign Economic and Trade Ministry promulgated the Opinions on the Problems of Listed Companies Involving Foreign Investment, which specifies the conditions and procedures for foreign investors to list on A and B stock markets. Foreign invested enterprises are allowed to purchase non-circulating shares issued by domestic listed companies in accordance with Temporary Regulations of Inbound Foreign Investment. The non-circulating stocks issued by corporations of foreign investment with B stocks are allowed to circulate in the B stock market. The Foreign Economic and Trade Ministry gives a Notice of Expansion of Business Rights in Export and Import to foreign investors. Foreign invested enterprises with an annual export value of 10 billion USD are allowed to purchase the products manufactured by domestic enterprises for export. Foreign investors are also allowed to import several hi-tech products for testing by foreign invested R&D centers. The regulations of Merger and Division of Foreign Invested Enterprises are modified so as to further manage the merging of foreign invested enterprises and domestic enterprises.

In June 2002, the Foreign Economic and Trade Ministry issued a notice of Competition Involving the Establishment of Branches by Foreign Invested Logistics Industry to allow the foreign invested logistics industry to run part or all of the international logistics business and the third party logistics industry.

In June 2003, the Commerce Ministry announced the Regulations for Establishment of Investing Companies by Foreigners to further expand on the business items run by investing companies. Such investing companies may engage in investments and reinvestments after approval from foreign economic and trade authorities in the province, autonomous region or municipal where they are located, and also pending further approval by the Commerce Ministry. The investing companies are allowed to finance the depreciated business and to run export and import trade, conditional export and import trade, and market the products manufactured by the enterprises invested by them in Mainland China.

The regulations and laws regarding the service trade industry have been expanded to attract foreign investment. The following regulations were promulgated: PRC Management Guidelines on Foreign Invested Insurance Companies, Administration of Foreign Attorneys Institutes in Mainland China, Temporary Regulations on Licensing of Foreign Invested Rental Business, Temporary Regulations of Chinese and Foreign Joint Invested Employment Agencies, Administration of Tourism Industry, Administration of Foreign Invested Telecommunication Business, Regulations of Foreign Invested Transportation Industry, PRC Regulations of Foreign Invested International Freight Service, Temporary Regulations of Foreign Invested Cinemas, Regulations of Chinese and Foreign Joint Invested Audio & Visual Products Marketing Industry, Temporary Regulations of Chinese and Foreign Joint Invested Medical Institutes, Establishment Regulations of Foreign Invested Securities Companies, Regulations of Foreign Investments in Civil Aviation, Licensing and Supervision Regulations of Accreditation Institutes, Training and Consultancy, Temporary Regulations of Foreign Invested Printing Business, etc.

6.4. *Adjustment of foreign investment policies*

Mainland China has been a member of the WTO for the past five years. The Chinese government has said that it is willing to promote liberty and convenience in trade and investment together with other member countries. It will promote a more open, fair and reasonable multilateral trade and investment system. The resulting international and domestic economic development has brought Mainland China unprecedented opportunities accompanied with severe challenges. So far, the Chinese government has been focusing on the soft investment environment in order to further strengthen comprehensive competitiveness and create a better environment for foreign investors. Mainland China's policies to attract foreign investment include the following:

6.4.1. Strengthen intellectual property rights and create a fair and open market environment

The Chinese central government insists on banning unreasonable charges, inspections and penalties placed on foreign invested enterprises. It also endeavors to break local protectionism and industrial monopoly. The government is increasingly executing laws in connection with intellectual property rights in its fight against piracy, fake goods and counterfeiting crimes. The Chinese government will also further modify the complaint placing system for foreign investors in order to protect foreign investors from violation against laws. After joining the WTO, the market will become even more competitive. Therefore, the Chinese government will continue to highly value protection of intellectual property rights and accelerate the progress of patent examination.

6.4.2. Further open the service trade industry

Mainland China will sequentially and positively open up its service trade industry. It will also announce new laws and regulations relating to mergers and acquisitions by foreign investment, venture investment, foreign trade, commerce, franchise management, etc. The government will also complete and make uniform the participation system for the foreign invested service industry in the Chinese market. Foreign investment in the logistics industry is encouraged, as is foreign investment in chain stores. Foreign invested purchase centers and commodity distribution centers have also been established. Modern ideas, advanced management experience and techniques, and market operation of the service industry in foreign countries are increasingly introduced to domestic enterprises. Therefore, the structure of the service industry in Mainland China will be improved and the level of the service industry will be enhanced.

6.4.3. Encourage foreign investment in western China

The twelve provinces in western China occupy 68.2% of the total national area and contain 28.6% of the total population in Mainland China. Therefore, developing western China is important. There are great prospects for the western and central areas of Mainland China, and the government plans to promote foreign investment policies to develop these areas. The government will further remove restrictions on the service industry and domestic financing conditions for foreign investment projects for these areas in a bid to attract foreign investments. By combining the modifications of Index of Industries for Foreign Investments, the related authorities are drawing up a new Index of Preferential Industries in Central and Western China for Foreign Investments to properly expand the range of industries for foreign investments. It is expected to facilitate foreign investments in infrastructure, mineral resources, tourism resources, ecological environment construction and protection, agricultural and livestock processed products, etc., in central and western China. Foreign invested enterprises established along the coast are encouraged to reinvest in central and western China. Foreign investors are also encouraged to participate in the projects in connection with natural gas delivery, power transmission, etc.

6.4.4. Encourage foreign investment in high-tech industries and satellite industries

Because there is great room for foreign investors to develop in Mainland China, they are expected to develop large-scale high-tech industries such as IC and biological engineering. Foreign investors are encouraged to introduce, develop and create new technologies. They are also encouraged to engage in capital and technical collective industries and establish advanced technology projects. The registered capital share owned by foreign investors and conditions for related investments will be properly restricted. The regulations in regards to venture investment will be further completed, and the environment and conditions built up for the establishment of the high-tech industry. Foreign investors are encouraged to use local raw materials. The government will promote domestic SMEs to strengthen their cooperation with foreign companies so as to introduce advanced technology. Packaging regulations for large-scale enterprises, and upgrading global production and sales network will also be carried out.

6.4.5. Positively explore new approaches of foreign investment

Transnational enterprises are encouraged to make investments in Mainland China. Studies on the policies for attracting transnational companies, establishing regional headquarters, and setting up international export purchase centers are being conducted. The international experience and approaches for mergers are taken into consideration by combining the economic features and concrete conditions of enterprises in Mainland China. Feasible regulations and laws relating to foreign investment in mergers will be completed. Transnational companies are encouraged to participate in the reform and reshuffle of domestic enterprises through the assignment of shares or purchase of assets. Transnational enterprises may transfer their advanced technology and advanced management experience so as to promote the reform and reshuffle of domestic enterprises in Mainland China and make possible the optimized distribution of social resources. The government will also select a group of reformed enterprises, especially large-scale public enterprises, to assign their shares to foreign investors. Further modifications will be made on the regulations regarding investing companies and corporations owned by foreigners. The promotion of BOT, franchise assignment, listing of foreign invested enterprises in domestic or overseas stock markets, etc., will be further specified and completed so as to create proper conditions for transnational enterprises to invest.

In short, enormous changes have occurred in the market environment in Mainland China. The market economy orders are significantly improved and a new environment open to foreigners has been formed. These changes will provide and create a more attractive investment environment for foreign investors in Mainland China.

Chapter 7

The Business and Investment Environment of the Information Technology and Electronics Industries in Taiwan

1. Evaluation of Taiwan's Business Environment

In September 2004, the report of *Countries Forecast — Taiwan* by the UK Economist Information Unit (EIU) pointed out that from 2004 to 2008, the Business Environment Forecast of Taiwan ranked 18th among the 60 evaluated countries, up two places compared to the ranking within the period of 1999 to 2003. The score of the Business Environment Forecast increased from 7.36 points within the historical period (1999–2003) to 8.05 points within the future forecast period (2004–2008). Another report by the EIU, *Country Forecast — Global Outlook,* pointed out that between 2004 and 2008, Taiwan will be ranked 18th in the Global Business Environment, just next to Singapore and Hong Kong, and ranked as 3rd in the Asian region (keeping its previous ranking), which is better than the ranking of South Korea, Japan, Malaysia, Thailand and Mainland China. That is, Taiwan's investment environment in the Business Environment Forecast is stable, and its excellence has been recognized once again. The evaluation of the Taiwan Business Environment Forecast ranges from "good" to "very good". In the Business Environment Forecast for 2004 to 2008, among the 10 indices of the Taiwan Business Environment Forecast, the performance of the macroeconomic environment is the best, ranked 3rd globally (eight places higher). The foreign trade and foreign exchange control is ranked 9th globally (21 places higher), the labor market is ranked 9th globally (four places higher), and the tax policy is ranked 7th (five places lower). As for other global rankings, the market opportunity is ranked 16th (four places higher), the infrastructure is ranked 17th (six places higher), the policy toward private enterprises and competitiveness is ranked 18th (one place higher), the foreign capital policy is ranked 20th (seven places higher), financial policy is ranked 20th (two places higher), and the political environment is ranked 22nd (unchanged). Even though there are many indices showing a continuously improving investment environment in Taiwan, as well as a common upgrading of labor productivity and quality, the indices also show advanced warning that structural unemployment will be a problem, and that insufficient labor force will be faced by the manufacturing industry in the near future. The EIU estimated that in 2004, Taiwan's economic growth rate would reach 5.8%, and the average economic growth rate will be 4.7% during 2005 to 2008. In addition, the EIU holds that the government will gradually relax restrictions on FDI in securities, telecommunications and infrastructure. Thus, the tariff will be further reduced, the quota policy will be gradually abolished, and the restriction on capital input/output will also be relaxed.

Furthermore, in the *2004/2005 Global Competitiveness Report* announced by the Switzerland World Economic Forum (WEF), the ranking of the Growing Competitiveness Index placed Taiwan in 4th place, allowing it to maintain the lead position within the Asian region. The advantageous items possessed by Taiwan such as technology still maintain leading positions, and as for the items with worse performance in the past, such as the macroeconomics performance, they are significantly improved now. Similarly, in June 2004, Switzerland IMD announced that in global competitiveness, Taiwan has advanced from 17th in 2003 to 12th place. Apart from the efficiency of enterprises possessing the most competitiveness, the rankings of the original weaker items, such as governmental efficiency and infrastructure, have all advanced. In addition, with the benefits of the prosperity recovery that motivated the performance of domestic macroeconomics, economic performance advanced nine places. Overall, whether viewing the evaluation of WEF or IMD, we can find that the national competitiveness of Taiwan is showing a positive development of upgrading. The reason Taiwan is ranked in leading positions is mainly due to its technology. In the evaluations of IMD and WEF, Taiwan's technology has occupied more than 1/3 of the most advantageous sub-indices. After stepping into the era of a knowledge-based economy, the WEF's evaluation made a great adjustment by heavily emphasizing the importance of technology and innovation, and even enhancing the weight of such sub-indices. From the change in each year's ranking, we can find that in the three indices of competitive growth — technology, public policy and macroeconomics — Taiwan's performance in technology is the most stable, is worse in public policy, and has its biggest fluctuation in ranking in the performance of macroeconomics.

IMD's *2004 National Competitiveness Report* pointed out that if Taiwan is able to upgrade the 20 sub-indices, especially raising its competitiveness to the average level with other countries, then the ranking of overall competitiveness for Taiwan can be upgraded to 10th globally. These items include the risk of political instability (54th place), the ratio of capital and property tax to GDP (48th place), social cohesion (49th place), and foreign inward investment (51st place), as well as lower-ranked items such as the international phone rate, hygiene and health expenditure, and wastewater disposition, which show great upgrading potential for Taiwan's competitiveness.

2. Competitive Strengths and Weaknesses in Taiwan's Information Technology and Electronics Industries

Table 7.1. SWOT Analysis of the Commercial and Investment Environment for the Taiwanese Information Technology and Electronics Industries

1. Science-based parks already possess economies of scale 2. Industrial chain is well-developed 3. Excellent manufacturing technology and experienced in business operation 4. Complete infrastructure 5. High quality of labor, good working attitude 6. HR possess internationalization experience 7. Good living environment and quality 8. Support of governmental policy	Internal Strengths	Internal Weaknesses	1. Limitation in cross-straits talent flows 2. Supply of high-tech talents is facing a bottleneck 3. Industrial water supply is seasonal and unstable 4. Unstable politics; lack of a common view of the economic and trading policy between Mainland China and Taiwan 5. Dependence on foreign advanced technology 6. No cross-straits direct transportation 7. Bad cooperation and contact of cross-straits financial control policy

(*Continued*)

Table 7.1. (*Continued*)

1. Financial reform is effectively opening up FDI in Taiwan's stock market and integrating the finance industries, in order to attract the European and US financial service industry to invest in Taiwan 2. Labor division of the international industrial chain for IT and electronic high-tech industries is becoming clearer 3. The technological transfer from US and Japanese IT and electronic high-tech industries is becoming active, and the trend of strategic alliances between Taiwanese and Japanese high-tech industries is significantly increasing	External Opportunities	External Threats	1. Global electronic high-tech technological progress is facing a bottleneck in excess production 2. Mainland China's lower labor and land costs 3. Mainland China's rapid economic development triggered price inflation of raw materials 4. Unknown Mainland China's macro-control policy; the tension of raising interest rates and the appreciation of the RMB still exists 5. Mainland China's banks have weak mechanisms, slow financial reform, and the potential risk of encountering financial instability 6. Mainland China has actively pursued investment recruiting

3. Development Strategy of Taiwanese High-Tech Industries

Currently, due to Taiwanese industries facing global labor disintegration, global layout, the vertically and horizontally competitive environment of new production, new markets and new technology, Taiwanese industries will start to establish an operation environment with knowledge-based production and application in the effort to develop high added-value industry. Taiwanese industries should go in the direction of upgrading its technological level, stepping into innovative R&D, and developing global logistics to increase added value. From the perspective of the "Smiling Curve" invented by the founder of Acer Group, Dr. Stan Shih, the leader of Taiwanese information industries, the curve's middle part is the lower priced manufacturing and assembly, which is the value chain Taiwanese industries are good at. The right-hand side of this curve is global logistics (logistics, marketing, branding); and the left-hand side of this curve is innovative

R&D (innovative R&D, design, key components). Thus, at both ends of the value chain are innovative R&D and global logistics, which is the direction in which Taiwanese industries should strive.

Basically, Taiwanese industries, after experiencing phases such as import substitution, export expansion, improvement of infrastructure, economic liberalization, industrial upgrade, industrial global layout, etc., do not possess many new advantages. However, with a superior geographic location situated in the hub of the Asia-Pacific Zone, and with the same language and race as Mainland China, Taiwan can be a gateway to entering the Mainland Chinese market. With its solid and accumulated manufacturing experience, the industrial structure is compact and complete. Taiwan has numerous SMEs that possess flexibility, plentiful high-quality manpower, strong capability of industrial R&D technological commoditization, sound venture industries and capital markets, international marketing skills, global logistics experience, and the capability of innovative R&D.

In addition, the Taiwanese government has entered the conventional industry, service industry, "Two-Trillion, Two-Star" and green industry to upgrade current industries and advance the emerging service industry, emerging industrial development, and strategic planning for establishing green industries. Industries that have been promoted are: high-tech, textile, health functional foods, high-level materials, light metal, optoelectronics and electrical chemicals, power-driven vehicles, sports and leisure, R&D services, information services, logistics services, resources, biotechnology, digital content, image display, and IC industries. With much strategic planning, the government hopes to invest a large amount of energy into creating a new peak for industrial innovation.

Because R&D and innovation in Taiwan are mainly oriented towards OEM, cost reducing, process innovation and the manufacturing industry, the average personnel adds value no matter whether the manufacturing or service industry is less developed than most advanced countries, which has resulted in a growth less that of the national income of Taiwan. Especially in the competitive era of globalization, when industries face the enormous pressure of transforming and upgrading, as well as global competitive pressure, they have to be innovative and break through the bottlenecks of talent, capital, technology, intellectual property and various service industries. Only then can they actively innovate and upgrade their competitive advantages. In order to upgrade competitive advantages, the government has created many consultancy plans to promote the innovation of technologies, products and business patterns, promotion of self-brands, and the establishment and layout of channels, with the strategy of equal emphasis on the manufacturing industry and service industry to build up the operating environment for supporting enterprise innovation, and upgrading the added value for each industry. In the past, conventional Taiwanese industries have used low-skilled and cheap labor as their foundation, which contributed to the so-called economic miracle. However, faced with the current competitiveness of the emerging industrial countries, Taiwanese industries now need to rebuild their industrial competitive advantages starting from the improvement of their constitution, and also use the strategy of industrial transformation in order to adapt to the new competitive environment. Industrial upgrade or transformation should be analyzed to check whether technology can be upgraded, if the material supply is sufficient, if the direction and scale of markets and the high-level R&D manpower is adequate, and whether there is a complete certification system and platform. In addition, global logistics, brand partnership, marketing depth, production transfer, key technology, strategic alliance, and the value of a knowledge-based economy are also key points that need to be considered.

As for the future internationalization strategy for Taiwanese industries, there needs to be emphasis on the R&D innovative international network layout, especially through the R&D conducted by the overseas branches of the industry and the usage of global innovative resources. To improve the level of industrial innovation, the industry, authority, academy and research institutes should use the international R&D innovative network and develop the usage of the leverage effect on international R&D innovative resources.

To expand global markets, Taiwan should establish a complete R&D design and application system which actively promotes industrial mergers and globalization layout, merges into the global supply chain, and enhances the international marketing channels. However, in order to assist enterprises in acquiring business opportunities and entering the target market, the government has to adopt the strategic solution of using investment to drive trading activities and establish logistic centers to assist SMEs entering the purchase circle of major international manufacturers. It also needs to upgrade the economic and trading relationship between Taiwan and major countries through the works of the free trade agreement, participation in multilateral economic and trading affairs, etc. Currently, Taiwan's economic development is facing many challenges from internal and external changes. According to the economic conditions and industrial developing trend in Taiwan, the government has selected key industries such as the "Two-Trillion, Two-Star" industry, communication, biotechnology, nano-technology, and knowledge-based services for rapid promotion in order to create new core competitive advantages.

The Asia-Pacific Zone plays a key role in the production and manufacturing of information and communication industries. Taiwan also displays powerful strength in the semiconductor and panel industries, and when the tide of digitalization comes into every industry, it is forecast that the technology industries will be rapidly transferred from the information and communication industries to the digital ones. Through the cooperation between industry, authority, and academy and research institutes, and the integration of the intelligence and experiences of Taiwanese up/downstream industries, there will be a common platform for Taiwanese industries, which will assist the information and electronics industry in controlling the global markets and technologies, and further consolidate Taiwan's advanced status in the global information electronic industry, as well as enhance the digitalization-related infrastructure.

4. Taiwan's Development Strategy to Become a Knowledge-Based Economy

Many specialists, scholars and core intelligence believe that among the four main facets of a knowledge-based economy, the performance in the infrastructure of its knowledge society is the worst for Taiwan, and needs to be improved first. Because the infrastructure of a knowledge-based society can be regarded as the foundation for the other three facets (i.e., innovative ability, the capacity and quality of science and technology, and social infrastructure), therefore, only with excellent infrastructure can the knowledge-based economy develop soundly. Otherwise, even the excellent performance of the other facets will be of no use. Therefore, Taiwan must start with improving the knowledge infrastructure to develop its knowledge-based economy.

For knowledge infrastructure, the common belief is that educational innovation, governmental policy, and mutual trust and respect are the items that need to be improved. Educational innovation needs to be emphasized for the incubation of basic education and science education. The educational policy needs to be well-planned and well-executed, and effectively use the educational resources to avoid unbalanced distribution from affecting the quality of education. The governmental administration must emphasize consistency and stability, and policy execution must be transparent and efficient. In addition, the government should be a model for its people, acting honestly and with foresight. It should also carefully execute its own policies and further establish mutual trust and respect in society.

Apart from improving knowledge infrastructure, the government also needs to improve the quality of talent, and the capacity and quality of science, technology and humanity. All this stems from education. If there is a comprehensive education policy and system, then the incubation and quality of talent will be upgraded. A knowledge-based economy also needs to emphasize accumulation, integration and continuous improvement of knowledge. Thus, during its development, it should not only focus on the development of high-tech innovation, but also effectively integrate and use the technological power that is contained in the conventional industries and SMEs, and enhance the exchange and cooperation of knowledge among industries, schools and research institutes, thereby establishing an effective knowledge management mechanism.

Chapter 8
Analysis of Each Index

1. Knowledge Capital

1.1. Quantity and quality of invention and design patent

1.1.1. Number of approved inventions and design patents in the US

Observing the sub-index of the number of approved invention and design patents in the US, we can find that since 1999 there has been a slight upward trend, but its growth has come to a standstill in the last three years, which indicates that the invention and design power in Taiwan has encountered a bottleneck and needs to break through it. This index lacks relevant data for these three regions in China, so we are unable to compare the relative quality among them.

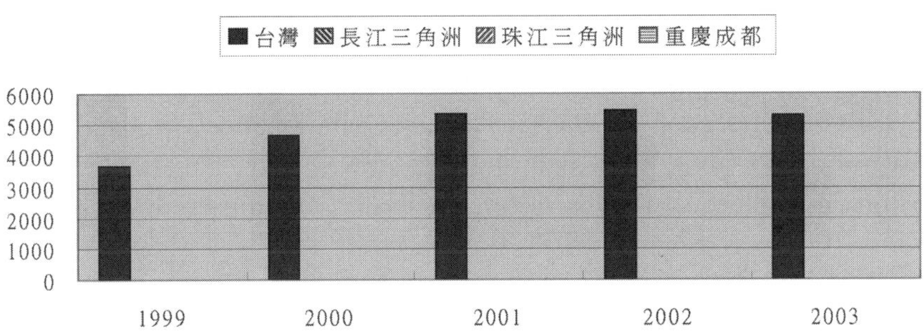

Figure 8.1. Number of Approved Inventions and Design Patents in the US

Legend (from left to right): Taiwan/Yanzhijiang Delta Region/Chongqing Chengdu Region/ Zhujiang Delta Region

1.1.2. Quantity of approved domestic patents

In terms of the quantity and quality of invention and design patents, observing this sub-index of the quantity of approved domestic patents, we can see that Zhujiang Delta Region took a big lead over

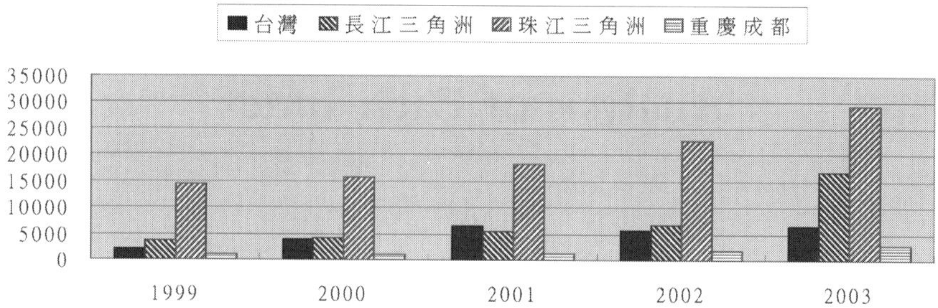

Figure 8.2. Quantity of Approved Domestic Patents

Legend (from left to right): Taiwan/Yanzhijiang Delta Region/Chongqing Chengdu Region/Zhujiang Delta Region

Taiwan, Yanzhijiang Delta Region, and Chongqing Chengdu Region in the absolute quantity, and displayed a gradual upward trend. However, the chasing situation in Yanzhijiang Delta Region is very clear, from 3,665 patents in 1999 up to 16,671 patents in 2003, which is almost five times more than before. In contrast, Taiwan displayed a standstill for this index, and it showed no change during 2001 to 2003.

1.1.3. Number of papers published in SCI

As for the number of papers that were published in SCI, due to the lack of individual data for the Yanzhijiang Delta Region, Zhujiang Delta Region, and Chongqing Chengdu Region, we use the situation in all of China to analyze this index. From the figure, we can see that whether

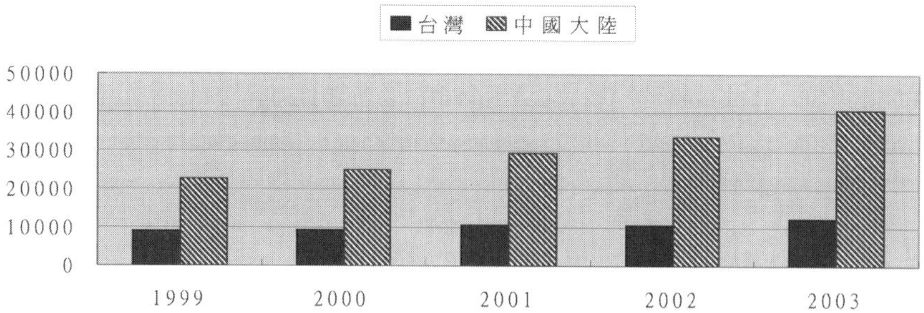

Figure 8.3. Number of Papers Published in SCI

Legend (from left to right): Taiwan/Mainland China

in Taiwan or China, the number of papers published in SCI have increased gradually in these last five years. However, for the absolute number or increase, Mainland China is significantly better than Taiwan.

1.1.4. Number of patents that have been cited: Current Impact Index (CII)

To observe this sub-index of the number of patents that have been cited, we use CII as an example. After the value reached 1.19 in 2000 for Taiwan, it displayed a decline, which indicated that the importance of Taiwan's recent patents or the impact has become weaker. Because this index lacks relevant data for the three regions in China, we are unable to compare the relative quality among these four regions.

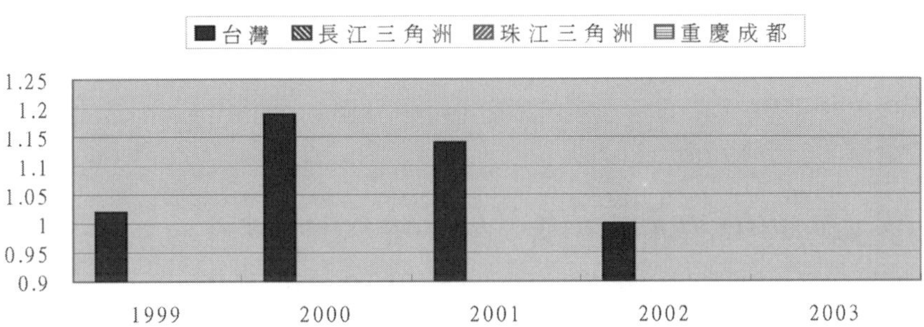

Figure 8.4. Number of Patents that have been Cited: Current Impact Index (CII)

Legend (from left to right): Taiwan/Yanzhijiang Delta Region/Chongqing Chengdu Region/Zhujiang Delta Region

1.1.5. Number of technical literature that each patent cited: Science Linkage Index (SLI)

Inspecting this index, Science Linkage Index (SLI; number of technical literature that each patent cited), we can see that Taiwan's value is around 0.15 to 0.25 for Taiwan, which indicates the linkage of scientific research and technical development is not very intensive in Taiwan. However, even though there are just two years of data for China, we can still find that its value is much higher than Taiwan's, therefore, Taiwan still has large room for expansion in this area.

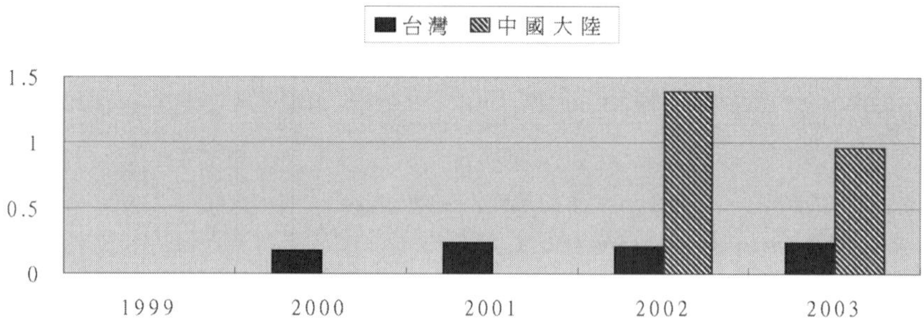

Figure 8.5. Number of Technical Literature that Each Patent Cited: Science Linkage Index (SLI)

Legend (from left to right): Taiwan/Mainland China

1.1.6. Number of invented patent rights possessed per 1 million inhabitants

Because this index lacks relevant data, no analysis is available at this moment.

1.2. Quantity and quality of human resources

1.2.1. Number of research personnel per 10,000 inhabitants

Observing this index, we can find that after 2000, Taiwan displayed an upward trend, which indicates that the number of research

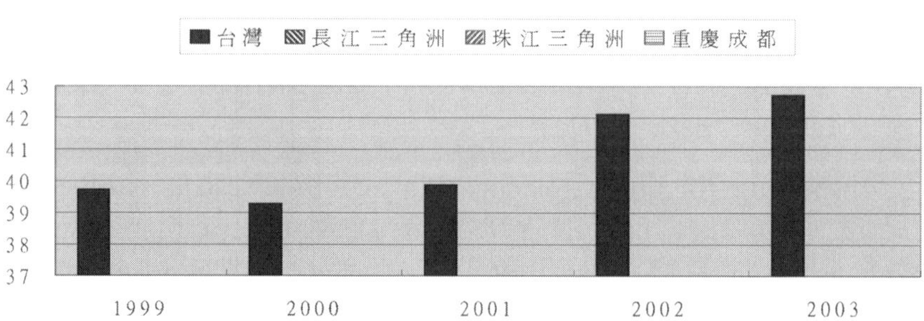

Figure 8.6. Number of Research Personnel per 10,000 Inhabitants

Legend (from left to right): Taiwan/Yanzhijiang Delta Region/Chongqing Chengdu Region/Zhujiang Delta Region

personnel in Taiwan increased gradually and advantageously to the country's R&D. Because this index lacks relevant data for the three regions in China, we are unable to compare the relative quality among them.

1.2.2. Number of research personnel per 10,000 labor inhabitants

Observing this index, we can see that the trend is similar to the index of number of research personnel per 10,000 inhabitants. It displays an upward trend, which indicates that the number of research personnel in Taiwan is increasing gradually and advantageously to Taiwan's R&D. Because this index lacks relevant data for the three regions in China, we are unable to compare the relative quality among them.

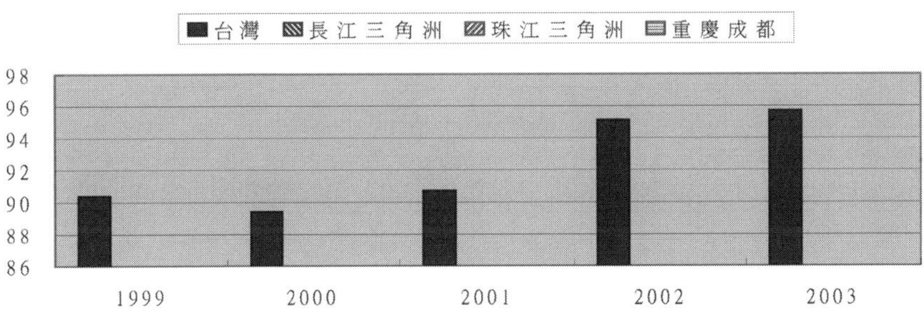

Figure 8.7. Number of Research Personnel per 10,000 Labor Inhabitants
Legend (from left to right): Taiwan/Yanzhijiang Delta Region/Chongqing Chengdu Region/Zhujiang Delta Region

1.2.3. Number of university students per 10,000 inhabitants

Observing this index, we can see that since 1999, the number of university students per 10,000 inhabitants in Taiwan is about 200 and increasing gradually, until 2003 when the index reached 400 and almost doubled its growth. In addition, it indicates that the recent manpower quality in Taiwan has risen to a certain level. Because this index lacks relevant data for the three regions in China, we are unable compare the relative quality among them.

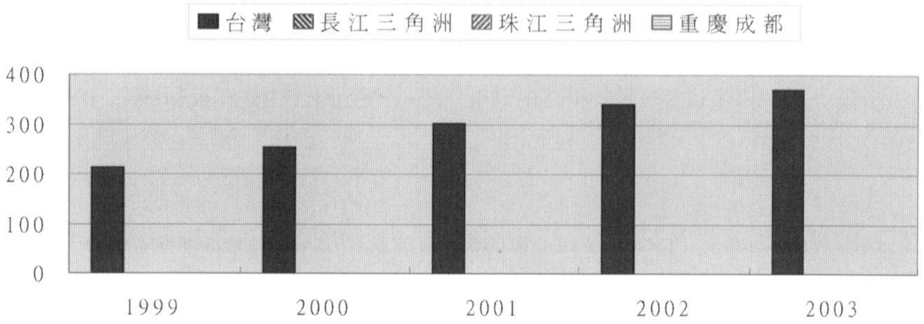

Figure 8.8. Number of University Students per 10,000 Inhabitants

Legend (from left to right): Taiwan/Yanzhijiang Delta Region/Chongqing Chengdu Region/Zhujiang Delta Region

1.2.4. Number of R&D personnel per 1,000 inhabitants

This index is the same as index 1.2.1, the number of research personnel per 10,000 inhabitants. Please refer to the analysis of 1.2.1.

1.2.5. Acquirability of high-skilled workers

Because this index lacks relevant data, no analysis is available at this moment.

1.3. Public basic technology, cultural capacity and development

1.3.1. Science education (0–10 points)

Observing this index, we can see that Taiwan took a significant lead over all of China and the Zhujiang Delta Region, but in 2003, the difference between Taiwan and these two regions was obviously reduced, which indicates that in recent years following China's economic development and prosperity, its level of science education is also improving. Because this index lacks relevant data for the Yanzhijiang Delta Region and Chongqing Chengdu Region, we are unable to compare them.

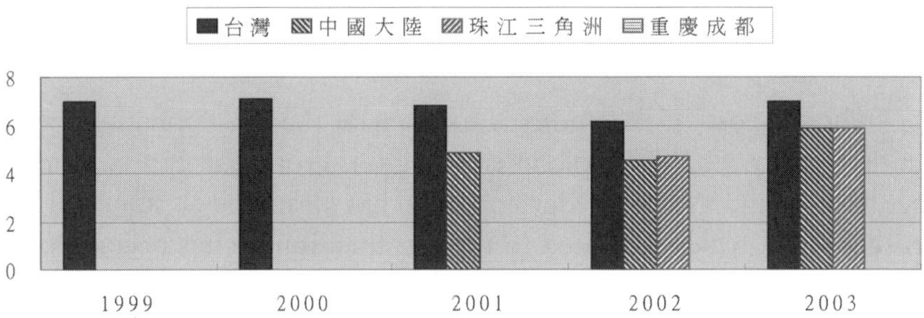

Figure 8.9. Science Education (0–10 points)

Legend (from left to right): Taiwan/Yanzhijiang Delta Region/Chongqing Chengdu Region/Zhujiang Delta Region

1.3.2. Interest of young people in science (0–10 points)

The trend of this index is similar to the trend of index 1.3.1, science education. In recent years, the differences between China, Zhujiang Delta Region, and Taiwan in this index has gradually narrowed. If we observe the global ranking, Taiwan's performance is not as good as in previous years. It ranked first place in 2000, but went down to 7th place in 2003. (For detailed rankings, please refer to the data organization of research indices for this study.)

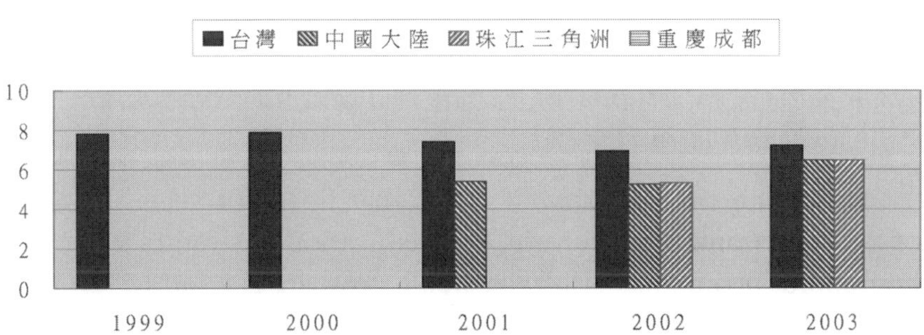

Figure 8.10. Interest of Young People in Science (0–10 points)

Legend (from left to right): Taiwan/Yanzhijiang Delta Region/Chongqing Chengdu Region/Zhujiang Delta Region

1.3.3. Amount of free time (per 1,000 persons) to participate in culture and art activities

The purpose of this index is to evaluate the development of the entire quality for the people in each region. From this figure, we can see that Taiwan's entire development has displayed a standstill in recent years, which is related to the fact that Taiwan has become one of the most developed countries. Therefore, its growth will slow down gradually. Because this index lacks relevant data for these three regions in China, we are unable to compare to the relative quality among them.

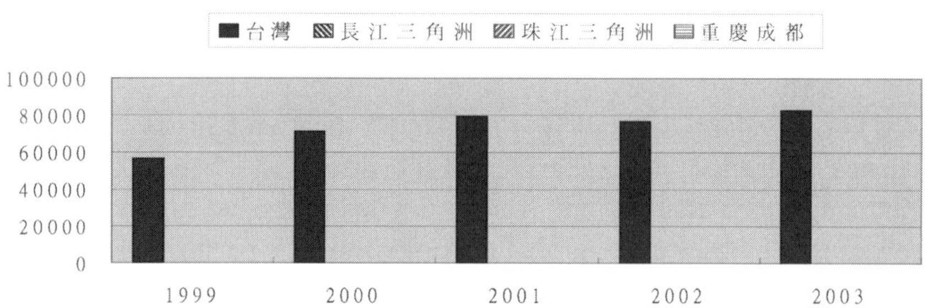

Figure 8.11. Amount of Free Time (per 1,000 persons) to Participate in Culture and Art Activities

Legend (from left to right): Taiwan/Yanzhijiang Delta Region/Chongqing Chengdu Region/Zhujiang Delta Region

1.3.4. Ratio of entertainment, culture and education expenditure to household income (%)

Observing this index, we can see that in recent years the ratio of the entertainment, culture, and education expenditure to the household income has displayed stable growth. The ratio is around 12.9% to 13.6%, which indicates that the entertainment, culture, and education expenditure has occupied a key position in households in Taiwan. Because this index lacks relevant data for the three regions in China, we are unable to compare the relative quality among them.

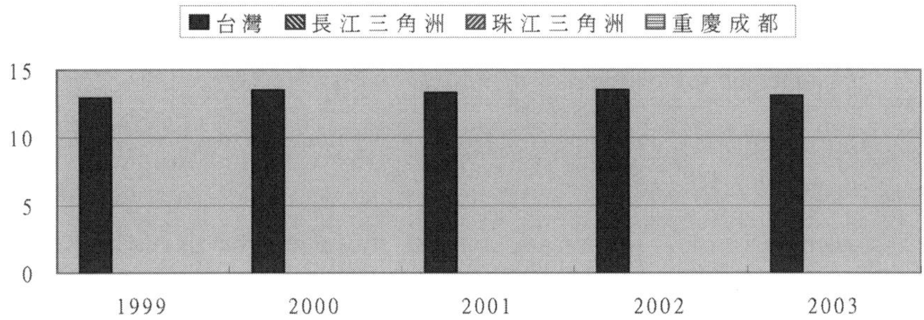

Figure 8.12. Ratio of Entertainment, Culture and Education Expenditure to Household Income (%)

Legend (from left to right): Taiwan/Yanzhijiang Delta Region/Chongqing Chengdu Region/ Zhujiang Delta Region

1.3.5. Ratio of government education expenditure to GDP (%)

Observing this index, we can see that from 1999 to 2002, the ratio of the three regions in China is not quite half the ratio in Taiwan, which indicates that Taiwan has more emphasis on the development of public education. In recent years, even though China has experienced rapid economic development, it has paid less attention to public education. For this index, Taiwan has a certain degree of advantage, which is also greatly aids the production of high-quality manpower.

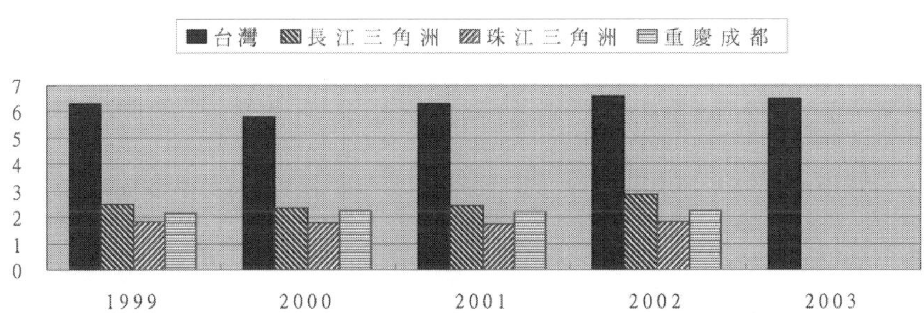

Figure 8.13. Ratio of Government Education Expenditure to GDP (%)

Legend (from left to right): Taiwan/Yanzhijiang Delta Region/Chongqing Chengdu Region/ Zhujiang Delta Region

2. Innovative Capability

2.1. Quality and quantity of research institutes

2.1.1. Technological research institutes' R&D budget/each researcher (million NTD)

From this index, we can see that the speed of growth of the R&D budget for each researcher in Taiwan is smooth, but the amount is significantly higher than that of the three regions in China. Even though the three regions in China have displayed growth, this growth has significant differences among them, among which, Yanzhijiang Delta Region has the fastest growth and seems to have caught up with Taiwan. The Chongqing Chengdu Region has fallen behind the other three regions, and there is still much room for growth.

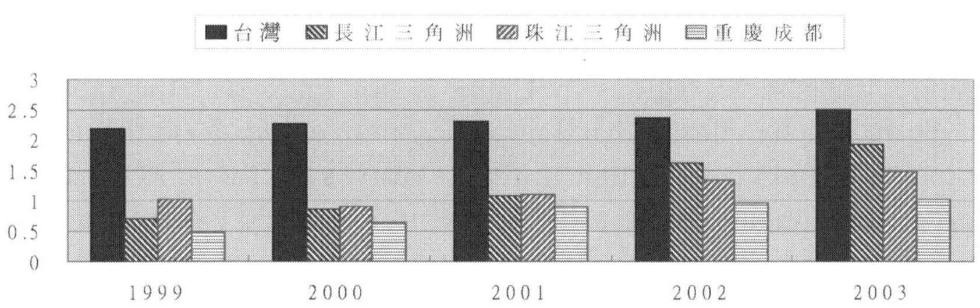

Figure 8.14. Technological Research Institutes' R&D Budget/Each Researcher (million NTD)

Legend (from left to right): Taiwan/Yanzhijiang Delta Region/Chongqing Chengdu Region/Zhujiang Delta Region

2.1.2. Ratio of research personnel with a master's degree or higher (%)

This index is similar to the trend of the previous index. Taiwan took a big lead over the three regions in China, and the main reason is that the promotion of higher education in Taiwan takes place far earlier than it does in China. In addition, with the policy of establishing universities and their own graduate schools within recent years, the number of students that have completed graduate school has increased greatly. As a result, due to the great number of graduates

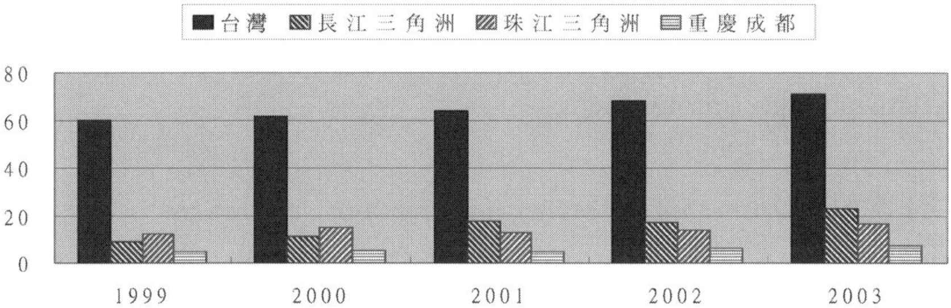

Figure 8.15. Ratio of Research Personnel with a Master's Degree or Higher (%)

Legend (from left to right): Taiwan/Yanzhijiang Delta Region/Chongqing Chengdu Region/ Zhujiang Delta Region

who have a master's degree, the research institutes have high standards in their employment criteria.

2.1.3. Number of approved foreign patent applications within the technology project (application)

This index indicates that within the last five years, the number of approved foreign patent applications within technology projects experienced stable growth in Taiwan. Apart from the outstanding performance in 2000 with nearly 500 approved applications, the number

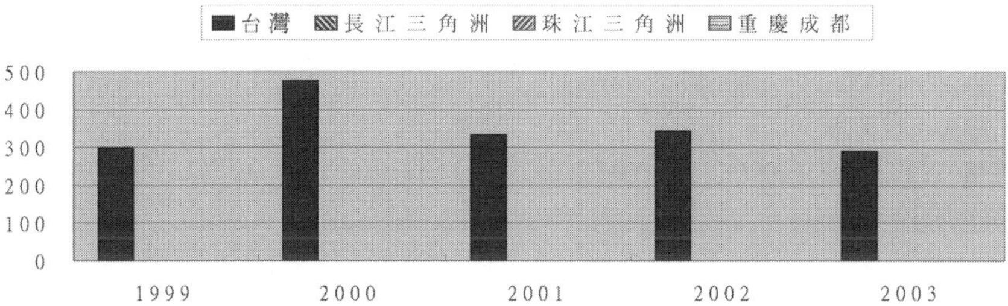

Figure 8.16. Number of Approved Foreign Patent Applications within the Technology Project (Application)

Legend (from left to right): Taiwan/Yanzhijiang Delta Region/Chongqing Chengdu Region/ Zhujiang Delta Region

is around 300 in the other four years. Because this index lacks relevant data for the three regions in China, we are unable to compare the relative quality among them.

2.1.4. Number of spin-off companies for technology projects

Because this index lacks relevant data, no analysis is available at this moment.

2.2. Quality and quantity of enterprise R&D

2.2.1. Number of enterprise R&D personnel per 1,000 inhabitants

Observing this index, we can see that the number of enterprise R&D personnel per 1,000 inhabitants in Taiwan is almost three people every year, but it is less than 0.5 in China, which indicates that Taiwan possesses a relative advantage in this aspect. Because this index lacks relevant data for Yanzhijiang Delta Region, Zhujiang Delta Region, and Chongqing Chengdu Region, the comparative analysis is done for the whole of China.

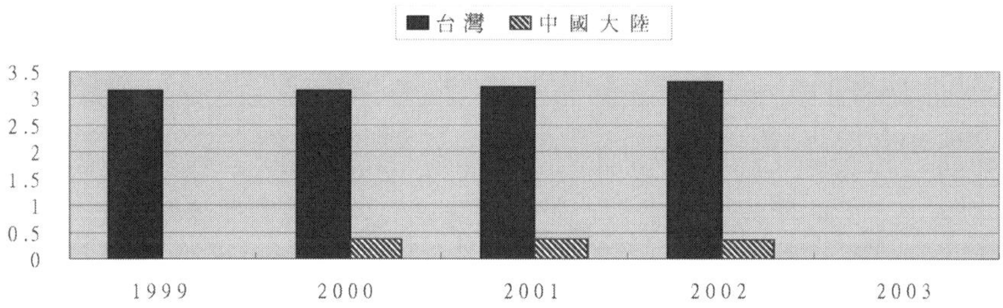

Figure 8.17. Number of Enterprise R&D Personnel per 1,000 Inhabitants
Legend (from left to right): Taiwan/Mainland China

2.2.2. Ratio of enterprise R&D to gross industrial production (%)

From the figures, we can see that in the last five years, the ratio of Taiwan's enterprise R&D to gross industrial production has declined. Except for its low point at 1.15 in 2001, it hovered around 1.25 for

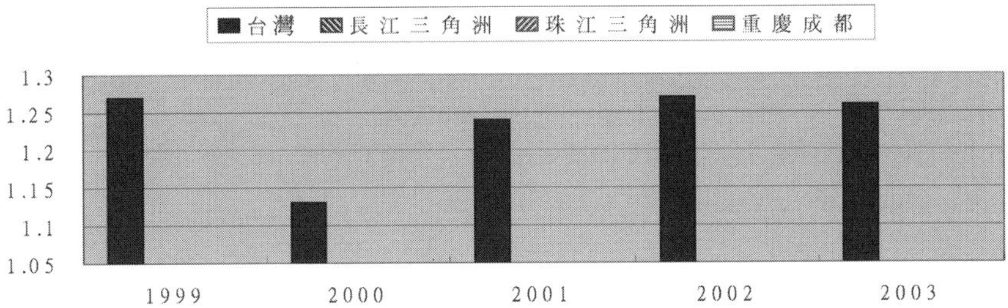

Figure 8.18. Ratio of Enterprise R&D to Gross Industrial Production (%)

Legend (from left to right): Taiwan/Yanzhijiang Delta Region/Chongqing Chengdu Region/ Zhujiang Delta Region

the next four years, which indicates that Taiwanese enterprises still emphasize on R&D expenditure. Because this index lacks relevant data for the three regions in China, we are unable to compare the relative quality among them.

2.2.3. Number of domestically approved patent applications for science-based parks (application)

Observing this index of the number of domestically approved patent applications for science-based parks, we can see that there has been growth in Taiwan from 1999 to 2001, but the country displayed a decline in 2002. However, generally speaking, it still maintained a

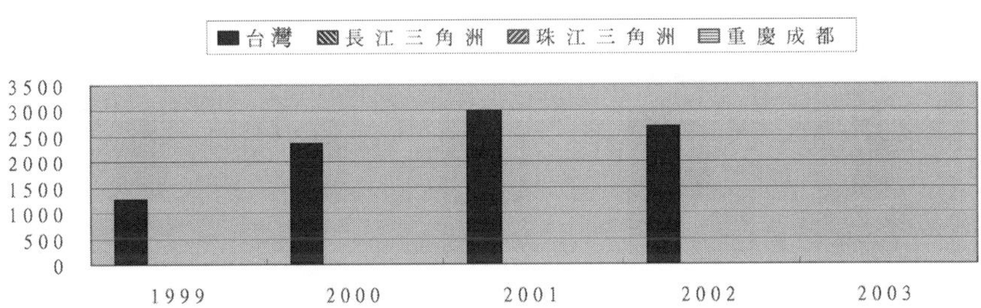

Figure 8.19. Number of Domestically Approved Patent Applications for Science-based Parks (Application)

Legend (from left to right): Taiwan/Yanzhijiang Delta Region/Chongqing Chengdu Region/ Zhujiang Delta Region

certain level. It indicates that the capability of innovative R&D for the companies in Taiwan's science-based park did not decline following the slowing economic prosperity in Taiwan. Because this index lacks relevant data for the three regions in China, we are unable to compare the relative quality among them.

2.3. Encouraged level of newly established enterprises and innovative activities

2.3.1. Ratio of the foreign capital within venture investment to the domestic capital (The data is the amount venture companies invested in the current year, 100 million NTD)

Because of the lack of data of foreign and domestic capital in venture investment, the index uses the investment that venture companies invested in the current year instead. From the data we can find that Taiwan's investment amount is up to 30 billion USD in 1999 and 2000, but in 2001, due to the international economic depression, the investment amount went down substantially to less than 10 billion USD. However, following the later recovery of prosperity, it displayed a slight growth in 2002 and 2003. Because this index lacks relevant

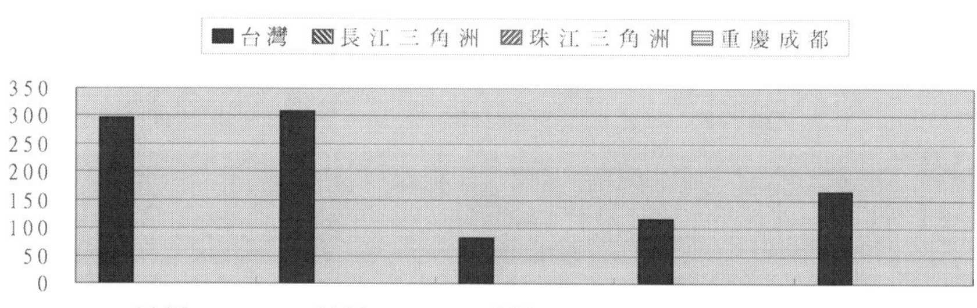

Figure 8.20. Ratio of the Foreign Capital within Venture Investment to the Domestic Capital

Legend (from left to right): Taiwan/Yanzhijiang Delta Region/Chongqing Chengdu Region/Zhujiang Delta Region

data for the three regions in China, we are unable to compare the relative quality among them.

2.3.2. Acquiring liability of venture capital (0–10 points)

According to the IMD data, we can see that in acquiring liability of venture capital, Taiwan is significantly better than China. Its global ranking went up from 14th in 2000 to fourth place in 2003. However, even though China's points increased, its ranking went from 43rd in 2001 down to 51st in 2003, which indicates that it still needs to be improved. Because this index lacks relevant data for Yanzhijiang Delta Region, Zhujiang Delta Region and Chongqing Chengdu Region, the comparative analysis is done for the whole of China.

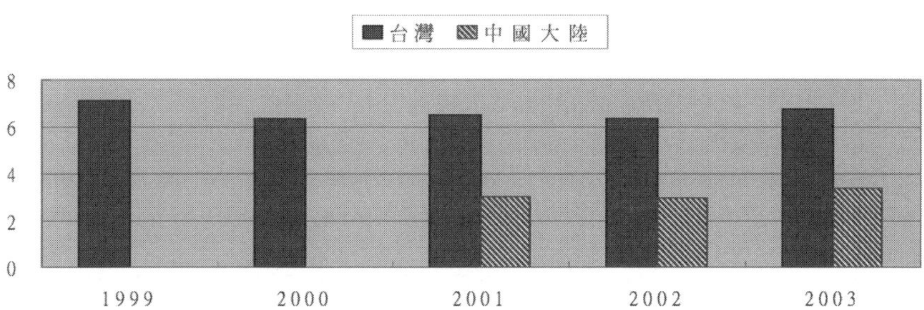

Figure 8.21. Acquiring Liability of Venture Capital (0–10 points)
Legend (from left to right): Taiwan/China

2.3.3. Ratio of R&D expenditure to GDP (%)

Observing the ratio of R&D expenditure to GDP, all the four regions displayed gradual growth. However, the historic ratio of Taiwan is the highest among them, over two every year, and even close to 2.5 in 2003, which indicates that Taiwan still emphasizes the input of R&D. However, Yanzhijiang Delta Region seems to have nearly caught up with Taiwan, but Chongqing Chengdu Region still needs to improve.

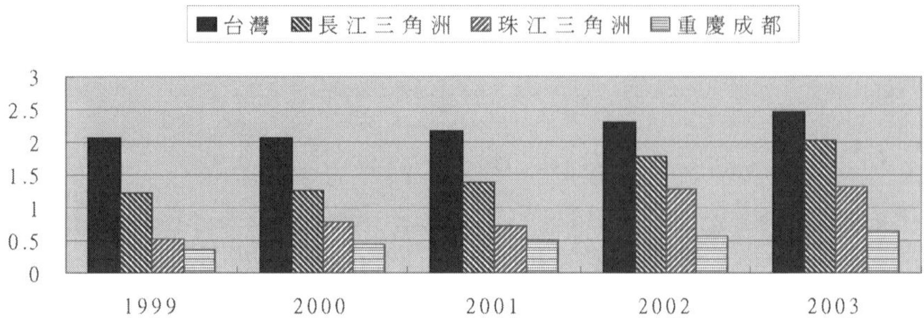

Figure 8.22. Ratio of R&D Expenditure to GDP (%)

Legend (from left to right): Taiwan/Yanzhijiang Delta Region/Chongqing Chengdu Region/Zhujiang Delta Region

2.3.4. Ratio of IT industry R&D expenditure to GDP (%)

Because this index lacks relevant data, no analysis is available at this moment.

2.3.5. Ratio of investment in IT and electronics industry to investment in total basic infrastructure and innovative reconstruction (%)

Because this index lacks relevant data, no analysis is available at this moment.

2.3.6. Ratio of the successful venture capital to the venture investment amount (%)

Because this index lacks relevant data, no analysis is available at this moment.

2.3.7. Ratio of the applied amount of the med/long-term funds that the government grants start-up businesses (%)

Because this index lacks relevant data, no analysis is available at this moment.

2.3.8. Ratio of the IT industry output value to GDP (%)

Because this index lacks relevant data, no analysis is available at this moment.

2.4. Knowledge circulation among industrial, academic, and research circles

2.4.1. Industry-academy cooperation (0–10 points)

Observing this industry-academy cooperation index, Taiwan's performance in the recent two or three years has fallen behind the whole of China's and Zhujiang Delta Region's. In 2002, even though Taiwan was ranked last among these three regions, the difference among them was not very big, Taiwan was 31st, China was 26th and Zhujiang Delta Region was 29th. But, in 2003, Taiwan's ranking was overtaken by Zhujiang Delta Region, whose global ranking went up to fourth, and the China's went up to 11th. Even though Taiwan's ranking went up to 18th, the difference among them has grown. This indicates that Taiwan still needs to improve the connection between industry and academics.

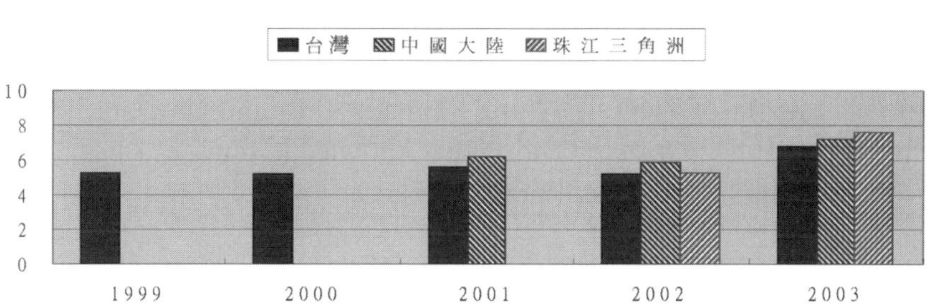

Figure 8.23. Industry-Academy Cooperation (0–10 points)
Legend (from left to right): Taiwan/China/Zhujiang Delta Region

2.4.2. Technology cooperation among enterprises (0–10 points)

Even though it fell behind China for the industry-academy cooperation index, the relationship of technology cooperation among Taiwanese enterprises is still very strong. Observing the performance

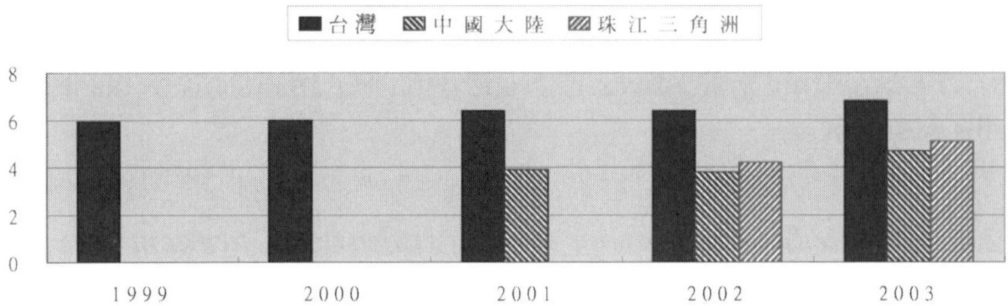

Figure 8.24. Technology Cooperation among Enterprises (0–10 points)
Legend (from left to right): Taiwan/China/Zhujiang Delta Region

over the years, we can see that it is significantly better than that of China's and Zhujiang Delta Region's. Taking 2003 as an example, the global ranking of Taiwan is 18th, which took a substantial lead over China's 51st place and Zhujiang Delta Region's 43rd. It is an indication that the technological cooperation among China's enterprises still needs to be improved.

2.4.3. Knowledge transfer sufficiency (0–10 points)

Observing this index, we can see that Taiwan is significantly ahead of China and Zhujiang Delta Region. Taking 2003 as an example, the global ranking of Taiwan is 13th, China is 46th and Zhujiang Delta Region is 49th. It is an indication that knowledge transfer is not practiced well in China and still needs to be improved.

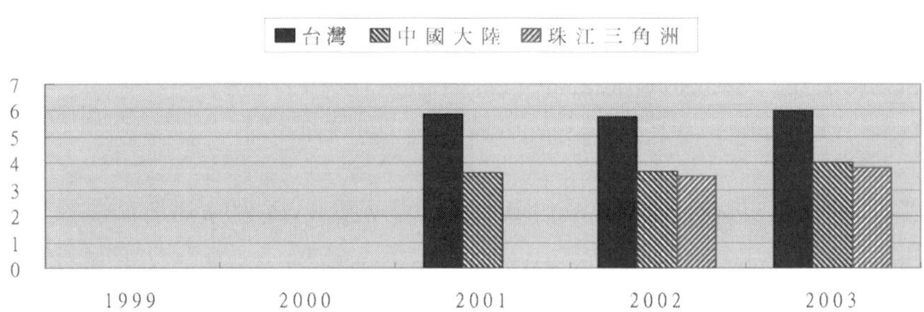

Figure 8.25. Knowledge Transfer Sufficiency (0–10 points)
Legend (from left to right): Taiwan/China/Zhujiang Delta Region

2.5. *Degree of international collaboration in innovative R&D*

2.5.1. *Purchase amount of foreign technology (100 million USD)*

In this index, we can see that Taiwan and Chongqing Chengdu Region fell way behind Yanzhijiang Delta Region and Zhujiang Delta Region, among which, Yanzhijiang Delta Region and Zhujiang Delta Region displayed a gradual increase. In addition, up to 2002, the purchase amount of foreign technology in Zhujiang Delta Region reached more than 30 billion USD. It indicates that Mainland China is very dependent on foreign technology sources for developing its economy and upgrading its industry. However, it is worth taking note that this index may be wrong because the data from China uses the export amount of high-tech products as its representative and it is different from the so-called technology, thus the amount might have been overestimated.

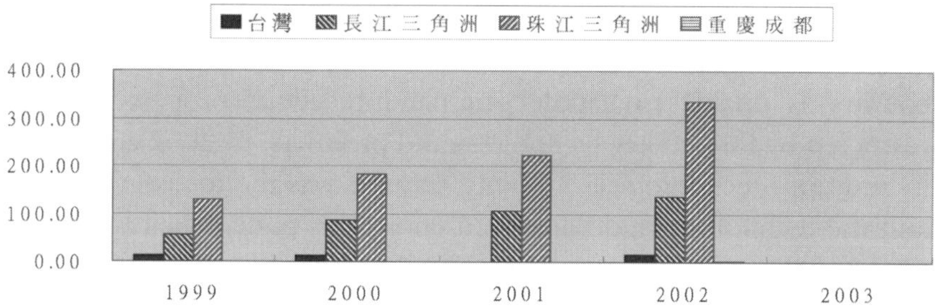

Figure 8.26. Purchase Amount of Foreign Technology (100 million USD)
Legend (from left to right): Taiwan/Yanzhijiang Delta Region/Chongqing Chengdu Region/Zhujiang Delta Region

2.5.2. *Amount of technology export (100 million USD)*

For this index, just as in the previous index, Taiwan and Chongqing Chengdu Region are significantly behind Yanzhijiang Delta Region and Zhujiang Delta Region, among which, the amount in Zhujiang Delta Region is the largest. However, it is worth taking note that this index may also be wrong because the data from China uses the export

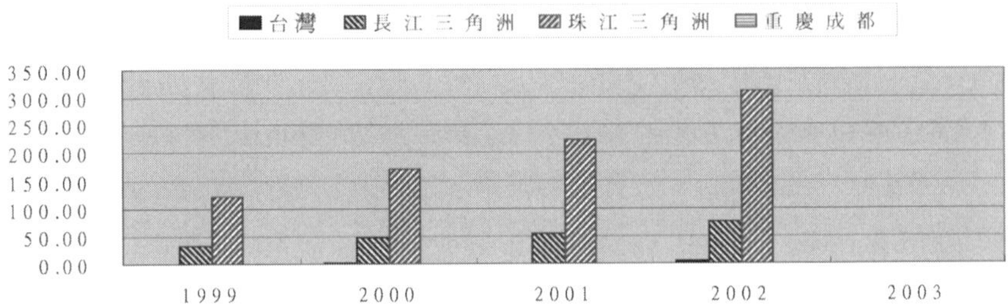

Figure 8.27. Amount of Technology Export (100 million USD)

Legend (from left to right): Taiwan/Yanzhijiang Delta Region/Chongqing Chengdu Region/Zhujiang Delta Region

amount of high-tech products as its representative and it is different from the so-called technology, thus the amount might have been overestimated.

2.5.3. Coverage ratio (%) of license fees and royalty

In terms of the coverage ratio of license fees and royalty, Taiwan grew before 2001, but it declined after 2001, which indicates that the country is unable to handle original intellectual property, and is restricted by foreign key technology and products. In 2002 and 2003, in addition, technological inbound transfer was the focus in Taiwan, and the technology trade displayed enormous trade deficits. Relative to the performance of the international patents, the massive amount of technology trade deficit shows even more the insufficient coverage

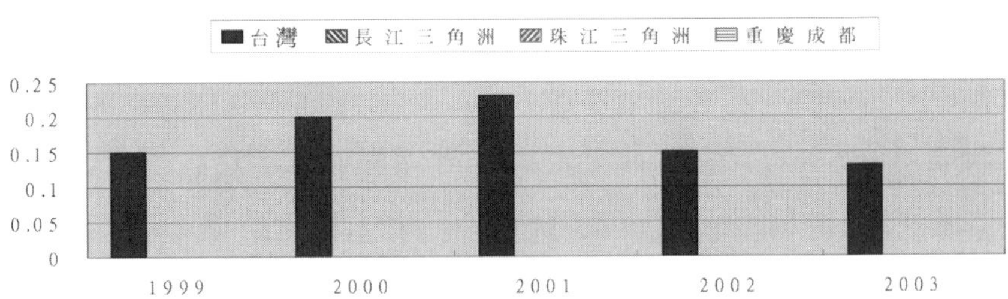

Figure 8.28. Coverage Ratio (%) of License Fees and Royalty

Legend (from left to right): Taiwan/Yanzhijiang Delta Region/Chongqing Chengdu Region/Zhujiang Delta Region

for the technological R&D value chain, and it needs a breakthrough immediately. Because this index lacks relevant data for Yanzhijiang Delta Region, Zhujiang Delta Region, and Chongqing Chengdu Region, the comparative analysis is done for the whole of China.

2.5.4. Amount of the joint venture investment with listed/OTC enterprises among OECD countries

Because this index lacks relevant data, no analysis is available at this moment.

2.6. Performance of innovative capability and entrepreneurship

2.6.1. Industrial cluster innovative environment

Because this index lacks relevant data, no analysis is available at this moment.

2.6.2. Connection between innovative infrastructure and industrial cluster environment

Because this index lacks relevant data, no analysis is available at this moment.

2.6.3. Research manpower of science and engineering

Because this index lacks relevant data, no analysis is available at this moment.

2.6.4. Innovation policies

Because this index lacks relevant data, no analysis is available at this moment.

2.6.5. Innovative orientation enterprises

Because this index lacks relevant data, no analysis is available at this moment.

2.6.6. Establishment of start-up businesses (0–10 points)

Observing this index, we can see that the establishment of start-up businesses in Taiwan has decreased recently, moving from the global ranking of 6th place in 2000 to 11th in 2002, and 10th in 2003. Zhujiang Delta Region overtook Taiwan in 2003, and this may be due to Taiwan's industrial migration in recent years.

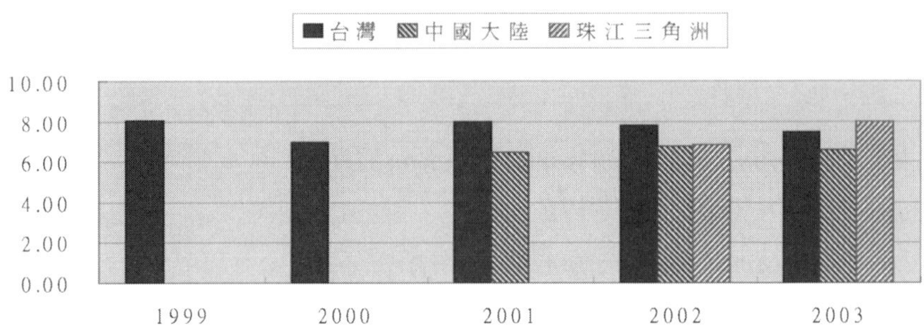

Figure 8.29. Establishment of Start-Up Businesses (0–10 points)

Legend (from left to right): Taiwan/Mainland China/Zhujiang Delta Region

2.6.7. Entrepreneurship (0–10 points)

Observing this index, we can see that Taiwan's performance is very stable. Its global ranking has been maintained within 10 places in these past five years. However, the performance of Zhujiang Delta Region within the previous year can be described as a soaring development

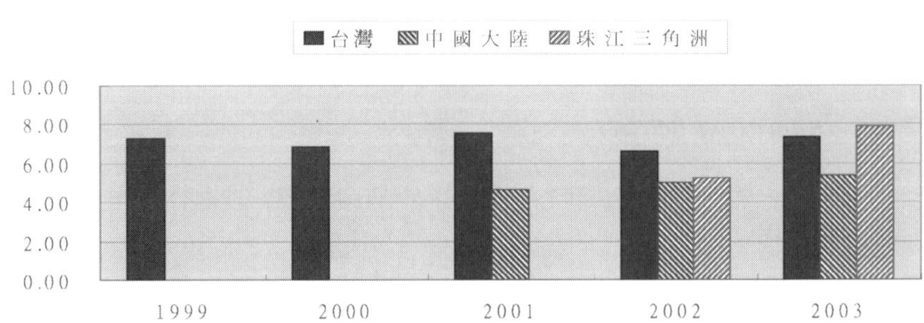

Figure 8.30. Entrepreneurship (0–10 points)

Legend (from left to right): Taiwan/Mainland China/Zhujiang Delta Region with its global ranking going up from 42nd in 2002 to the first in 2003.

2.6.8. Positiveness of introducing new technology to enterprises

Because this index lacks relevant data, no analysis is available at this moment.

3. Application of Information Technology

3.1. Degree of e-industry

3.1.1. Ratio of enterprises that engaged in relevant e-commerce activities

Because this index lacks relevant data, no analysis is available at this moment.

3.1.2. New information technology application that meets enterprises' demand (0–10 points)

Observing this index, we can see that Taiwan is significantly better than China and Zhujiang Delta Region. It has been ranked around 20th to 30th place globally in the last five years, and China and Zhujiang Delta Region are ranked around 50th to 60th place.

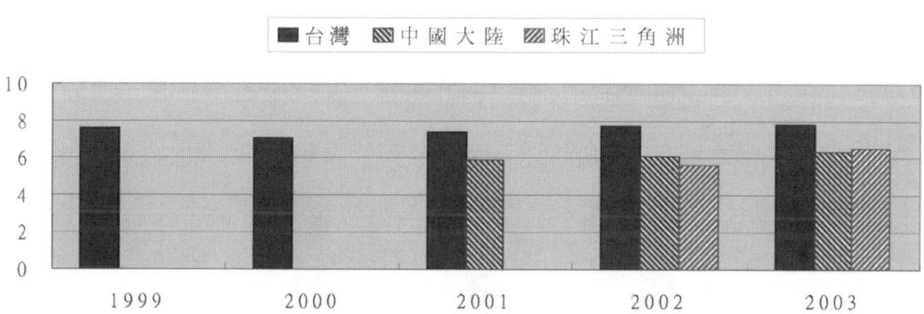

Figure 8.31. New Information Technology Application that Meets Enterprises' Demand (0–10 points)

Legend (from left to right): Taiwan/Mainland China/Zhujiang Delta Region

3.1.3. Electronic degree (points)

Because this index lacks relevant data, no analysis is available at this moment.

3.1.4. Maturity of industrial electronics (points)

Because this index lacks relevant data, no analysis is available at this moment.

3.2. Abundance of digitalization contents

3.2.1. Popularization of governmental websites (1–7 points)

Because this index lacks relevant data, no analysis is available at this moment.

3.2.2. Evaluation of e-government (1–7 points)

Because this index lacks relevant data, no analysis is available at this moment.

3.2.3. Evaluation of e-government websites (points)

There is a considerable lack of data for this field, and there is only data from the last three years in Taiwan which can be analyzed. In 2000, Taiwan was ranked as second globally, and first in 2001 and 2002,

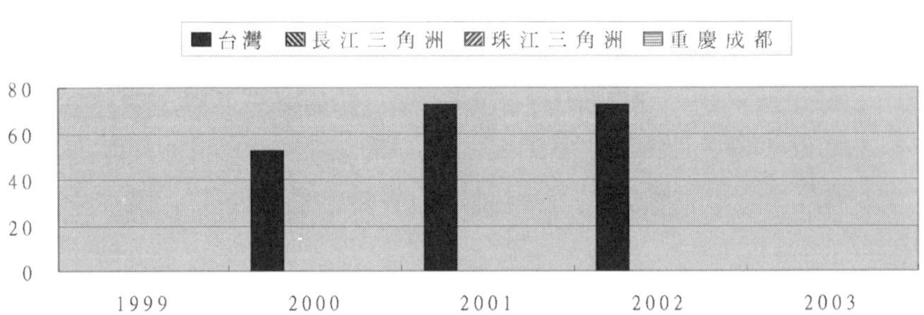

Figure 8.32. Evaluation of e-Government Websites (points)

Legend (from left to right): Taiwan/Yanzhijiang Delta Region/Chongqing Chengdu Region/Zhujiang Delta Region

which indicate that the electronic degree for the government sectors is very pronounced in Taiwan and helped it take a big lead over other countries. Because this index lacks relevant data for Yanzhijiang Delta Region, Zhujiang Delta Region, and Chongqing Chengdu Region, it is not possible to conduct any comparison at this moment.

3.3. e-learning popularization (household/school/other)

3.3.1. Household on-line popularization (%)

There is a considerable lack of data for this field, and there is only data from the last three years in Taiwan which can be analyzed. Viewing the figures, we can find that households with on-line access in Taiwan has gradually increased from 14.5% in 1999 to 27.4% in 2001, when it almost doubled its growth. This will be very helpful to communication and the development of information in Taiwan. Because this index lacks relevant data for Yanzhijiang Delta Region, Zhujiang Delta Region, and Chongqing Chengdu Region, it is not possible to conduct any comparison at this moment.

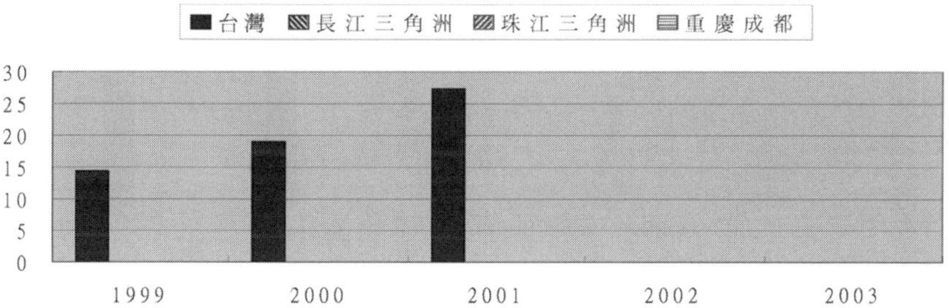

Figure 8.33. Household On-line Popularization (%)

Legend (from left to right): Taiwan/Yanzhijiang Delta Region/Chongqing Chengdu Region/Zhujiang Delta Region

3.3.2. Schools' on-line rate (%)

There is a considerable lack of data for this field, and there is only data for the last three years in Taiwan which can be analyzed. The developing trend of on-line usage in Taiwan's schools is similar to the household on-line popularization in Taiwan, and both of these have

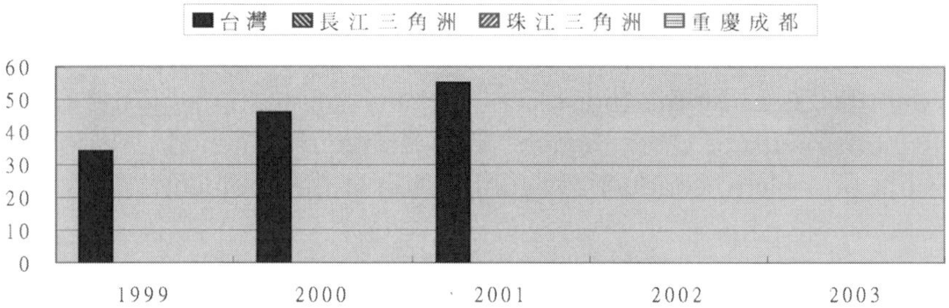

Figure 8.34. Schools' On-line Rate (%)

Legend (from left to right): Taiwan/Yanzhijiang Delta Region/Chongqing Chengdu Region/Zhujiang Delta Region

displayed much growth. The on-line rate in Taiwan's schools grew from 34% in 1999 to 55% in 2001, which indicates that 50% of Taiwan's schools already possess network capability, and this will be helpful in information searches. Because this index lacks relevant data for Yanzhijiang Delta Region, Zhujiang Delta Region, and Chongqing Chengdu Region, it is not possible to conduct any comparison at this moment.

4. Social Infrastructure

4.1. Measures for encouraging foreign investment and financial environment and capital markets

4.1.1. Statute for upgrading industries and relevant tax deduction laws and regulations

This index is descriptive data, so it cannot be analyzed using figures and tables. Please refer to the data organization of research indices for this study.

4.1.2. High-tech business tax exemption plan or tax preference policy

This index is descriptive data, so it cannot be analyzed using figures and tables. Please refer to the data organization of research indices for this study.

4.1.3. Tax exemption of venture capital investment

This index is descriptive data, so it cannot be analyzed using figures and tables. Please refer to the data organization of research indices for this study.

4.1.4. Taxation burden rate of foreign enterprises

This index is descriptive data, so it cannot be analyzed using figures and tables. Please refer to the data organization of research indices for this study.

4.1.5. Amount of international capital transfer and time of outward remittance

This index is descriptive data, so it cannot be analyzed using figures and tables. Please refer to the data organization of research indices for this study.

4.1.6. Credit rating of banks

This index is descriptive data, so it cannot be analyzed using figures and tables. Please refer to the data organization of research indices for this study.

4.1.7. Volatility of exchange rate

This index is descriptive data, so it cannot be analyzed using figures and tables. Please refer to the data organization of research indices for this study.

4.1.8. Bank management transparency (systematization and standardization)

This index is descriptive data, so it cannot be analyzed using figures and tables. Please refer to the data organization of research indices for this study.

4.1.9. Enterprise directors and supervisors management transparency (systematization and standardization)

This index is descriptive data, so it cannot be analyzed using figures and tables. Please refer to the data organization of research indices for this study.

4.1.10. Liability to accquire huge amount of capital for foreign enterprises

This index is descriptive data, so it cannot be analyzed using figures and tables. Please refer to the data organization of research indices for this study.

4.1.11. Ratio for the amount of absorbing foreign capital to total investment

Because only Taiwan possesses such data of the ratio for the amount of absorbing foreign capital to the total investment, in order to uniformly compare criteria, we use the overseas Chinese and foreign investment amount for the data for Taiwan, and for China, we use the amount of attracted foreign capital. From the figures, we can see that, since 2000, the overseas Chinese and foreign investment amount in Taiwan has declined. However, it has increased in Yanzhijiang Delta Region and Zhujiang Delta Region. Among these

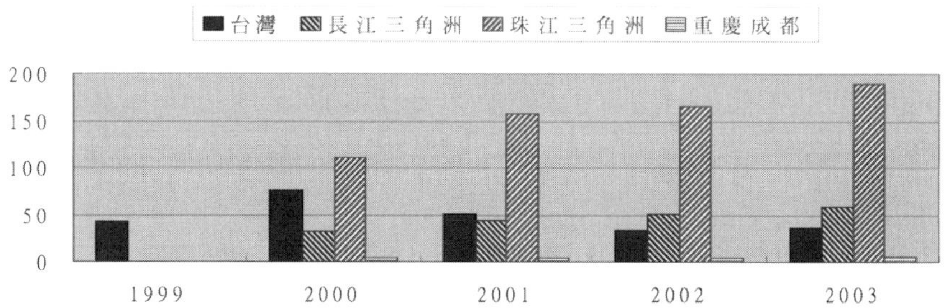

Figure 8.35. Ratio for the Amount of Absorbing Foreign Capital to Total Investment

Legend (from left to right): Taiwan/Yanzhijiang Delta Region/Chongqing Chengdu Region/Zhujiang Delta Region

four regions, the absolute amount of Zhujiang Delta Region is ranked in first place, which indicates that the capability of attracting foreign capital in Zhujiang Delta Region is significantly better than that of the other three regions.

4.1.12. Tariff

This index is descriptive data, so it cannot be analyzed using figures and tables. Please refer to the data organization of research indices for this study.

4.1.13. Personal income tax

This index is descriptive data, so it cannot be analyzed using figures and tables. Please refer to the data organization of research indices for this study.

4.1.14. Condition of business credit and business ethics for enterprises in investing regions

This index is descriptive data, so it cannot be analyzed using figures and tables. Please refer to the data organization of research indices for this study.

4.1.15. Whether foreign banks are able to engage in local currency business

This index is descriptive data, so it cannot be analyzed using figures and tables. Please refer to the data organization of research indices for this study.

4.2. Degree of social internationalization and liberalization

4.2.1. Ratio of outbound visitors to regional population (%)

Observing this index, we can see that the ratio of the number of outbound visitors to the regional population in Taiwan has seldom

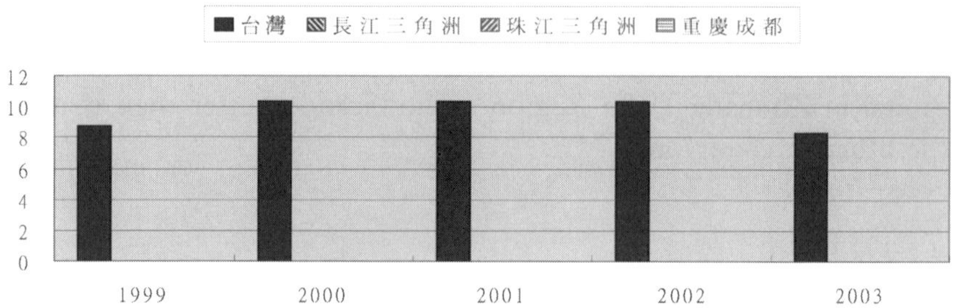

Figure 8.36. Ratio of Outbound Visitors to Regional Population (%)

Legend (from left to right): Taiwan/Yanzhijiang Delta Region/Chongqing Chengdu Region/ Zhujiang Delta Region

changed in recent years. It was around 10% during 2000 to 2002, but declined to 8% in 2003. Because this index lacks relevant data for Yanzhijiang Delta Region, Zhujiang Delta Region, and Chongqing Chengdu Region, it is not possible to conduct any comparison at this moment.

4.2.2. Ratio of inbound visitors to regional population (%)

Observing this index, we can see that the ratio in Taiwan is much higher than that of the three regions in China, and is mainly affected

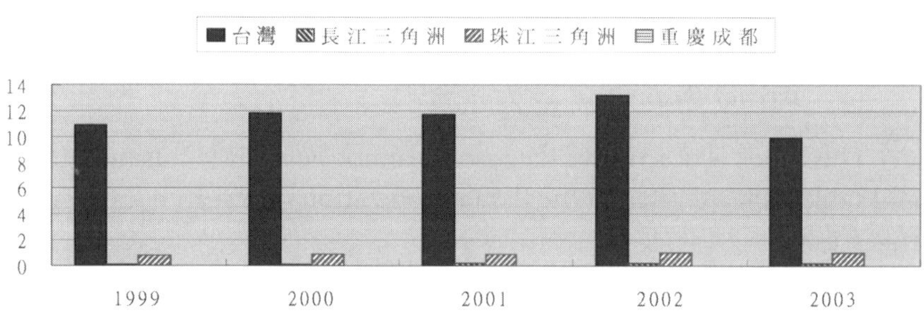

Figure 8.37. Ratio of Inbound Visitors to Regional Population (%)

Legend (from left to right): Taiwan/Yanzhijiang Delta Region/Chongqing Chengdu Region/ Zhujiang Delta Region

by the regional population. Therefore, attention needs to be paid to determine the social internationalization and liberalization level when using this index.

4.2.3. Ratio of legal foreign residents to regional population (%)

Observing this index, we can see that, recently, the ratio for the number of legal foreign residents to the regional population in Taiwan has displayed a slight growth from only 1.55% in 1999 to 1.79% in 2003. Because this index lacks relevant data for Yanzhijiang Delta Region, Zhujiang Delta Region, and Chongqing Chengdu Region, it is not possible to conduct any comparison at this moment.

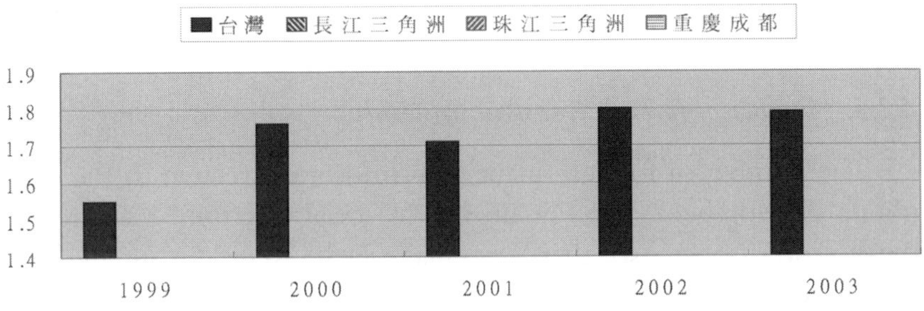

Figure 8.38. Ratio of Legal Foreign Residents to Regional Population (%)
Legend (from left to right): Taiwan/Yanzhijiang Delta Region/Chongqing Chengdu Region/ Zhujiang Delta Region

4.2.4. Number of international conferences/exhibitions and their participants

The data of this index is mainly the number of foreign participants that attended the international seminars in Taiwan. Significantly, after 2000, it declined from 35,000 people in 2000 to 31,542 in 2003. Because this index lacks relevant data for Yanzhijiang Delta Region, Zhujiang Delta Region, and Chongqing

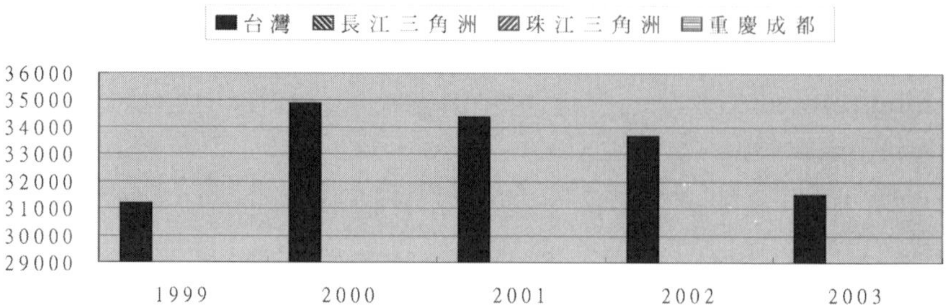

Figure 8.39. Number of International Conferences/Exhibitions and Their Participants

Legend (from left to right): Taiwan/Yanzhijiang Delta Region/Chongqing Chengdu Region/Zhujiang Delta Region

Chengdu Region, it is not possible to conduct any comparison at this moment.

4.2.5. Number of foreign financial institutions

The number of foreign financial institutions is related to the economic development and the degree of social freedom. Looking at these figures, we can see that Yanzhijiang Delta Region is significantly in the lead over Taiwan and displayed growth as well. Taiwan displayed a slight decline, which indicates the capability of attracting foreign

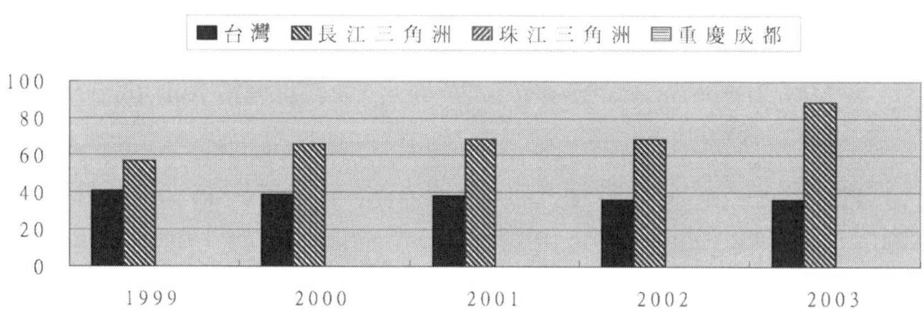

Figure 8.40. Number of Foreign Financial Institutions

Legend (from left to right): Taiwan/Yanzhijiang Delta Region/Chongqing Chengdu Region/Zhujiang Delta Region

financial institutions in Yanzhijiang Delta Region is better than in Taiwan. Because this index lacks relevant data for Zhujiang Delta Region and Chongqing Chengdu Region, it is not possible to conduct any comparison at this moment.

4.2.6. Ratio of study-abroad students (with student visas) to graduate (and above) students (%)

Observing this index, we can see that the number of study-abroad students (with student visas) to the number of graduate (and above) students in Taiwan has declined from nearly 6% in 1999 to 3% in 2003. The decrease almost doubled, which indicates that the number of study-abroad students is declining in Taiwan, but the ratio of university students is growing. In addition, in China, there was a slight increase from 0.58% in 1999 to 1.39% in 2003. Because this index lacks relevant data for Yanzhijiang Delta Region, Zhujiang Delta Region, and Chongqing Chengdu Region, it is not possible to conduct any comparison at this moment.

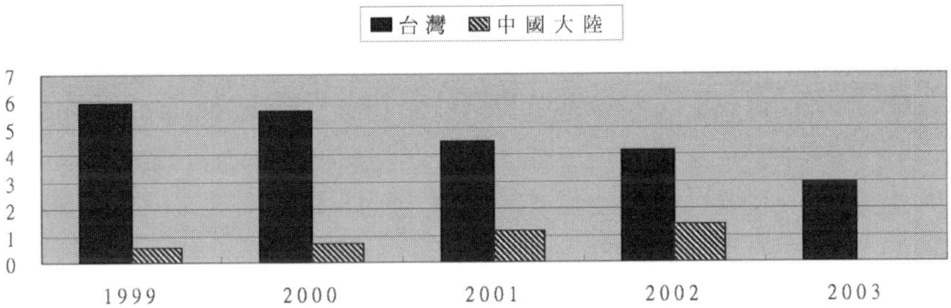

Figure 8.41. Ratio of Study-abroad Students (with Student Visas) to Graduate (and above) Students (%)

Legend (from left to right): Taiwan/Mainland China

4.2.7. Ratio of foreign students to total regional students (%)

Looking at this index, we can see that Taiwan has maintained 0.1% for the last five years with little change. Because this index lacks relevant data for Yanzhijiang Delta Region, Zhujiang Delta Region,

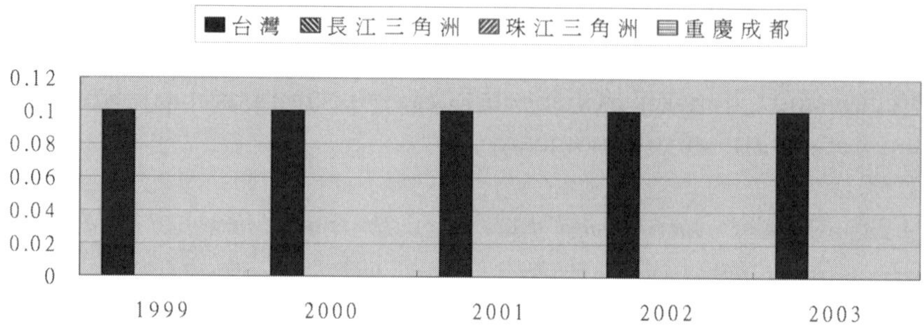

Figure 8.42. Ratio of Foreign Students to Total Regional Students (%)

Legend (from left to right): Taiwan/Yanzhijiang Delta Region/Chongqing Chengdu Region/ Zhujiang Delta Region

and Chongqing Chengdu Region, it is not possible to conduct any comparison at this moment.

4.2.8. Foreign labor ratio (%)

Observing this index, we can see that the foreign labor ratio of Taiwan displayed a decline after 2000, then reached 2.98% in 2003. It is mainly affected by the current government's policies. Because the index lacks the relevant foreign labor data for China, we are unable to compare them.

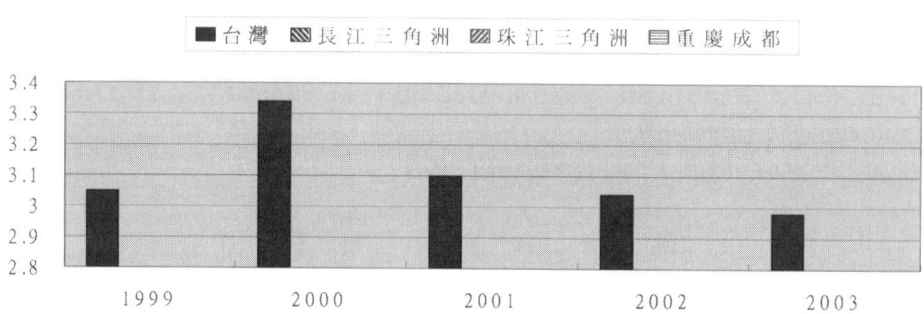

Figure 8.43. Foreign Labor Ratio (%)

Legend (from left to right): Taiwan/Yanzhijiang Delta Region/Chongqing Chengdu Region/ Zhujiang Delta Region

4.2.9. Exit and entry policy

This index is descriptive data, so it cannot be analyzed using figures and tables. Please refer to the data organization of research indices for this study.

4.2.10. Traffic cost of exit and entry

This index is descriptive data, so it cannot be analyzed using figures and tables. Please refer to the data organization of research indices for this study.

4.2.11. Foreign language (English) popularization (GEPT — General English Placement Test)

(China used the number of passed Advanced Level (Level 4) to represent the number of university students.)

From the data, we can see that the popularization of foreign languages has risen in Taiwan, which will be helpful in a certain level of internationalization for Taiwan society. Because this index lacks relevant data for Yanzhijiang Delta Region, Zhujiang Delta Region, and Chongqing Chengdu Region, it is not possible to conduct any comparison at this moment.

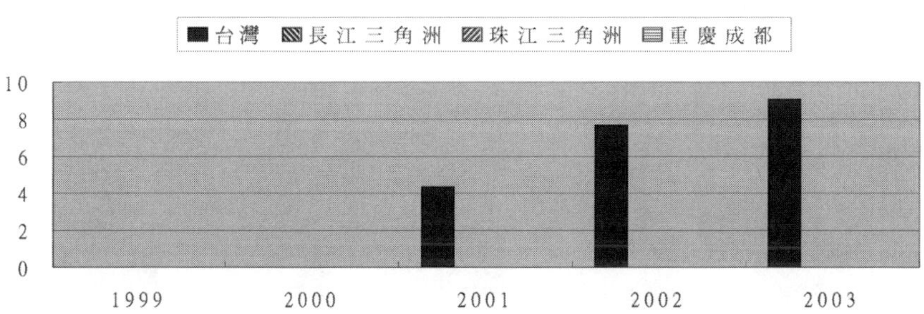

Figure 8.44. Foreign Language (English) Popularization (GEPT — General English Placement Test)

Legend (from left to right): Taiwan/Yanzhijiang Delta Region/Chongqing Chengdu Region/Zhujiang Delta Region

4.2.12. Acceptance level of local to foreigner

Because this index lacks relevant data, no analysis is available at this moment.

4.2.13. Satisfaction to international schools

Because this index lacks relevant data, no analysis is available at this moment.

4.2.14. Fees of studying in international schools

Because this index lacks relevant data, no analysis is available at this moment.

4.3. Habitations, medical care, and social security

4.3.1. Number of western medical doctors per 10,000 inhabitants

Observing this index, we can see that the number of medical personnel in Yanzhijiang Delta Region is larger than those of the other three regions, and there is almost no difference between Taiwan, Zhujiang Delta Region, and Chongqing Chengdu (except

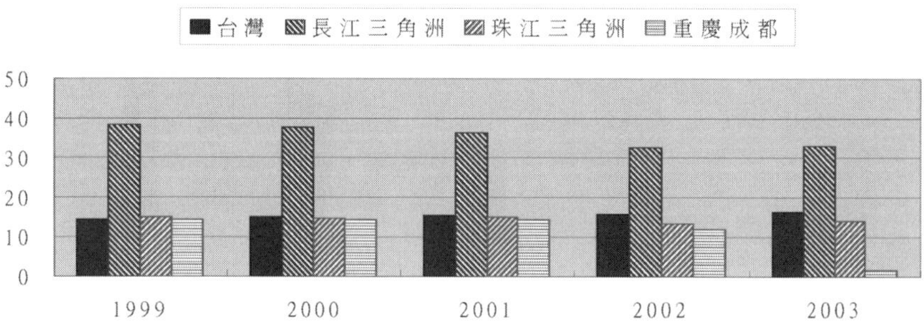

Figure 8.45. Number of Western Medical Doctors per 10,000 Inhabitants

Legend (from left to right): Taiwan/Yanzhijiang Delta Region/Chongqing Chengdu Region/Zhujiang Delta Region

for Chongqing Chengdu Region in 2003, which was significantly reduced).

4.3.2. Number of hospital beds per 10,000 inhabitants

This index displayed a significant contrast. Taiwan and Yanzhijiang Delta Region can be regarded as a set, and Zhujiang Delta Region and Chongqing Chengdu Region can be regarded as another set. In the last five years, the number of hospital beds per 10,000 inhabitants has been around 50 to 60 in Taiwan and Yanzhijiang Delta Region, which is significantly better than the number of around 20 in Zhujiang Delta Region and Chongqing Chengdu Region. That indicates that the medical care environment in Taiwan and Yanzhijiang Delta Region is better than those of Zhujiang Delta Region and Chongqing Chengdu Region.

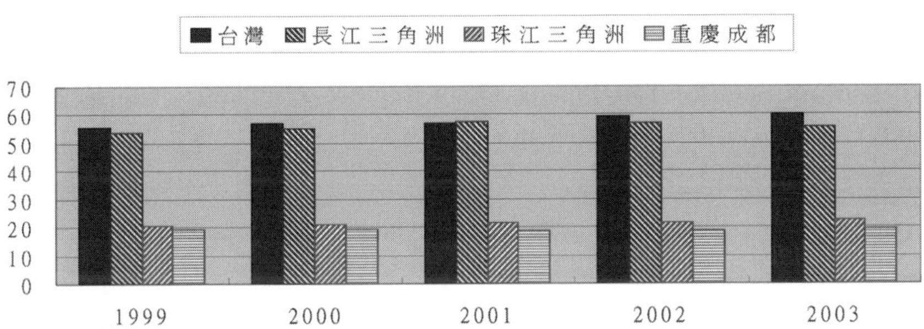

Figure 8.46. Number of Hospital Beds per 10,000 Inhabitants

Legend (from left to right): Taiwan/Yanzhijiang Delta Region/Chongqing Chengdu Region/Zhujiang Delta Region

4.3.3. Completion rate of regular epidemic inoculation

There is no difference among the four regions for this index. The inoculation completion rate is 100% for all three regions in China, but is 95% in Taiwan region for 1999, and lacks the data for the last four years. However, the difference is not expected to be large.

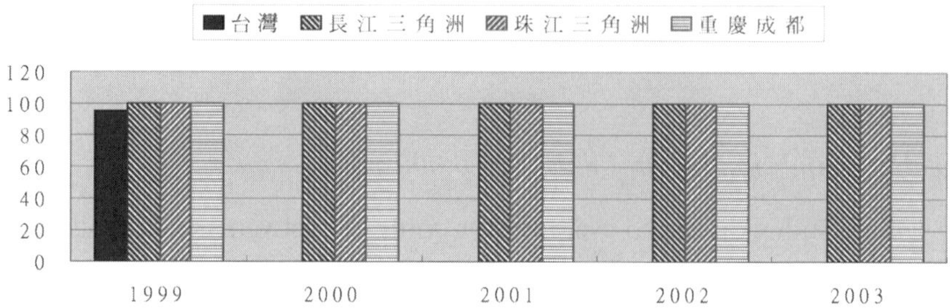

Figure 8.47. Completion Rate of Regular Epidemic Inoculation

Legend (from left to right): Taiwan/Yanzhijiang Delta Region/Chongqing Chengdu Region/Zhujiang Delta Region

4.3.4. Ratio of medical care expenditure to GDP

Observing this index, we can see that the ratio of medical care expenditure to GDP in Taiwan has displayed gradual growth. It formerly fell behind China, but after 2001, Taiwan took the lead, which indicates that Taiwan has emphasized medical care, and this might be also linked to the establishment of National Health Insurance. Because of the lack of relevant data for Yanzhijiang Delta Region, Zhujiang Delta Region, and Chongqing Chengdu Region, the whole of China is used.

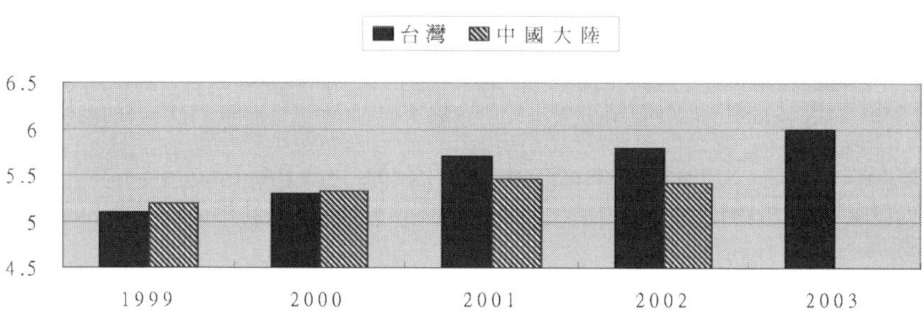

Figure 8.48. Ratio of Medical Care Expenditure to GDP

Legend (from left to right): Taiwan/China

4.3.5. Crime occurrence rate per 100,000 inhabitants

Observing this index, we can see that the crime occurrence rate per 100,000 inhabitants in China is significantly higher than in Taiwan, which indicates that Taiwan is relatively safer than China in

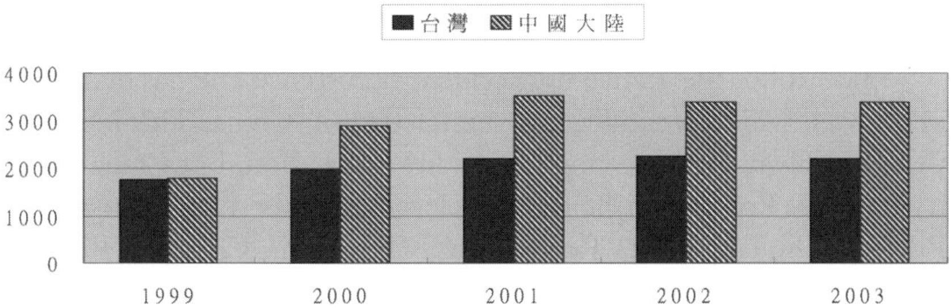

Figure 8.49. Crime Occurrence Rate per 100,000 Inhabitants

Legend (from left to right): Taiwan/China

public security. However, the rate has slightly increased in both regions. Because of the lack of relevant data for Yanzhijiang Delta Region, Zhujiang Delta Region, and Chongqing Chengdu Region, the whole of China is used.

4.3.6. Larceny occurrence rate per 100,000 inhabitants

Observing this index, we can see that the larceny occurrence rate per 100,000 inhabitants in China was lower than that of Taiwan's in 1999, however, since 2000, the rate has risen considerably from 1,150.7 cases in 1999 to 2,253.3 cases in 2001. The rate has almost doubled since and is also higher than Taiwan's in the past two or three years, which indicates that Taiwan is still relatively safer than China. Because of the lack of relevant data for Yanzhijiang Delta Region, Zhujiang Delta Region, and Chongqing Chengdu Region, the whole of China is used.

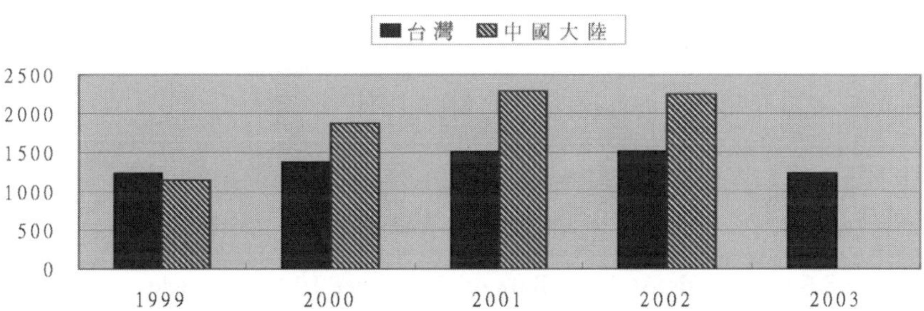

Figure 8.50. Larceny Occurrence Rate per 100,000 Inhabitants

Legend (from left to right): Taiwan/China

4.3.7. Annual fire accident occurrence number increase rate

Since 2000, the annual fire accident occurrence rate has decreased in Taiwan, which is significantly better than in China. This indicates that the Taiwanese have emphasized fire prevention, but China needs to improve. Because of the lack of relevant data for Yanzhijiang Delta Region, Zhujiang Delta Region, and Chongqing Chengdu Region, the whole of China is used.

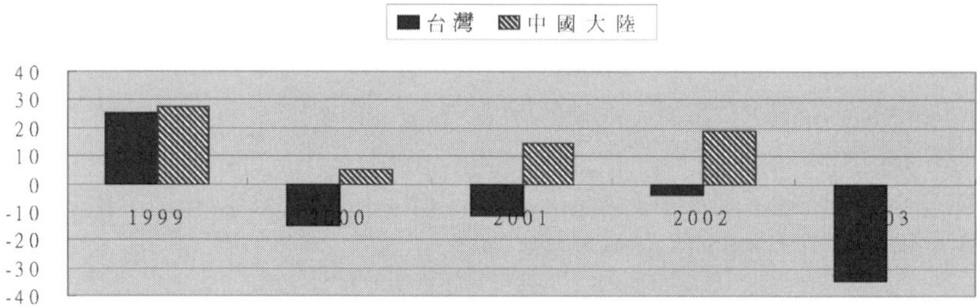

Figure 8.51. Annual Fire Accident Occurrence Number Increase Rate

Legend (from left to right): Taiwan/China

4.3.8. Traffic accident rate on roads per 10,000 vehicles

Taiwan's performance in this index is not impressive, rising significantly from only 1.54% in 1999 to 48.77% in 2003, which indicates the frequent occurrence of traffic accidents. Because this index lacks relevant data for Yanzhijiang Delta Region, Zhujiang Delta

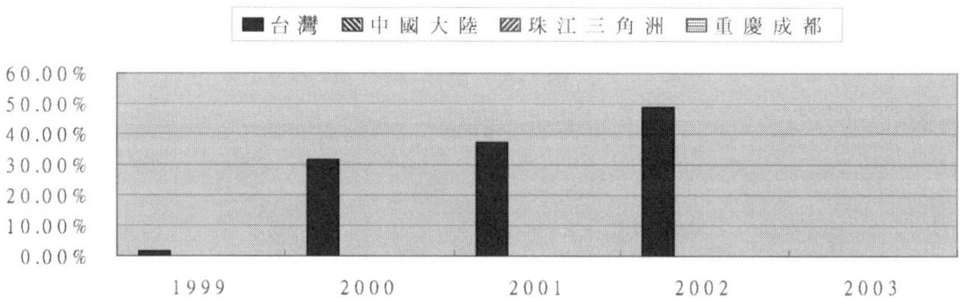

Figure 8.52. Traffic Accident Rate on Roads per 10,000 Vehicles

Legend (from left to right): Taiwan/Yanzhijiang Delta Region/Chongqing Chengdu Region/Zhujiang Delta Region

Region, and Chongqing Chengdu Region, it is not possible to conduct any comparison at this moment.

4.4. Governmental performance transparency and administrative efficiency

4.4.1. Performance transparency (0–10 points)

Observing this index we can see that the leading comparison is between Taiwan and China. In 2001 and 2003, Taiwan was better than China, but in 2002, China was better than Taiwan. However, looking at the latest year, 2003, we can see the amazing growth in Zhujiang Delta Region, which rapidly rose from a global ranking of 38th in 2002 to 19th in 2003.

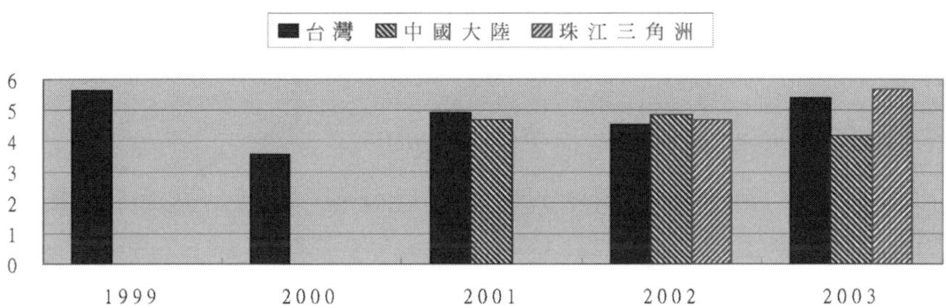

Figure 8.53. Performance Transparency (0–10 points)

Legend (from left to right): Taiwan/China/Zhujiang Delta Region

4.4.2. Government administrative efficiency

Observing this index, we can see that Taiwan declined from its global ranking of 18th place in 1999 to 20th in 2000, and it further declined to 21st in 2001. Because this index lacks relevant data for Yanzhijiang Delta Region, Zhujiang Delta Region, and Chongqing Chengdu Region, it is not possible to conduct any comparison at this moment.

Taiwan has performed better than China in the past three years, but its overall ranking has significantly declined, from 29th place globally in 2000 to 36th place in 2003. In contrast, within the past

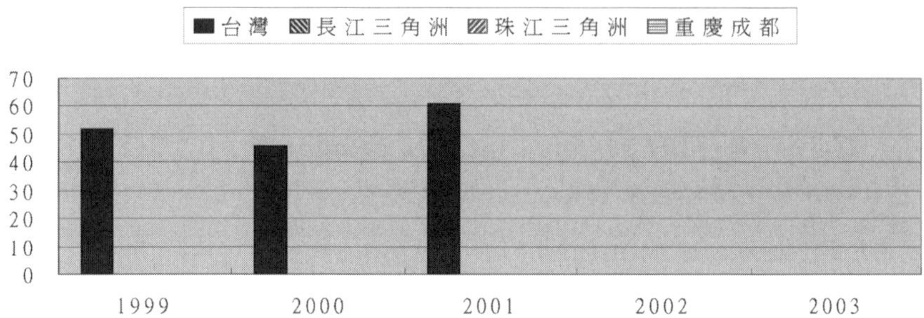

Figure 8.54. Government Administrative Efficiency

Legend (from left to right): Taiwan/Yanzhijiang Delta Region/Chongqing Chengdu Region/Zhujiang Delta Region

two years, the change in Zhujiang Delta Region is considerable, and its ranking rose from 37th globally in 2002 to 20th in 2003 with a respectably improvement.

4.4.3. Complexity of administrative program

Because this index lacks relevant data, no analysis is available at this moment.

4.4.4. FDI transparency of verification

This index is descriptive data, so it cannot be analyzed using figures and tables. Please refer to the data organization of research indices for this study.

4.4.5. Communication mechanism between government and foreign companies

This index is descriptive data, so it cannot be analyzed using figures and tables. Please refer to the data organization of research indices for this study.

4.5. *Integrity and execution level of intellectual property rights and relevant scientific legal systems*

4.5.1. Level of patents and intellectual property rights protection

Observing this index we can see that Taiwan is significantly better than the whole of China and Zhujiang Delta Region. Taking 2003 as an example, Taiwan's global ranking was 27th, but China and Zhujiang Delta Region were 41st and 42nd respectively, which indicates less protection of intellectual property rights in China.

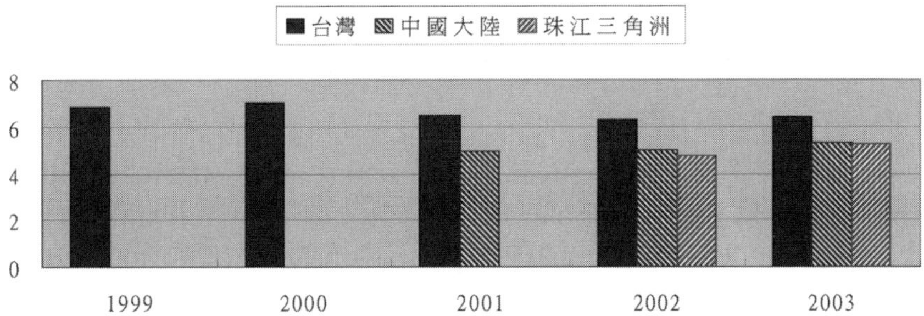

Figure 8.55. Level of Patents and Intellectual Property Rights Protection
Legend (from left to right): Taiwan/China/Zhujiang Delta Region

4.5.2. Level of intellectual property rights trade and circulation

Observing this index, we can see that the trading and circulating level of intellectual property rights in Yanzhijiang Delta Region is significantly higher than those of Zhujiang Delta Region and Chongqing Chengdu Region. From this trend, we can see growth in Yanzhijiang Delta Region and Zhujiang Delta Region, and stability in Chongqing Chengdu Region. Because of a lack of relevant data for Taiwan, it is not possible to conduct any comparison at this moment.

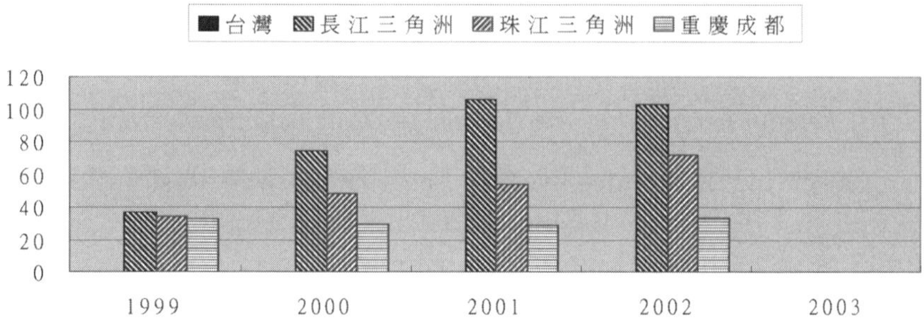

Figure 8.56. Level of Intellectual Property Rights Trade and Circulation

Legend (from left to right): Taiwan/Yanzhijiang Delta Region/Chongqing Chengdu Region/Zhujiang Delta Region

4.6. Social trust and mutual respect level

4.6.1. Equity

Observing this index, we can see that Taiwan was better than the whole of China in the past three years, but its overall ranking has significantly declined, from 29th place globally in 2000 to 36th in 2003, and its performance is still declining. In contrast, within the past two years, the level in Zhujiang Delta Region is rising considerably. Its ranking rose from 37th place globally in 2002 to 20th in 2003.

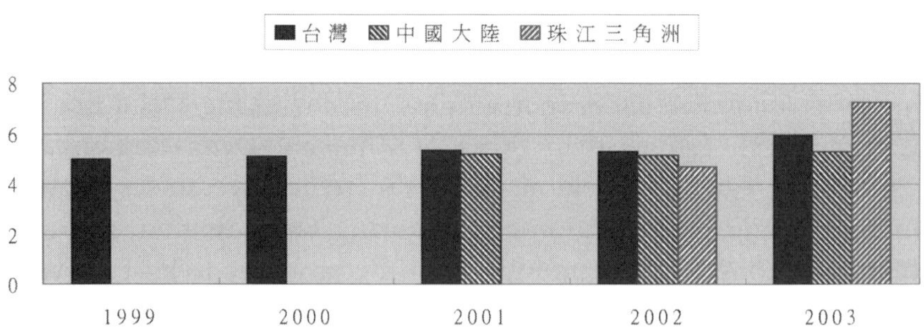

Figure 8.57. Equity

Legend (from left to right): Taiwan/China/Zhujiang Delta Region

4.6.2. Unitization

Observing this index, we can see that Taiwan significantly fell behind China and Zhujiang Delta Region. The ranking of Taiwan in 2002 and 2003 was 50th and 49th respectively, however, the rankings of China and Zhujiang Delta Region rose from 25th and 33rd to 19th and 17th.

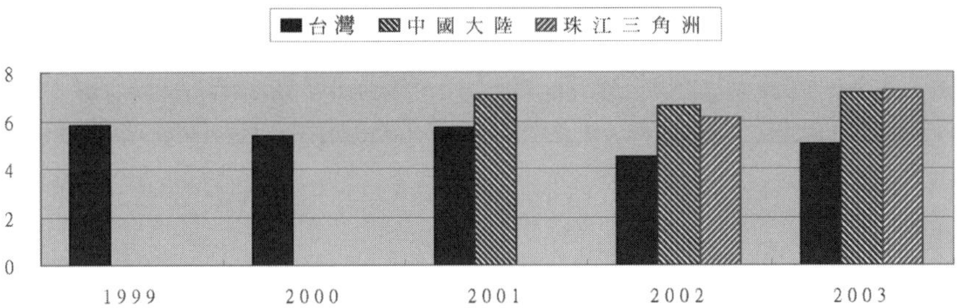

Figure 8.58. Unitization

Legend (from left to right): Taiwan/China/Zhujiang Delta Region

4.6.3. Privacy protection

Observing this index, we can see that the emphasis on privacy in Taiwan is significantly better than in China and Zhujiang Delta Region, but the Zhujiang Delta Region is showing improvement and should continue to be watched.

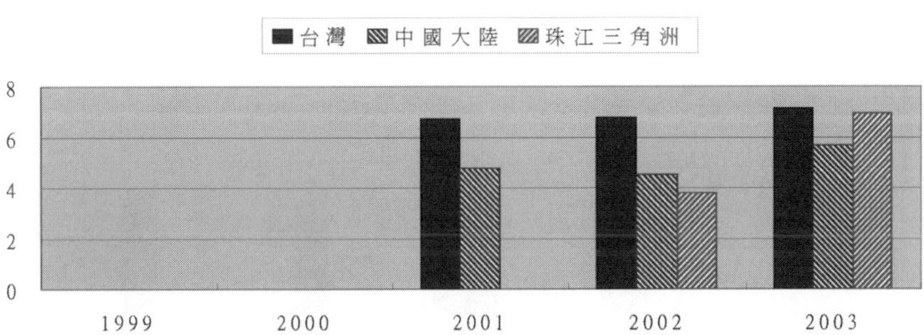

Figure 8.59. Privacy Protection

Legend (from left to right): Taiwan/Yanzhijiang Delta Region/Chongqing Chengdu Region/ Zhujiang Delta Region

4.6.4. Culture diversification and compatibility

4.6.5. Harmonious labor-capital relations

This index is descriptive data, so it cannot be analyzed using figures and tables. Please refer to the data organization of research indices for this study.

4.7. Positiveness of innovative teaching and education

4.7.1. Number of prizewinners that participated in international scientific contests and its ratio to the total student population

Because this index lacks relevant data, no analysis is available at this moment.

4.7.2. Creative teaching budget and its ratio to the total government expenditure

Because this index lacks relevant data, no analysis is available at this moment.

4.8. Convenience and quality of communications and transportation facilities

4.8.1. Quantity of international routes (passenger and cargo)

Observing this index, we can see that the quantity of international routes for Taiwan is higher than China's, which indicates that the air traffic in Taiwan is better.

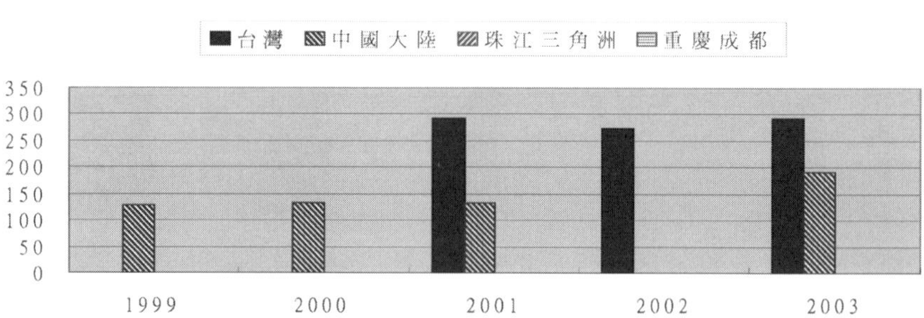

Figure 8.60. Quantity of International Routes (Passenger and Cargo)

Legend (from left to right): Taiwan/Yanzhijiang Delta Region/Chongqing Chengdu Region/ Zhujiang Delta Region

4.8.2. Frequency of international flights per week

From 1999 to 2000, the frequency of international flights gradually increased in Taiwan, but it showed a large decline in 2003. Because this index lacks relevant data for Yanzhijiang Delta Region, Zhujiang Delta Region, and Chongqing Chengdu Region, it is not possible to conduct any comparison at this moment.

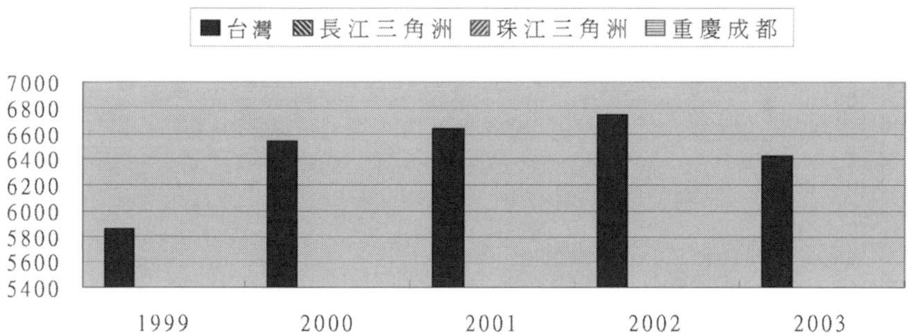

Figure 8.61. Frequency of International Flights per Week

Legend (from left to right): Taiwan/Yanzhijiang Delta Region/Chongqing Chengdu Region/Zhujiang Delta Region

4.8.3. On-time rate of international flights

In this index, Taiwan is significantly better than China. The on-time rate of international flights in Taiwan is almost 100%, and only 80% in China, which indicates that Taiwan possesses a more precise

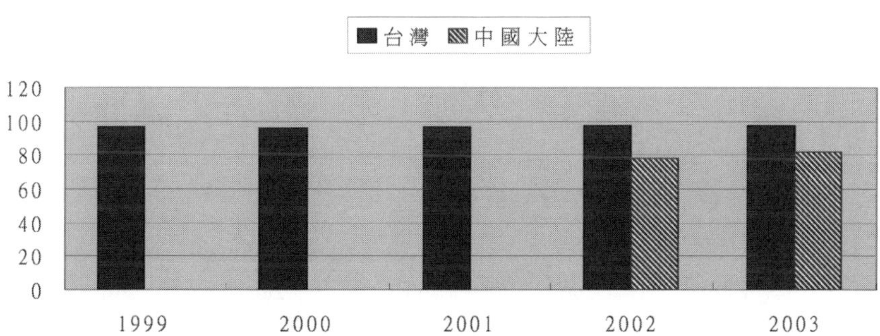

Figure 8.62. On-time Rate of International Flights

Legend (from left to right): Taiwan/China

time-handling capability for international flights, and China still needs to improve this aspect.

4.8.4. Aircraft movement in airports

Observing this index, we can see that the number for Taiwan is significantly less than that for China, and it is absolutely related to the number of airports.

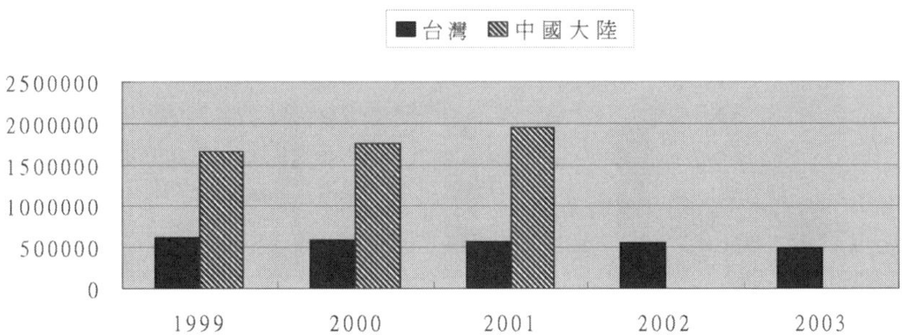

Figure 8.63. Aircraft Movement in Airports

Legend (from left to right): Taiwan/China

4.8.5. Accident rate of aircrafts

We can see that the accident rate of aircrafts is not generally high for Taiwan, and no accident occurred from 2000 to 2001. However, because of the lack of relevant data for China, it is not possible to conduct any comparison at this moment.

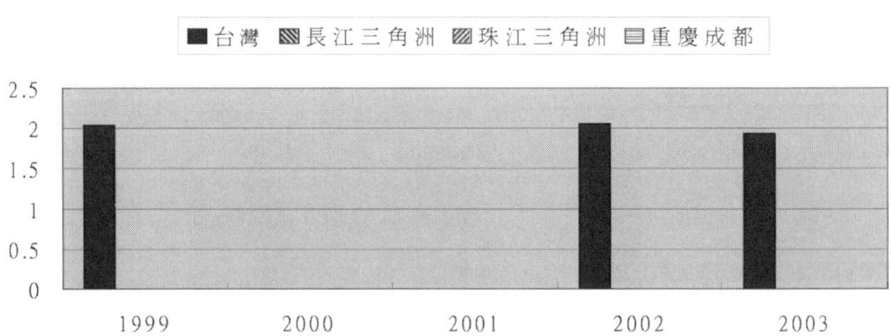

Figure 8.64. Accident Rate of Aircrafts

Legend (from left to right): Taiwan/Yanzhijiang Delta Region/Chongqing Chengdu Region/ Zhujiang Delta Region

4.8.6. Rate of increase of highway mileage

This index is related to the developing degree of national infrastructure. Due to the enormous size of China and the vigorous development of the economy recently, highway construction is very active, with an explosive increase in 2001.

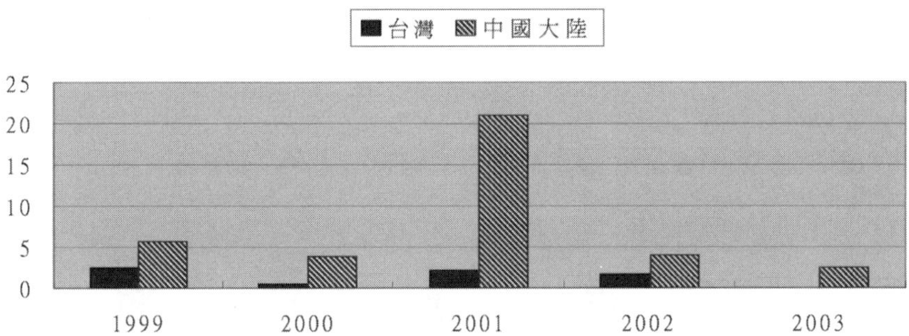

Figure 8.65. Rate of Increase of Highway Mileage

Legend (from left to right): Taiwan/Mainland China

4.8.7. Port/dock cargo loading volume

We can see that Taiwan's loading volume is significant higher than those of Yanzhijiang Delta Region, Zhujiang Delta Region, and Chongqing Chengdu Region. This is strongly related to the fact that Taiwan is an island with comprehensive cargo forwarding functions.

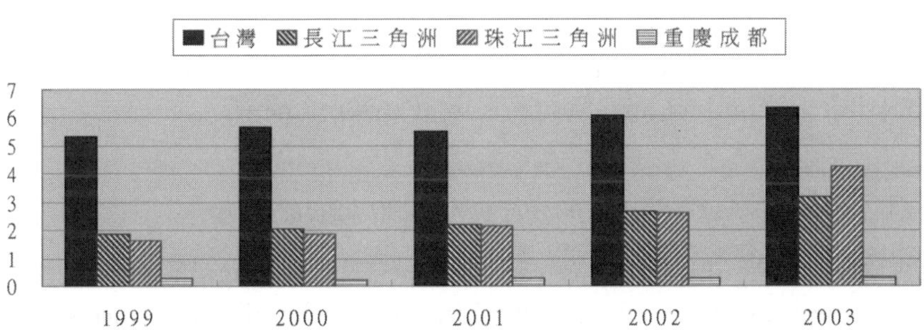

Figure 8.66. Port/Dock Cargo Loading Volume

Legend (from left to right): Taiwan/Yanzhijiang Delta Region/Chongqing Chengdu Region/Zhujiang Delta Region

However, in recent years, the growth in Zhujiang Delta Region has increased significantly, and shows a trend of catching up with Taiwan.

4.8.8. Rapid transit system (subway and light-rail train)

Observing this index, we can see a gradual increase, which indicates that Taiwan has emphasized construction of public transportation in recent years. However, because of the lack of relevant data for China, it is not possible to conduct any comparison at this moment.

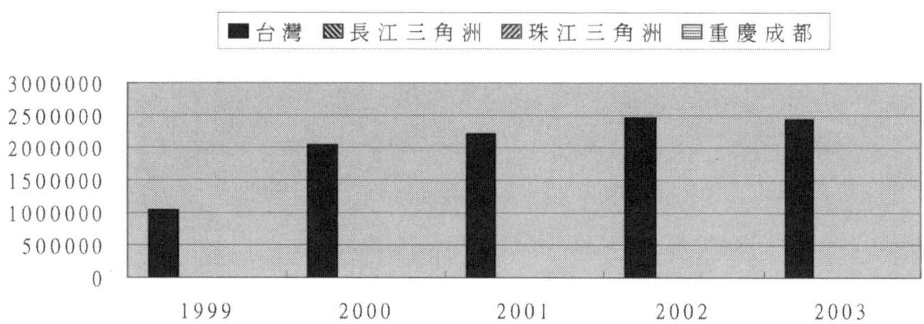

Figure 8.67. Rapid Transit System (Subway and Light-rail Train)
Legend (from left to right): Taiwan/Yanzhijiang Delta Region/Chongqing Chengdu Region/ Zhujiang Delta Region

4.9. Convenience and quality of communication facilities

4.9.1. Quantity of telephone switchboards

Due to different calculating criteria of the data between Taiwan and China, and due to the lack of relevant data for China, it is not possible to conduct any comparison at this moment.

4.9.2. Number of cellular phones per 1,000 inhabitants

Observing this index we can see that Taiwan is significantly better than the other three regions in China. However, all four regions displayed a gradual increase, and the Yanzhijiang Delta Region seems to be gradually catching up with Taiwan.

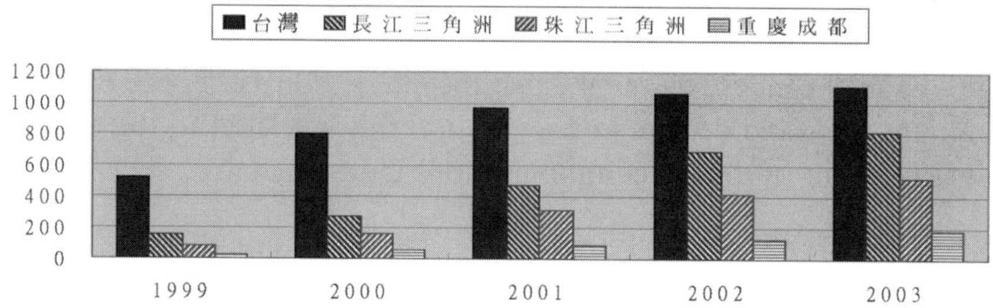

Figure 8.68. Number of Cellular Phones per 1,000 Inhabitants

Legend (from left to right): Taiwan/Yanzhijiang Delta Region/Chongqing Chengdu Region/ Zhujiang Delta Region

4.9.3. Number of local phone sets per 1,000 inhabitants

Observing this index, we can see that the difference between Taiwan and Yanzhijiang Delta Region is not significant, but Zhujiang Delta Region and Chongqing Chengdu Region are falling behind significantly.

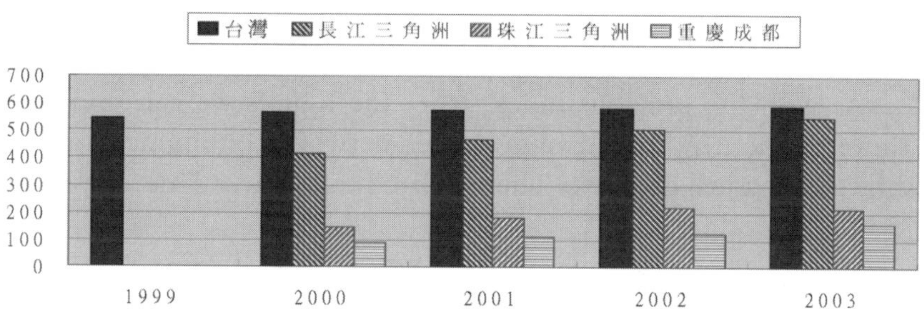

Figure 8.69. Number of Local Phone Sets per 1,000 Inhabitants

Legend (from left to right): Taiwan/Yanzhijiang Delta Region/Chongqing Chengdu Region/ Zhujiang Delta Region

4.9.4. Cellular/communication price rate per minute

Because the quantity of cellular/phone businesses is enormous for each area, and the difference among various pricing plans is considerable, it is difficult to make comparisons with this index. However, a decreasing price rate is the overall trend.

4.9.5. International call price rate per minute

From the index, we can see that the international call price rate per minute showed a decline in Taiwan, and this is related to the overall development of the telecommunications industry.

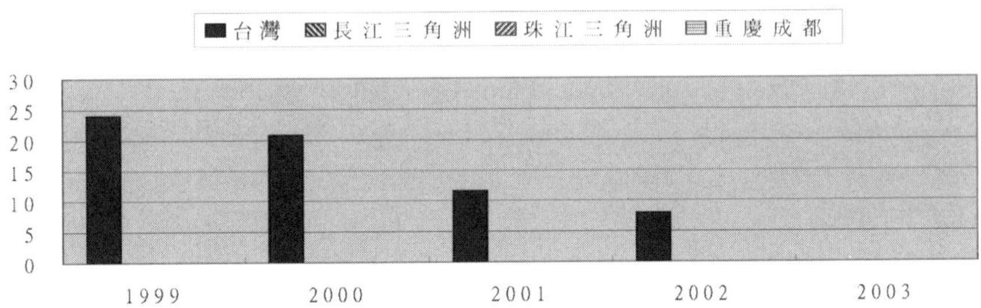

Figure 8.70. International Call Price Rate per Minute

Legend (from left to right): Taiwan/Yanzhijiang Delta Region/Chongqing Chengdu Region/Zhujiang Delta Region

4.9.6. Monthly broadband (cable) network communication fee (512k/64k)

As with the previous index, from the figures we can find that Taiwan's monthly broadband (cable) network communication fee showed a gradual decrease. This is related to the overall development

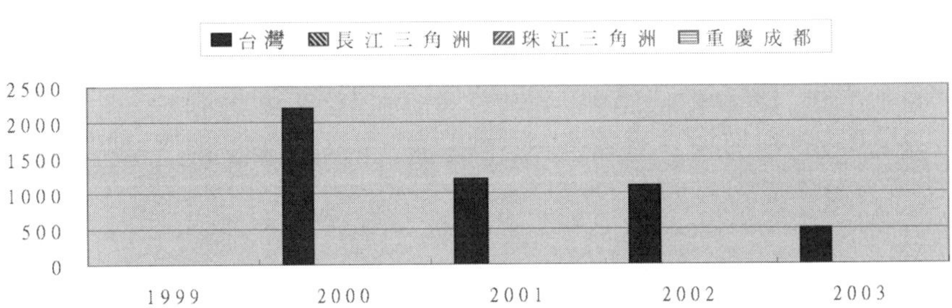

Figure 8.71. Monthly Broadband (Cable) Network Communication Fee (512k/64k)

Legend (from left to right): Taiwan/Yanzhijiang Delta Region/Chongqing Chengdu Region/Zhujiang Delta Region

of the telecommunications industry and the popularization of broadband.

4.9.7. Network stability

This index is descriptive data, so it cannot be analyzed using figures and tables. Please refer to the data organization of research indices for this study.

4.10. Public environment and facilities and environmental protection

4.10.1. Popularization of power consumption

From the index, we can see that the trend of Taiwan's popularization of power consumption has increased, which indicates that Taiwan's power network is evenly distributed island-wide.

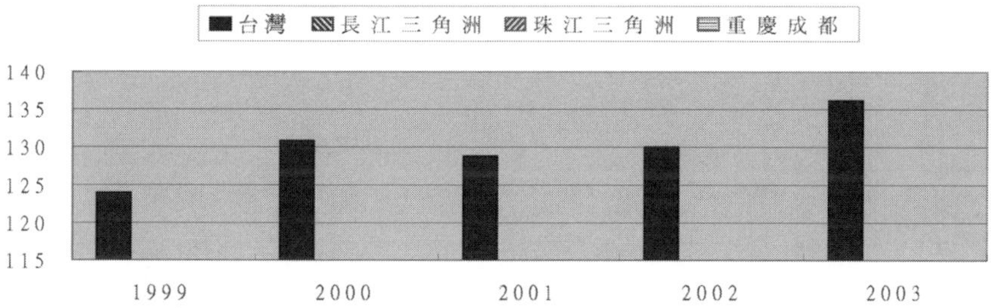

Figure 8.72. Popularization of Power Consumption

Legend (from left to right): Taiwan/Yanzhijiang Delta Region/Chongqing Chengdu Region/Zhujiang Delta Region

4.10.2. Average electricity price

Since 2001, Taiwan's electricity price started to gradually decrease from NTD$2.11 per KWH in 2001 to NTD$2.07 per KWH currently, and this is greatly related to the government's policy.

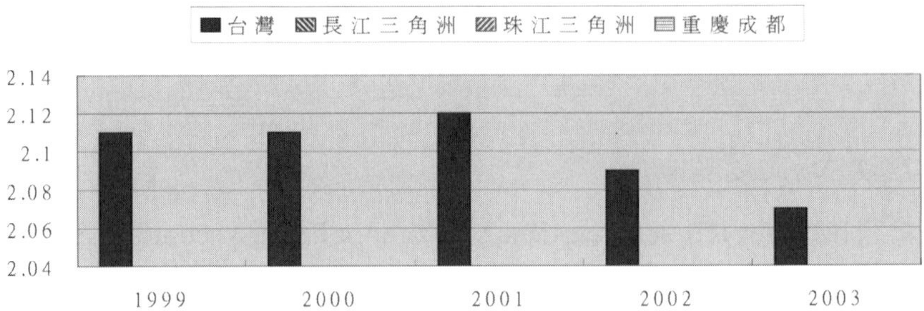

Figure 8.73. Average Electricity Price

Legend (from left to right): Taiwan/Yanzhijiang Delta Region/Chongqing Chengdu Region/Zhujiang Delta Region

4.10.3. Peak load and percent reserve margin

Observing this index, we can see that due to incredible household and industrial electricity consumption in Taiwan, the peak load rate frequently breaks the highest record indicating that the demand for electricity consumption is considerable in Taiwan. However, because of the lack of relevant data for China, it is not possible to conduct any comparison at this moment.

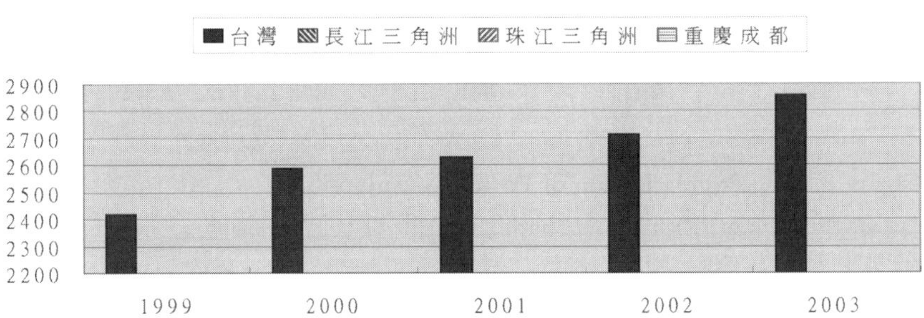

Figure 8.74. Peak Load and Percent Reserve Margin

Legend (from left to right): Taiwan/Yanzhijiang Delta Region/Chongqing Chengdu Region/Zhujiang Delta Region

4.10.4. Electrical power equipment operating efficiency: including the capacity utilization rate, the net heating efficiency of boiler combustion air mechanism, the line loss rate, the automatic emergency stop frequency of nuclear mechanism

(i) Capacity utilization rate

Observing this index, we can see a slight difference of the utilization rate for the electrical power equipment between Taiwan and China, and the capacity utilization rate is also fully loaded in Taiwan, indicating that the requirement of power consumption is very large.

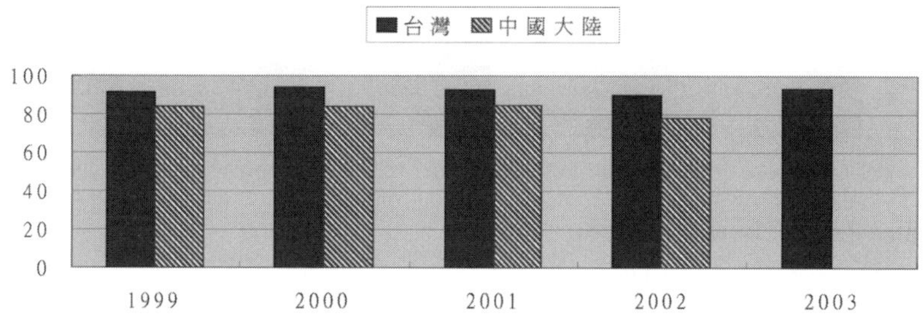

Figure 8.75. Capacity Utilization Rate

Legend (from left to right): Taiwan/China

(ii) Net heating efficiency of boiler combustion air mechanism

Observing this index, we can see that the net heating efficiency for Taiwan and China is about 35% and 33% respectively; there is little difference between them. This indicates that the equipment operating efficiency is stable for both locations.

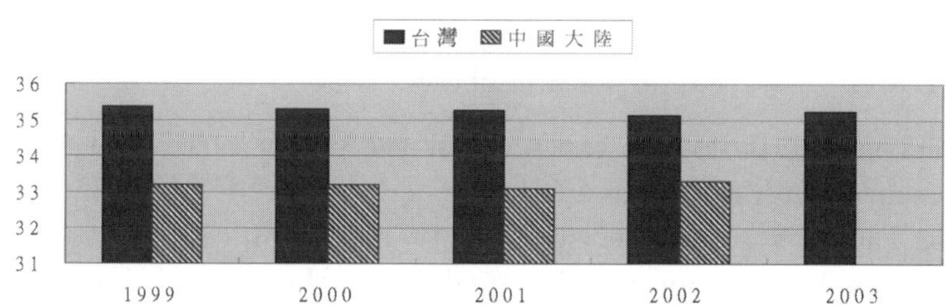

Figure 8.76. Net Heating Efficiency of Boiler Combustion Air Mechanism

Legend (from left to right): Taiwan/Mainland China

(iii) Line loss rate

Observing this index, we can see that, within the past five years, Taiwan's line loss rate is about 4% to 6%, which is slightly lower than China's 6%~8%, and this indicates that Taiwan's equipment operating efficiency is slightly higher than China's.

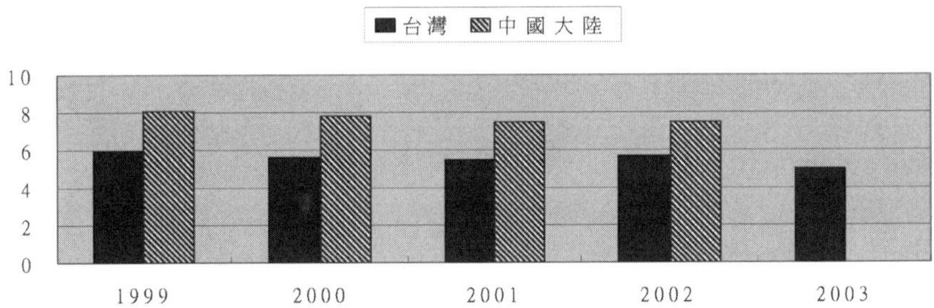

Figure 8.77. Line Loss Rate

Legend (from left to right): Taiwan/China

(iv) Automatic emergency stop frequency of nuclear mechanism

Because this index lacks relevant data, no analysis is available at this moment.

4.10.5. Power supply quality: including the power recovery time per household, the power recovery time from accidents, and the power recovery time from construction

(i) Power recovery time per household

From the index, we can see that the power recovery time per household in Taiwan has gradually been reduced, which indicates that Taiwan's power service quality has significantly improved, and this helps improve the quality of the overall investment environment.

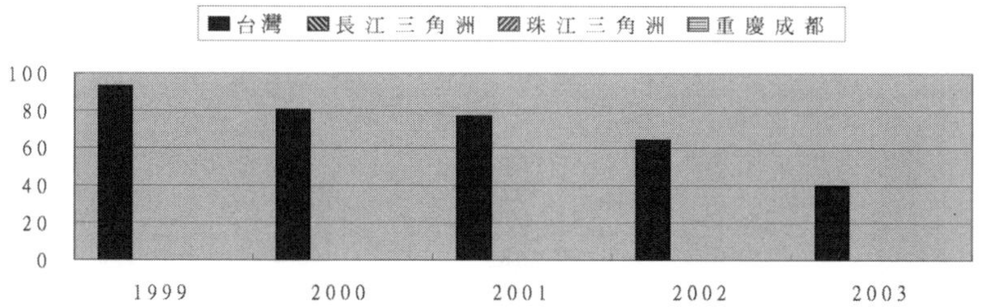

Figure 8.78. Power Recovery Time per Household

Legend (from left to right): Taiwan/Yanzhijiang Delta Region/Chongqing Chengdu Region/Zhujiang Delta Region

(ii) Power recovery time from accidents

As with the above-mentioned index and figures, we can see that the power recovery time from accidents in Taiwan has gradually been reduced, which indicates that Taiwan's power service quality has significantly improved, and this helps improve the quality of the overall investment environment.

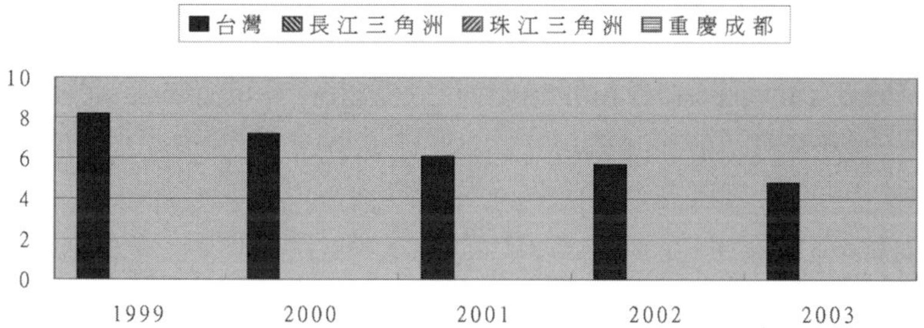

Figure 8.79. Power Recovery Time from Accidents

Legend (from left to right): Taiwan/Yanzhijiang Delta Region/Chongqing Chengdu Region/Zhujiang Delta Region

(iii) Power recovery time from construction

As with the above-mentioned index and figures, we can see that the power recovery time from construction in Taiwan has gradually been reduced, which indicates that Taiwan's power service quality has

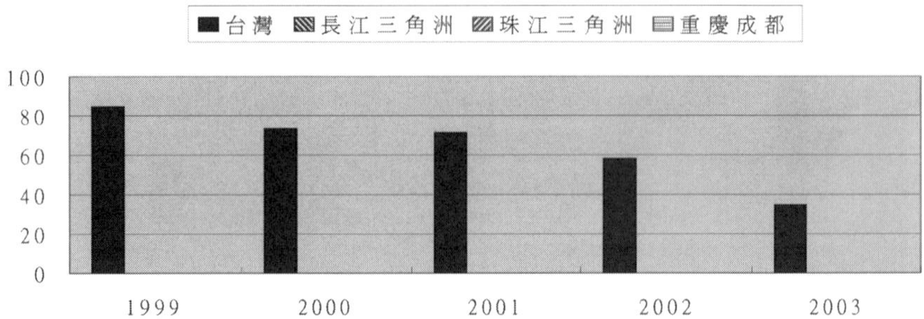

Figure 8.80. Power Recovery Time from Construction

Legend (from left to right): Taiwan/Yanzhijiang Delta Region/Chongqing Chengdu Region/ Zhujiang Delta Region

significantly improved, and this helps improve the quality of the overall investment environment.

4.10.6. Tap water popularization

Observing this index, we can see that tap water popularization has reached 100% in Yanzhijiang Delta Region, which is better than the other three regions, even though the difference of popularization between Taiwan and Zhujiang Delta Region is not large at around 90%. However, Chongqing Chengdu Region is the one with the lowest popularization of tap water and needs to be improved.

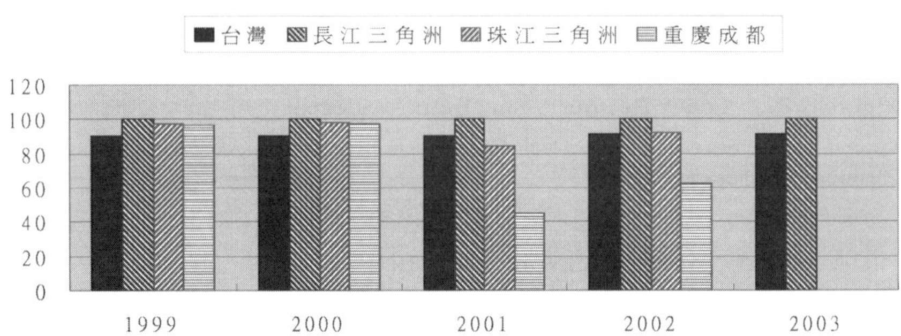

Figure 8.81. Tap Water Popularization

Legend (from left to right): Taiwan/Yanzhijiang Delta Region/Chongqing Chengdu Region/ Zhujiang Delta Region

4.10.7. Average monthly water bill per person

The monthly water bill in China is significantly higher than that of Taiwan's, and the difference is large and unreasonable. It may be caused by factors in the pricing unit, and still needs to be examined.

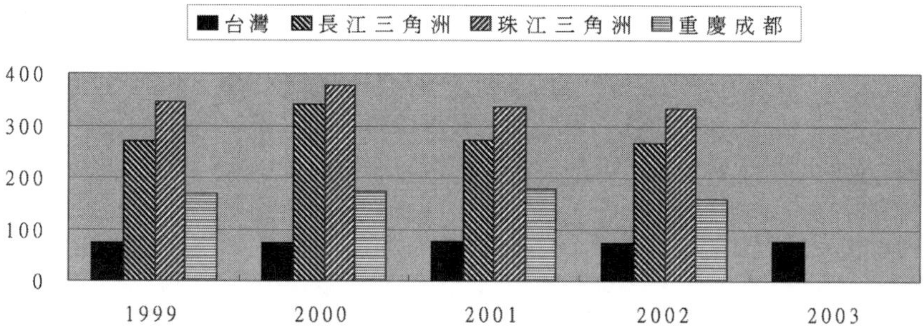

Figure 8.82. Average Monthly Water Bill per Person

Legend (from left to right): Taiwan/Yanzhijiang Delta Region/Chongqing Chengdu Region/ Zhujiang Delta Region

4.10.8. Annual water supply volume for cities

Observing this index, we can see that the Yanzhijiang Delta Region is far ahead of the other three regions with the maximum annual water supply volume. Zhujiang Delta Region is next, and the difference between Taiwan and Chongqing Chengdu Region is not large.

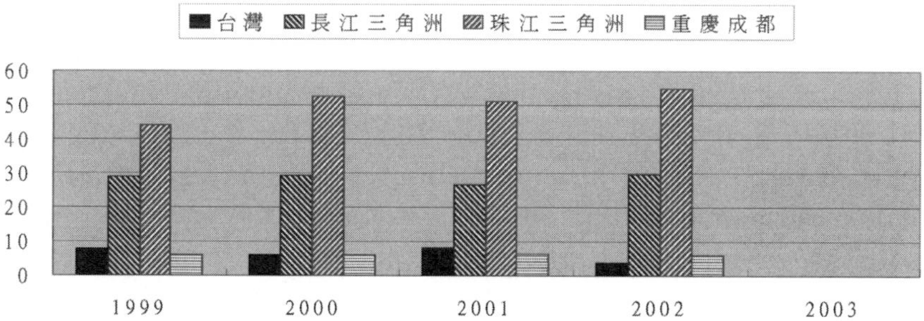

Figure 8.83. Annual Water Supply Volume for Cities

Legend (from left to right): Taiwan/Yanzhijiang Delta Region/Chongqing Chengdu Region/ Zhujiang Delta Region

4.10.9. Popularization of household sewer connections

Observing this index, we can see that the popularization of household sewer connections in Taiwan has gradually increased, from 45.63% in 1999 to 68.13% in 2003, which indicates the significant improvement in sewage management in Taiwan. Because of different calculating criteria in China, a comparison is not possible, so please refer to the data organization of research indices for this study.

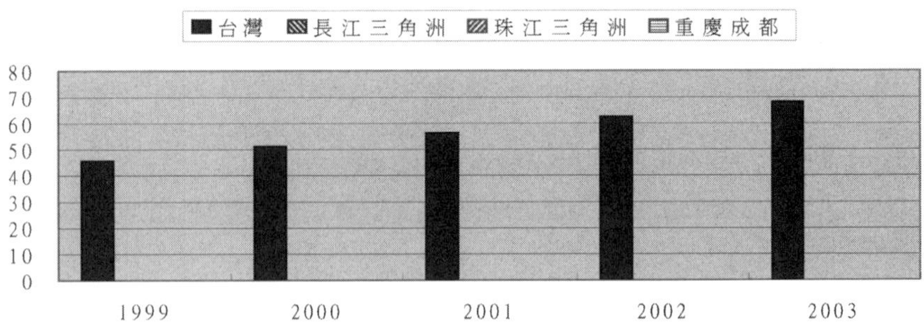

Figure 8.84. Popularization of Household Sewer Connections

Legend (from left to right): Taiwan/Yanzhijiang Delta Region/Chongqing Chengdu Region/Zhujiang Delta Region

4.10.10. Rainwater drainage construction rate

Observing this index, we can see that the popularization of rainwater drainage construction in Taiwan has gradually increased, from 49.3% in 1999 to 55.3% in 2002, which indicates the significant improvement in construction in Taiwan. Because of different calculating criteria in China, a comparison is not possible, so please refer to the data organization of research indices for this study.

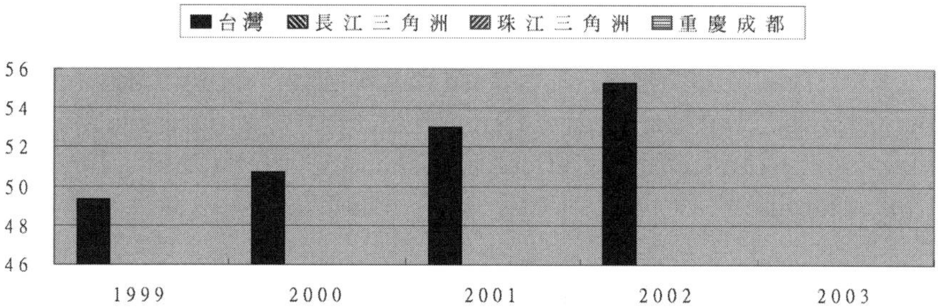

Figure 8.85. Rainwater Drainage Construction Rate

Legend (from left to right): Taiwan/Yanzhijiang Delta Region/Chongqing Chengdu Region/ Zhujiang Delta Region

4.10.11. Popularization of cable TV

Observing this index, we can see that the popularization of cable TV in Taiwan has gradually increased, from 67.9% in 1999 to 76.18% in 2002, which indicates the significant improvement in the layout of cable TV in Taiwan. Because of different calculating criteria in China, a comparison is not possible, so please refer to the data organization of research indices for this study.

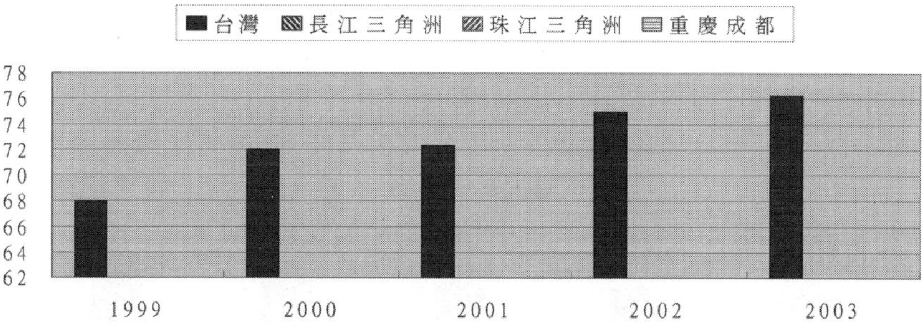

Figure 8.86. Popularization of Cable TV

Legend (from left to right): Taiwan/Yanzhijiang Delta Region/Chongqing Chengdu Region/ Zhujiang Delta Region

4.10.12. Recycling rate

Observing this index, we can see that China is significantly better than Taiwan, but the large difference seems to be unreasonable. It may result from factors in different calculating criteria, and still needs to be examined.

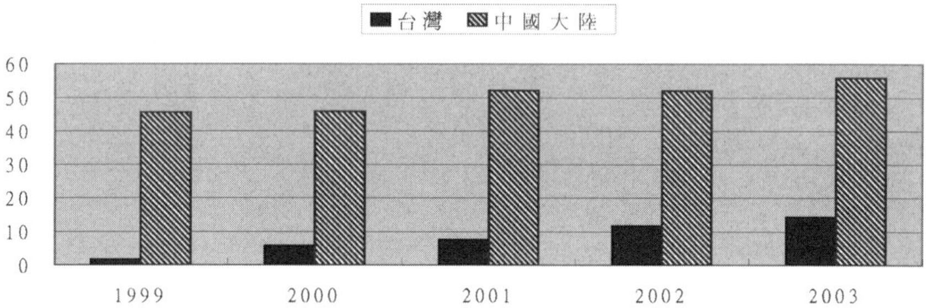

Figure 8.87. Recycling Rate

Legend (from left to right): Taiwan/China

4.10.13. Garbage disposal rate

Observing this index, we can see that the garbage disposal rate in Taiwan increased from 86.7% in 1999 to 98.2% in 2003, which indicates significant improvement in garbage disposal in Taiwan. However, it is only about 50% in China, and there is still considerable room for improvement.

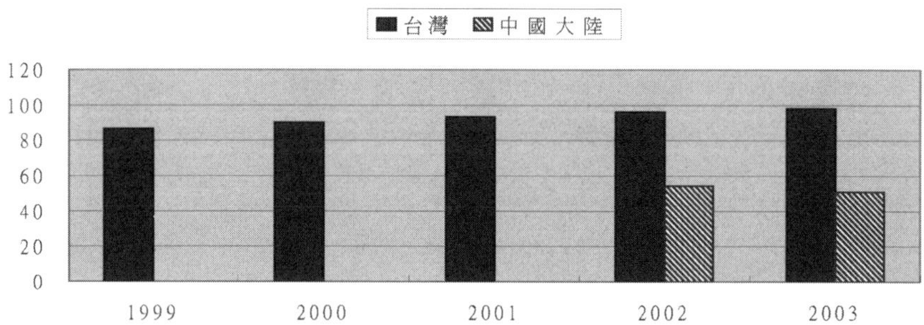

Figure 8.88. Garbage Disposal Rate

Legend (from left to right): Taiwan/China

4.10.14. Annual frequency and loss from natural disasters (including typhoons, earthquakes, and floods)

Observing this index, we can see that there are natural disasters in Taiwan nearly every year, especially unavoidable typhoons. The Taiwanese government should complete relevant protective and preventative measures for such disasters, which will be helpful to enhance investment confidence for foreign capital.

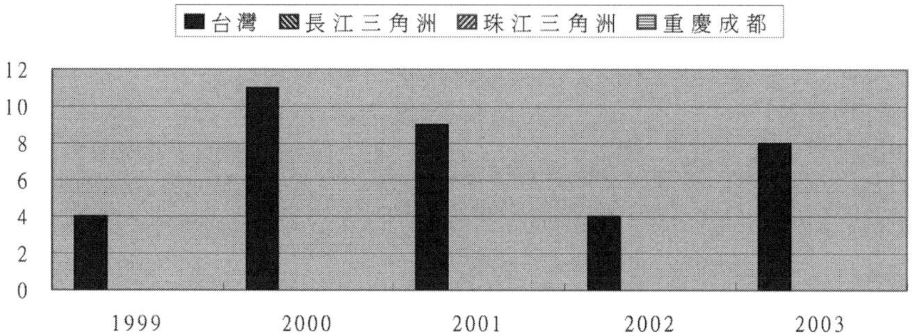

Figure 8.89. Annual Frequency and Loss from Natural Disasters

Legend (from left to right): Taiwan/Yanzhijiang Delta Region/Chongqing Chengdu Region/ Zhujiang Delta Region

4.10.15. Quantity and quality of city environment

Because this index lacks relevant data, no analysis is available at this moment.

Appendix A

1. Knowledge Capital

Main Index	Detailed Indices
1. Quantity and Quality of Invention and Design Patents	1. Number of accepted USPTO applications 2. Number of accepted domestic patent applications (four areas) 3. Number of SCI papers (National Science Council, NSC) 4. Current Impact Index (CII) for patents (Taiwan Institute for Economic Research) 5. Interrelatedness of Patent and Science (TIER) (average papers cited) 6. Invention Patents per 100 people (WEF)
2. Quantity and Quality of Human Resources	1. Number of researchers per 10,000 people (NSC) (four areas) 2. Number of researchers per 10,000 working adults (NSC) (four areas) 3. Number of university degrees per 10,000 people (Ministry of Science and Technology, MOST) 4. R&D Personnel (FTE) per 10,000 population (IMD) 5. Availability of highly skilled workers (European Chamber of Commerce)
3. Basic Science and Humanities Preparation and Development of the Population	1. Science education (IMD) 2. Young people's level of interest in science (IMD) 3. Attendance at cultural and arts events (Council for Cultural Affairs) 4. Educational expenditures as a proportion of household income 5. Government expenditure on education as a proportion of GDP

2. Innovative Ability

Main Index	Detailed Indices
1. Quality and Quantity of Research Institutions	1. R&D budget per researcher for sci-tech institutions (NSC) 2. Proportion of research institute researchers holding Masters or higher degrees (NSC) 3. Approved foreign patents from Sci-Tech Development Plan (other than technology) 4. Spin-off enterprises from Sci-Tech Development Plan (other than technology)
2. Quality and Quantity of Corporate R&D	1. R&D personnel per 1,000 employees (IMD) 2. R&D as a proportion of enterprise gross production (NSC) 3. Number of domestic patents from science and technical parks and proportion of companies in such parks (Science and Technical Parks)
3. Incentive Programs for Start-ups and Innovative Activities	1. Foreign investment as a proportion of venture capital (Venture Capital Association) (Science and Technical Venture Capital Company) 2. Ease of obtaining venture capital (IMD) 3. R&D Budget as a proportion of GDP (NSC) (MOST) 4. IT R&D expenditures as a proportion of GDP 5. Investment in IT and electronics as a proportion of all basic infrastructure and upgrade investment 6. Successful venture capital amount as a proportion of venture capital investment (Venture Capital Association) 7. Proportion of start-ups receiving medium- or long-term government operating monies (CEPD) 8. IT and high-tech electronic industry enterprise value as a proportion of GDP

(*Continued*)

(*Continued*)

Main Index	Detailed Indices
4. Flow of Knowledge between Industry and Academy	1. Industry-academy cooperation (IMD) 2. Technical cooperation between enterprises (IMD) 3. Completeness of knowledge transfer (IMD)
5. Innovative International R&D Cooperation Projects	1. Budget for purchase of foreign technology (Ministry of Economic Affairs, Directorate-General of Budget, Accounting and Statistics — MOEA DGBAS) 2. Value of technology sold abroad (MOEA–DGBAS) 3. Balance of payments on licensing fees and royalties (Central Bank) 4. Dollar amount of joint ventures with listed OECD nation companies
6. Indication of Ability to Innovate and Entrepreneurial Spirit	1. Industry cluster innovation environment (WEF) 2. Correlation between innovation infrastructure and industry cluster innovation environment (WEF) 3. Science and engineering research personnel (WEF) 4. Innovation policies (WEF) 5. Enterprise innovation orientation (WEF) 6. Frequency of start-ups (IMD) 7. Entrepreneurial spirit (IMD) 8. Enterprise proactivity in bringing in new technology (WEF)

3. Utilization of Information Technology

Main Index	Detailed Indices
1. Degree of Digitization of Industry	1. Proportion of factories and companies engaged in e-commerce (DGBAS survey) 2. New applications of IT in line with enterprise demands (IMD) 3. Degree of digitization (IDC) 4. Degree of industry digitization (EIU)
2. Extent of Digitized Content	1. Extensiveness of government Web sites (NRI) 2. Assessment of e-government services (NRI) 3. Assessment of e-government Web sites (Brown University)
3. Extent of E-learning (Households/Schools/ Other)	1. Household Internet Connection Rate (Statistical Report on the National Condition) 2. School Internet Connection Rate (Ministry of Education Computation Center)

4. Social Infrastructure

Main Index	Detailed Indices
1. Incentive Measures and Financial Environment for Foreign Investment; Capital Market	1. Laws and regulations on tax and other incentives to promote industry upgrades (as current taxes, deductions, and post-deduction taxes) 2. Tax relief plans or preferential treatment strategies for high-tech industry 3. Rental tax relief for start-ups 4. Foreign enterprise tax burden (current taxes, deductions, and post-deduction taxes) 5. Controls on international capital flow and time required for incoming and outgoing transfers 6. Bank credit rating (S&P) 7. Exchange rate fluctuations (descriptive) 8. Transparency of bank management (systematization and standardization) (S&P) 9. Transparency of corporate board management (systematization and standardization) (descriptive) 10. Ease with which foreign firms can obtain large sums of capital (listing, insofar as possible, numbers, or using descriptions) 11. Foreign investment as a proportion of all investment 12. Tariffs 13. Individual income taxes 14. Corporate business credit and ethics in investment target area 15. Foreign banks' ability to engage in local monetary business
2. Internationalization and Liberalization of Society	1. Number and proportion of people going abroad (Ministry of Transportation and Communications) 2. Number and proportion of people entering the area (Ministry of Transportation and Communications)

(*Continued*)

(Continued)

Main Index	Detailed Indices
	3. Number and proportion of legal foreign residents (Ministry of the Interior)
	4. Number of international conferences and exhibitions held and number of attendees
	5. Number of foreign financial institutions
	6. Visas granted for students studying abroad — number of students and proportion of all university students (Ministry of Education)
	7. Number and proportion of foreign students studying in the area (Ministry of Education)
	8. Proportion of foreign laborers (IMD)
	9. Entry and exit policies
	10. Transportation cost for entering or leaving the area
	11. Extent of foreign language ability (English) (General English Proficiency Test, GEPT) (for China: proportion of Level 4 passes among university students)
	12. Degree of acceptance of foreigners by locals (European Chamber of Commerce)
	13. Satisfaction with international schools (European Chamber of Commerce)
	14. Tuition for international schools (European Chamber of Commerce)
3. Living Environment, Medical Care, and Social Order	1. Doctors of Western medicine/doctors per 10,000 (for China: please note if no solid figures for doctors of Western medicine)
	2. Hospital beds per 10,000 population
	3. Prevalence of ordinary immunizations
	4. Medical expenditures as proportion of GDP
	5. Criminal offenses per 100,000 population
	6. Thefts per 100,000 population
	7. Fire accidents per year
	8. Vehicular accidents per 10,000 vehicles

4. Government Implementation of Transparency and Administrative Efficiency	1. Degree of implementation of transparency (IMD) 2. Government administrative efficiency (IMD) 3. Complexity of administrative procedures (WEF) 4. Transparency of approval process for foreign investment applications (descriptive) 5. Communication mechanisms between government and foreign firms (descriptive)
5. Completeness and Degree of Application of IP and Related Technology Laws	1. Degree of protection of patents and intellectual property rights (IMD) 2. Selling and buying of intellectual property rights
6. Degree of Trust and Mutual Respect in Society	1. Fairness (IMD) 2. Cohesion (IMD) 3. Protection of privacy (IMD) 4. Cultural diversity and tolerance (descriptive) 5. Harmony of labor-management relations (descriptive)
7. Proactivity in Innovative Education and Educational Innovation	1. Number and proportion of students participating in international science competitions 2. Budget for college and above education and proportion of total government education expenditures (Ministry of Education) (If no concrete figures available, describe related policies)
8. Convenience and Quality of Transportation and Shipping Infrastructure	1. Number of international airlines 2. Number of international flights weekly 3. On-time performance of international flights (MOTC) 4. Air cargo throughput (international and domestic) 5. Aviation accidents (MOTC)

(Continued)

(Continued)

Main Index	Detailed Indices
	6. Increase in highway mileage (DGBAS) 7. Port cargo loaded and unloaded (DGBAS) 8. Mass rapid transit (subways and light rail systems)
9. Convenience and Quality of Communications Infrastructure	1. Number of telephone exchanges (DGBAS) 2. Cellular phones per 1,000 population (DGBAS) 3. Fixed-line telephones per 1,000 population (DGBAS) 4. Cellular telephone tariff per minute (Directorate General of Telecommunications) 5. International telephone tariff per minute (Directorate General of Telecommunications) 6. Monthly broadband Internet fee for 512k/64k connection (Directorate General of Telecommunications) 7. Internet openness and stability (descriptive)
10. Public Environment and Infrastructure; Environmental Protection	1. Electrification (DGBAS) 2. Average price of electricity (Taipower) 3. Peak load and reserve capacity (Taipower) 4. Efficiency of operation for electrical equipment (including capacity utilization ratio, steam turbine net thermal efficiency, transmission line loss, number of automated shutdown events for nuclear reactors) (Taipower) 5. Quality of electric supply (including distribution lines per household, time taken to restore power after outage, workers available to restore power after outage) (Taipower) 6. Black-outs and suspended water supply per year 7. Extent of connection to running water (DGBAS) 8. Average unit cost for water (household and industrial use, respectively) 9. Annual city water supply

(Continued)

Main Index	Detailed Indices
	10. Sanitary sewer connection rate (DGBAS)/Wastewater treatment capacity (China) 11. Cable TV penetration (DGBAS) 12. Extent of resource recycling (DGBAS) 13. Extent of appropriate handling of trash (DGBAS)/Neutralization of trash (China) 14. Annual natural disasters and damages caused (including typhoons, earthquakes, flooding) (Ministry of the Interior) 15. Quality of city environment

5. Business Environment

Main Index	Detailed Indices
1. Business Cost	
2. Market potential and growth	1. Market size 2. Accessible market 3. Market growth

Appendix B

1. Knowledge Capital

Main Index	Detailed Indices (Sources)	Trend for Recent Five Years (World Ranking)					Description of Indices
		1999	2000	2001	2002	2003	
1. Quantity and Quality of Invention and Design Patents	1. Number of accepted USPTO applications (Taiwan)	3,693(5)	4,667(4)	5,371(4)	5,431(4)	5,298(4)	
	Number of accepted USPTO applications (Mainland China)						
	a. Number of accepted USPTO applications (Yangtze River Delta)						
	b. Number of accepted USPTO applications (Pearl River Delta)						
	c. Number of accepted USPTO applications (Chongqing & Chengdu)						

(*Continued*)

(Continued)

Main Index	Detailed Indices (Sources)	Trend for Recent Five Years (World Ranking)					Description of Indices
		1999	2000	2001	2002	2003	
	2. Number of accepted domestic patent applications (Taiwan)	2,139	3,834	6,477	5,683	6,399	
	Number of accepted domestic patent applications (Mainland China)	92,112	95,236	99,278	112,103	149,588	
	a. Number of accepted domestic patent applications (Yangtze River Delta)	3,665	4,050	5,371	6,695	16,671	
	b. Number of accepted domestic patent applications (Pearl River Delta)	14,328	15,799	18,259	22,761	29,235	
	c. Number of accepted domestic patent applications (Chongqing & Chengdu)	1,078	1,158	1,197	1,761	2,883	
	3. Number of SCI papers (National Science Council, NSC) (Taiwan)	8,944(19)	9,203(19)	10,635(17)	10,831(18)	12,392(18)	

Number of SCI papers (National Science Council, NSC) (Mainland China)	22,743(9)	24,923(9)	29,381(8)	33,561(6)	40,709(6)
a. Number of SCI papers (National Science Council, NSC) (Yangtze River Delta)					
b. Number of SCI papers (National Science Council, NSC) (Pearl River Delta)					
c. Number of SCI papers (National Science Council, NSC) (Chongqing & Chengdu)					
4. Current Impact Index (CII) for patents (Taiwan)	1.02	1.19	1.14	1.00	—
Current Impact Index (CII) for patents (Mainland China)					
a. Current Impact Index (CII) for patents (Yangtze River Delta)					
b. Current Impact Index (CII) for patents (Pearl River Delta)					

(*Continued*)

(Continued)

Main Index	Detailed Indices (Sources)	Trend for Recent Five Years (World Ranking)					Description of Indices
		1999	2000	2001	2002	2003	
	c. Current Impact Index (CII) for patents (Chongqing & Chengdu)						
	5. Interrelatedness of patent and science (Taiwan) (average papers cited)	—	0.18	0.24	0.21	0.24	
	Interrelatedness of patent and science (Mainland China)				1.39	0.96	
	a. Interrelatedness of patent and science (Yangtze River Delta)						
	b. Interrelatedness of patent and science (Pearl River Delta)						
	c. Interrelatedness of patent and science (Chongqing & Chengdu)						
	6. Invention patents per 100 people (WEF) (Taiwan, WEF)					234.4	
	Invention patents per 100 people (WEF) (Mainland China)	0.73	0.75	0.78	0.87		

	a. Invention patents per 100 people (Yangtze River Delta)					
	b. Invention patents per 100 people (Pearl River Delta)					
	c. Invention patents per 100 people (Chongqing & Chengdu)					
2. Quantity and Quality of Human Resources	1. Number of researchers per 10,000 people (NSC) (Taiwan)	39.7	39.3	39.9	42.1	42.7
	Number of researchers per 10,000 people (Mainland China)					
	a. Number of researchers per 10,000 people (Yangtze River Delta)					
	b. Number of researchers per 10,000 people (Pearl River Delta)					
	c. Number of researchers per 10,000 people (Chongqing & Chengdu)					

(*Continued*)

(Continued)

Main Index	Detailed Indices (Sources)	Trend for Recent Five Years (World Ranking)					Description of Indices
		1999	2000	2001	2002	2003	
	2. Number of researchers per 10,000 working adults (NSC) (Taiwan)	90.4	89.4	90.7	95.1	95.7	
	Number of researchers per 10,000 working adults (Mainland China)						
	a. Number of researchers per 10,000 working adults (Yangtze River Delta)						
	b. Number of researchers per 10,000 working adults (Pearl River Delta)						
	c. Number of researchers per 10,000 working adults (Chongqing & Chengdu)						
	3. Number of university degrees per 10,000 people (Ministry of Science and Technology, MOST)	212.8	253.2	302.2	342.3	370.5	

	Number of university degrees per 10,000 people (Mainland China)					
	a. Number of university degrees per 10,000 people (Yangtze River Delta)					
	b. Number of university degrees per 10,000 people (Pearl River Delta)					
	c. Number of university degrees per 10,000 people (Chongqing & Chengdu)					
4.	R&D Personnel (FTE) per 10,000 population (IMD) **(TSDA)** (Taiwan)	3.97	3.93	3.99	4.21	4.27
	R&D Personnel (FTE) per 10,000 population (Mainland China)					
	a. R&D Personnel (FTE) per 10,000 population (Yangtze River Delta)					

(*Continued*)

(Continued)

Main Index	Detailed Indices (Sources)	Trend for Recent Five Years (World Ranking)					Description of Indices
		1999	2000	2001	2002	2003	
	b. R&D Personnel (FTE) per 10,000 population (Pearl River Delta)						
	c. R&D Personnel (FTE) per 10,000 population (Chongqing & Chengdu)						
	5. Availability of highly skilled workers (European Chamber of Commerce)						
	Availability of highly skilled workers (Mainland China)						
	a. Availability of highly skilled workers (Yangtze River Delta)						
	b. Availability of highly skilled workers (Pearl River Delta)						
	c. Availability of highly skilled workers (Chongqing & Chengdu)						

3. Basic Science and Humanities Preparation and Development of the Population	1. Science education (IMD) (Taiwan, IMD, 0-10分)	7.00	7.114(3)	6.82(6)	6.139(18)	7.01(6)
	Science education (IMD) (Mainland China)			4.88(27)	4.56(37)	5.85(19)
	a. Science education (IMD) (Yangtze River Delta)					
	b. Science education (IMD) (Pearl River Delta)				4.720(34)	5.85(19)
	c. Science education (IMD) (Chongqing & Chengdu)					
	2. Young people's level of interest in science (IMD) (Taiwan, IMD, 0-10)	7.80	7.897(1)	7.39(2)	6.944(5)	7.21(7)
	Young people's level of interest in science (Mainland China)			5.38(23)	5.253(32)	6.50(12)
	a. Young people's level of interest in science (Yangtze River Delta)					
	b. Young people's level of interest in science (Pearl River Delta)				5.32(31)	6.44(13)
	c. Young people's level of interest in science (Chongqing & Chengdu)					

(*Continued*)

(Continued)

Main Index	Detailed Indices (Sources)	Trend for Recent Five Years (World Ranking)					Description of Indices
		1999	2000	2001	2002	2003	
	3. Attendance at cultural and arts events (Council for Cultural Affairs) (Taiwan)	56,653	71,547	79,624	76,543	82,455	
	Attendance at cultural and arts events (Mainland China)						
	a. Attendance at cultural and arts events (Yangtze River Delta)						
	b. Attendance at cultural and arts events (Pearl River Delta)						
	c. Attendance at cultural and arts events (Chongqing & Chengdu)						
	4. Educational expenditures as a proportion of household income (DGBAS) (Taiwan)	— 12.96	7.5 13.51	7.5 13.32	— 13.52	— 13.12	

Educational expenditures as a proportion of household income (Mainland China)					
a. Educational expenditures as a proportion of household income (Yangtze River Delta)					
b. Educational expenditures as a proportion of household income (Pearl River Delta)					
c. Educational expenditures as a proportion of household income (Chongqing & Chengdu)					
5. Government expenditure on education as a proportion of GDP (Taiwan)	6.3	5.8	6.3	6.6	6.5
Government expenditure on education as a proportion of GDP (Mainland China)	2.79	2.86	3.14	2.18	

(*Continued*)

(Continued)

Main Index	Detailed Indices (Sources)	Trend for Recent Five Years (World Ranking)					Description of Indices
		1999	2000	2001	2002	2003	
	a. Government expenditure on education as a proportion of GDP (Yangtze River Delta)	2.49	2.33	2.45	2.84		
	b. Government expenditure on education as a proportion of GDP (Pearl River Delta)	1.82	1.78	1.75	1.83		
	c. Government expenditure on education as a proportion of GDP (Chongqing & Chengdu)	2.15	2.25	2.21	2.23		

2. Innovative Ability

Main Index	Detailed Indices (Sources)	Trend for Recent Five Years (World Ranking)					Description of Indices
		1999	2000	2001	2002	2003	
1. Quality and Quantity of Research Institutions	1. R&D budget per researcher for sci-tech institutions (NSC) (Taiwan)	2.18	2.26	2.30	2.37	2.50	
	R&D budget per researcher for sci-tech institutions (Mainland China)	14.2	15.4	18.6	22	24.6	
	a. R&D budget per researcher for sci-tech institutions (Yangtze River Delta)	18.2	22.6	26.5	38.8	46.3	
	b. R&D budget per researcher for sci-tech institutions (Pearl River Delta)	26	24	27	32.3	35.7	
	c. R&D budget per researcher for sci-tech institutions (Chongqing & Chengdu)	12.4	17.1	22.1	23.2	24.4	

(*Continued*)

(Continued)

Main Index	Detailed Indices (Sources)	Trend for Recent Five Years (World Ranking)					Description of Indices
		1999	2000	2001	2002	2003	
	2. Proportion of research institute researchers holding master's or higher degrees (NSC) (Taiwan)	59.5	61.4	63.9	68.2	70.9	
	Proportion of research institute researchers holding master's or higher degrees (NSC) (Mainland China)	8.4	9.7	11	11	13	
	a. Proportion of research institute researchers holding master's or higher degrees (Yangtze River Delta)	9.0	11.3	17.5	16.9	22.9	
	b. Proportion of research institute researchers holding master's or higher degrees (Pearl River Delta)	12.1	14.7	12.9	14.1	16.5	
	c. Proportion of research institute researchers holding master's or higher degrees (Chongqing & Chengdu)	4.8	5.3	5	6.3	7.4	

3. Approved foreign patents from Sci-Tech Development Plan (Taiwan)	300	476	333	344	289
Foreign patents approved from Sci-Tech Development Plan (Mainland China)					
a. Foreign patents approved from Sci-Tech Development Plan (Yangtze River Delta)					
b. Foreign patents approved from Sci-Tech Development Plan (Pearl River Delta)					
c. Foreign patents approved from Sci-Tech Development Plan (Chongqing & Chengdu)					
4. Spin-off enterprises from Sci-Tech Development Plan (Taiwan)	17	23	15		

(Continued)

(Continued)

Main Index	Detailed Indices (Sources)	Trend for Recent Five Years (World Ranking)					Description of Indices
		1999	2000	2001	2002	2003	
	Spin-off enterprises from Sci-Tech Development Plan (Mainland China) a. Spin-off enterprises from Sci-Tech Development Plan (Yangtze River Delta) b. Spin-off enterprises from Sci-Tech Development Plan (Pearl River Delta) c. Spin-off enterprises from Sci-Tech Development Plan (Chongqing & Chengdu)						

2. Quality and Quantity of Corporate R&D	1. R&D personnel per 1,000 employees (IMD) (IMD) (Taiwan)	3.146(9)	3.151(10)	3.213(14)	3.312(15)	—
	R&D personnel per 1,000 employees (IMD) (Mainland China)		0.376(31)	0.379(39)	0.346(43)	—
	a. R&D personnel per 1,000 employees (IMD) (Yangtze River Delta)					
	b. R&D personnel per 1,000 employees (IMD) (Pearl River Delta)					
	c. R&D personnel per 1,000 employees (IMD) (Chongqing & Chengdu)					
	2. R&D as a proportion of enterprise gross production (NSC) (Taiwan)	1.27	1.13	1.24	1.27	1.26
	R&D as a proportion of enterprise gross production (NSC) (Mainland China)					
	a. R&D as a proportion of enterprise gross production (Yangtze River Delta)					0.86

(*Continued*)

(Continued)

Main Index	Detailed Indices (Sources)	Trend for Recent Five Years (World Ranking)					Description of Indices
		1999	2000	2001	2002	2003	
	b. R&D as a proportion of enterprise gross production (Pearl River Delta)					1.12	
	c. R&D as a proportion of enterprise gross production (Chongqing & Chengdu)					0.76	
	3. Number of domestic patents from science and technical parks and proportion of companies in such parks (Science and Technical Parks) (Taiwan)	1,260/	2,366/	2,991/	2,688/	—	
	Number of domestic patents from science and technical parks and proportion of companies in such parks (Science and Technical Parks) (Mainland China)						

	a. Number of domestic patents from science and technical parks and proportion of companies in such parks (Yangtze River Delta)					
	b. Number of domestic patents from science and technical parks and proportion of companies in such parks (Pearl River Delta)					
	c. Number of domestic patents from science and technical parks and proportion of companies in such parks (Chongqing & Chengdu)					
3. Incentive Programs for Start-ups and Innovative Activities	1. Foreign investment as a proportion of venture capital (Venture Capital Association) (Science and Technical Venture Capital Company) (Taiwan)	295.92	308.03	81.46	117.4	165.4

(Continued)

(Continued)

Main Index	Detailed Indices (Sources)	Trend for Recent Five Years (World Ranking)					Description of Indices
		1999	2000	2001	2002	2003	
	Foreign investment as a proportion of venture capital (Mainland China)						
	a. Foreign investment as a proportion of venture capital (Yangtze River Delta)						
	b. Foreign investment as a proportion of venture capital (Pearl River Delta)						
	c. Foreign investment as a proportion of venture capital (Chongqing & Chengdu)						
	2. Ease of venture capital (IMD) (IMD, 0-10) (Taiwan)	7.133	6.354(14)	6.47(9)	6.36(6)	6.76(4)	
	Ease of venture capital (IMD) (Mainland China)			2.99(43)	2.95(52)	3.37(51)	
	a. Ease of venture capital (IMD) (Yangtze River Delta)						

b. Ease of venture capital (IMD) (Pearl River Delta)				3.68(42)	2.93(54)
c. Ease of venture capital (IMD) (Chongqing & Chengdu)					
3. R&D Budget as a proportion of GDP (NSC) (MOST) (Taiwan)	2.06	2.06	2.17	2.31	2.45
R&D Budget as a proportion of GDP (Mainland China)	0.69	0.83	1.0	1.1	1.23
a. R&D Budget as a proportion of GDP (Yangtze River Delta)	1.23	1.27	1.38	1.78	2.03
b. R&D Budget as a proportion of GDP (Pearl River Delta)	0.52	0.78	0.73	1.29	1.33
c. R&D Budget as a proportion of GDP (Chongqing & Chengdu)	0.37	0.44	0.5	0.57	0.64
4. IT R&D expenditures as a proportion of GDP (Taiwan)					

(*Continued*)

(Continued)

Main Index	Detailed Indices (Sources)	Trend for Recent Five Years (World Ranking)					Description of Indices
		1999	2000	2001	2002	2003	
	IT R&D expenditures as a proportion of GDP (Mainland China)						
	a. IT R&D expenditures as a proportion of GDP (Yangtze River Delta)						
	b. IT R&D expenditures as a proportion of GDP (Pearl River Delta)						
	c. IT R&D expenditures as a proportion of GDP (Chongqing & Chengdu)						
	5. Investment in IT and electronics as a proportion of all basic infrastructure and upgraded investment (Taiwan) Investment in IT and electronics as a proportion of all basic infrastructure and upgraded investment (Mainland China)						

a. Investment in IT and electronics as a proportion of all basic infrastructure and upgraded investment (Yangtze River Delta)	
b. Investment in IT and electronics as a proportion of all basic infrastructure and upgraded investment (Pearl River Delta)	
c. Investment in IT and electronics as a proportion of all basic infrastructure and upgraded investment (Chongqing & Chengdu)	
6. Successful venture capital amount as a proportion of venture capital investment (Venture Capital Association) (Taiwan)	

(*Continued*)

(Continued)

Main Index	Detailed Indices (Sources)	Trend for Recent Five Years (World Ranking)					Description of Indices
		1999	2000	2001	2002	2003	
	Successful venture capital amount as a proportion of venture capital investment (Mainland China)						
	a. Successful venture capital amount as a proportion of venture capital investment (Yangtze River Delta)						
	b. Successful venture capital amount as a proportion of venture capital investment (Pearl River Delta)						
	c. Successful venture capital amount as a proportion of venture capital investment (Chongqing & Chengdu)						
	7. Proportion of start-ups receiving medium- or long-term government operating monies (CEPD) (Taiwan)						

	Proportion of start-ups receiving medium- or long-term government operating monies (Mainland China)
	a. Proportion of start-ups receiving medium- or long-term government operating monies (Yangtze River Delta)
	b. Proportion of start-ups receiving medium- or long-term government operating monies (Pearl River Delta)
	c. Proportion of start-ups receiving medium- or long-term government operating monies (Chongqing & Chengdu)
8	IT production (high-tech electronics) as a proportion of GDP (Taiwan)
	IT production (high-tech electronics) as a proportion of GDP (Mainland China)

(*Continued*)

(Continued)

Main Index	Detailed Indices (Sources)	Trend for Recent Five Years (World Ranking)					Description of Indices
		1999	2000	2001	2002	2003	
	a. IT production (high-tech electronics) as a proportion of GDP (Yangtze River Delta)						
	b. IT production (high-tech electronics) as a proportion of GDP (Pearl River Delta)						
	c. IT and high-tech electronic industry enterprise value as a proportion of GDP (Chongqing & Chengdu)						
4. Flow of Knowledge between Industry and Academy	1. Industry-academic cooperation (IMD) (IMD, 0-10) (Taiwan)	5.267	5.18(12)	5.58(24)	5.194(31)	6.79(18)	
	Industry-academic cooperation (IMD) (Mainland China)			6.17(20)	5.899(26)	7.19(11)	
	a. Industry-academic cooperation (IMD) (Yangtze River Delta)						

b. Industry-academic cooperation (IMD) (Pearl River Delta)				5.265(29)	7.63(4)
c. Industry-academic cooperation (IMD) (Chongqing & Chengdu)					
2. Technical cooperation between enterprises (IMD) (IMD, 0-10) (Taiwan)	5.900	6.00(9)	6.38(13)	6.389(17)	6.79(18)
Technical cooperation between enterprises (IMD) (Mainland China)	523.45	650.75	782.75 3.88(47)	884.17 3.8(57)	1,084.67 4.69(51)
a. Technical cooperation between enterprises (IMD) (Yangtze River Delta)	36.63	73.9	106.16	120.22	142.7
b. Technical cooperation between enterprises (IMD) (Pearl River Delta)	34.45	48.21	53.97	68.45 4.2(54)	80.57 5.07(43)
c. Technical cooperation between enterprises (IMD) (Chongqing & Chengdu)	32.53	29.66	28.95	–40.94	55.5

(Continued)

(Continued)

Main Index	Detailed Indices (Sources)	Trend for Recent Five Years (World Ranking)					Description of Indices
		1999	2000	2001	2002	2003	
	3. Completeness of knowledge transfer (IMD) (IMD, 0-10) (Taiwan)			5.82(8)	5.746(14)	5.98(13)	
	Completeness of knowledge transfer (IMD) (Mainland China)			3.64(32)	3.697(47)	4.02(46)	
	a. Completeness of knowledge transfer (IMD) (Yangtze River Delta)						
	b. Completeness of knowledge transfer (IMD) (Pearl River Delta)				3.48(52)	3.81(49)	
	c. Completeness of knowledge transfer (IMD) (Chongqing & Chengdu)						
5. Innovative International R&D Cooperation Projects	1. Private budget for purchase of foreign technology (Ministry of Economic Affairs) (Taiwan)	419.5	431	—	493.9	—	

Private budget for purchase of foreign technology (Mainland China)	376.0	525.1	641.1	828.4	
a. Private budget for purchase of foreign technology (Yangtze River Delta)	56.47	84.93	105.34	135.92	
b. Private budget for purchase of foreign technology (Pearl River Delta)	129.79	183.15	221.91	337.12	
c. Private budget for purchase of foreign technology (Chongqing & Chengdu)	0.83	0.74	0.8	2.98	
2. Government budget for technology sold abroad (MOEA) (Taiwan)	31.1	63.2	—	150.8	—
Government budget for technology sold abroad (Mainland China)	247.0	370.4	464.5	678.6	
a. Government budget for technology sold abroad (Yangtze River Delta)	33.04	48.08	54.06	74.94	

(*Continued*)

(Continued)

Main Index	Detailed Indices (Sources)	Trend for Recent Five Years (World Ranking)					Description of Indices
		1999	2000	2001	2002	2003	
	b. Government budget for technology sold abroad (Pearl River Delta)	120.3	170.2	222.88	310.69		
	c. Government budget for technology sold abroad (Chongqing & Chengdu)	0.28	0.3	0.25	0.61		
	3. Balance of payments on the ratios of licensing fees and royalties (Central Bank) (Taiwan)	0.15	0.20	0.23	0.15	0.13	
	Balance of payments on the ratios of licensing fees and royalties (Mainland China)						
	a. Balance of payments on the ratios of licensing fees and royalties (Yangtze River Delta)						
	b. Balance of payments on the ratios of licensing fees and royalties (Pearl River Delta)						

	c. Balance of payments on the ratios of licensing fees and royalties (Central Bank) (Chongqing & Chengdu)
4.	Dollar amount of joint ventures with listed OECD nation companies (Taiwan) Dollar amount of joint ventures with listed OECD nation companies (China)
	a. Dollar amount of joint ventures with listed OECD nation companies (Yangtze River Delta)
	b. Dollar amount of joint ventures with listed OECD nation companies (Pearl River Delta)
	c. Dollar amount of joint ventures with listed OECD nation companies (Chongqing & Chengdu)

(*Continued*)

(Continued)

Main Index	Detailed Indices (Sources)	Trend for Recent Five Years (World Ranking)					Description of Indices
		1999	2000	2001	2002	2003	
6. Indication of Ability to Innovate and Entrepreneurial Spirit	1. Industry cluster innovation environment (WEF) (Taiwan) Industry cluster innovation environment (Mainland China) a. Industry cluster innovation environment (Yangtze River Delta) b. Industry cluster innovation environment (Pearl River Delta) c. Industry cluster innovation environment (Chongqing & Chengdu)				5.98(4)		
	2. Correlation between innovation infrastructure and industry cluster innovation environment (WEF) (Taiwan) Correlation between innovation infrastructure and industry cluster innovation environment (Mainland China)				4.43(15)		

a. Correlation between innovation infrastructure and industry cluster innovation environment (Yangtze River Delta)	
b. Correlation between innovation infrastructure and industry cluster innovation environment (Pearl River Delta)	
c. Correlation between innovation infrastructure and industry cluster innovation environment (Chongqing & Chengdu)	
3. Science and engineering research personnel (WEF) (Taiwan)	7.87(16)
Science and engineering research personnel (Mainland China)	
a. Science and engineering research personnel (Yangtze River Delta)	
b. Science and engineering research personnel (Pearl River Delta)	

(*Continued*)

(Continued)

Main Index	Detailed Indices (Sources)	Trend for Recent Five Years (World Ranking)					Description of Indices
		1999	2000	2001	2002	2003	
	c. Science and engineering research personnel (Chongqing & Chengdu)						
	4. Innovation policies (WEF) (Taiwan)				4.00(5)		
	Innovation policies (WEF) (Mainland China)						
	a. Innovation policies (WEF) (Yangtze River Delta)						
	b. Innovation policies (WEF) (Pearl River Delta)						
	c. Innovation policies (WEF) (Chongqing & Chengdu)						
	5. Enterprise innovation orientation (WEF) (Taiwan)				5.49(14)		
	Enterprise innovation orientation (Mainland China)						
	a. Enterprise innovation orientation (Yangtze River Delta)						
	b. Enterprise innovation orientation (Pearl River Delta)						

c. Enterprise innovation orientation (Chongqing & Chengdu)					
6. Frequency of start-ups (IMD) (IMD, 0-10) (Taiwan)	8.067	6.949(6)	7.97(6)	7.803(11)	7.45(10)
Frequency of start-ups (IMD) (Mainland China)			6.48(30)	6.76(26)	6.57(23)
a. Frequency of start-ups (IMD) (Yangtze River Delta)					
b. Frequency of start-ups (IMD) (Pearl River Delta)				6.84(23)	7.89(4)
c. Frequency of start-ups (IMD) (Chongqing & Chengdu)					
7. Entrepreneurial spirit (IMD) (IMD, 0-10分)	7.300	6.861(7)	7.53(7)	6.611(9)	7.36(5)
Entrepreneurial spirit (China)			4.61(47)	5.02(49)	5.35(45)
a. Entrepreneurial spirit (Yangtze River Delta)					
b. Entrepreneurial spirit (Pearl River Delta)				5.24(42)	7.93(1)
c. Entrepreneurial spirit (Chongqing & Chengdu)					

(*Continued*)

(Continued)

Main Index	Detailed Indices (Sources)	Trend for Recent Five Years (World Ranking)					Description of Indices
		1999	2000	2001	2002	2003	
	8. Enterprise proactivity in bringing in new technology (WEF) (Taiwan) Enterprise proactivity in bringing in new technology (Mainland China) a. Enterprise proactivity in bringing in new technology (Yangtze River Delta) b. Enterprise proactivity in ringing in new technology (Pearl River Delta) c. Enterprise proactivity in bringing in new technology (Chongqing & Chengdu)						

3. Utilization of Information Technology

Main Index	Detailed Indices (Sources)	Trend for Recent Five Years (World Ranking)					Description of Indices
		1999	2000	2001	2002	2003	
1. Degree of Digitization of Industry	1. Proportion of factories and companies engaged in e-commerce (DGBAS survey) (Taiwan)	7.60	11.48				
	Proportion of factories and companies engaged in e-commerce (Mainland China)						
	a. Proportion of factories and companies engaged in e-commerce (Yangtze River Delta)						
	b. Proportion of factories and companies engaged in e-commerce (Pearl River Delta)						
	c. Proportion of factories and companies engaged in e-commerce (Chongqing & Chengdu)						
	2. New applications of IT in line with enterprise demands (IMD) (IMD, 0-10) (Taiwan)	7.567	7.09(20)	7.37(25)	7.718(23)	7.82(27)	

(*Continued*)

(Continued)

Main Index	Detailed Indices (Sources)	Trend for Recent Five Years (World Ranking)					Description of Indices
		1999	2000	2001	2002	2003	
	New applications of IT in line with enterprise demands (Mainland China)			5.84(44)	6.06(51)	6.35(54)	
	a. New applications of IT in line with enterprise demands (Yangtze River Delta)						
	b. New applications of IT in line with enterprise demands (Pearl River Delta)				5.6(55)	6.52(52)	
	c. New applications of IT in line with enterprise demands (Chongqing & Chengdu)						
	3. Degree of digitization (IDC) (Taiwan)		4,296(18)	6,292(10)			
	Degree of digitization (IDC) (Mainland China)						
	a. Degree of digitization (IDC) (Yangtze River Delta)						

	b. Degree of digitization (IDC) (Pearl River Delta)			
	c. Degree of digitization (IDC) (Chongqing & Chengdu)			
	4. Degree of industry digitization (EIU) (Taiwan) Degree of industry digitization (EIU) (Mainland China)	7.22(16)	7.26(20)	
	a. Degree of industry digitization (EIU) (Yangtze River Delta)			
	b. Degree of industry digitization (EIU) (Pearl River Delta)			
	c. Degree of industry digitization (EIU) (Chongqing & Chengdu)			
2. Extent of Digitized Content	1. Extensiveness of government websites (NRI) (Taiwan, NRI, 1-7) Extensiveness of government websites (NRI) (Mainland China)		6.0(15)	5.18(15)

(*Continued*)

(Continued)

Main Index	Detailed Indices (Sources)	Trend for Recent Five Years (World Ranking)					Description of Indices
		1999	2000	2001	2002	2003	
	a. Extensiveness of government websites (NRI) (Yangtze River Delta)						
	b. Extensiveness of government websites (NRI) (Pearl River Delta)						
	c. Extensiveness of government websites (NRI) (Chongqing & Chengdu)						
	2. Assessment of e-government services (NRI) (Taiwan)			4.90(7)			
	Assessment of e-government services (NRI) (Mainland China)						
	a. Assessment of e-government services (NRI) (Yangtze River Delta)						

b. Assessment of e-government services (NRI) (Pearl River Delta)			
c. Assessment of e-government services (NRI) (Chongqing & Chengdu)			
3. Assessment of e-government websites (Brown University) (Taiwan)	52.5(2)	72.5(1)	72.5(1)
Assessment of e-government websites (Mainland China)			
a. Assessment of e-government websites (Yangtze River Delta)			
b. Assessment of e-government websites (Pearl River Delta)			
c. Assessment of e-government websites (Chongqing & Chengdu)			

(*Continued*)

(Continued)

Main Index	Detailed Indices (Sources)	Trend for Recent Five Years (World Ranking)					Description of Indices
		1999	2000	2001	2002	2003	
3. Extent of E-learning (Households/ Schools/ Other)	1. Household Internet connection rate (Statistical Report on the National Condition, %) (Taiwan)	14.5	19.1	27.4			
	Household Internet connection rate (Mainland China)	88.24	202	310	419.6	431.6	
	a. Household Internet connection rate (Yangtze River Delta)			432.92	968.08	1,107.37	
	b. Household Internet connection rate (Pearl River Delta)						
	c. Household Internet connection rate (Chongqing & Chengdu)		10	22.6	57	99.2	

2. School Internet connection rate (Ministry of Education Computation Center, %) (Taiwan)	34.0	46.0	55.0
School Internet connection rate (Mainland China)			
a. School Internet connection rate (Yangtze River Delta)			
b. School Internet connection rate (Pearl River Delta)			
c. School Internet connection rate (Chongqing & Chengdu)			

4. Social Infrastructure

Main Index	Detailed Indices (Sources)	Trend for Recent Five Years (World Ranking)					Description of Indices
		1999	2000	2001	2002	2003	
1. Incentive Measures and Financial Environment for Foreign Investment; Capital Market	1. Laws and regulations on tax and other incentives to promote industry upgrades (as current taxes, deductions, and post-deduction taxes) (Taiwan) Laws and regulations on tax and other incentives to promote industry upgrades (Mainland China) a. Laws and regulations on tax and other incentives to promote industry upgrades (Yangtze River Delta) b. Laws and regulations on tax and other incentives to promote industry upgrades (Pearl River Delta)						

	c. Laws and regulations on tax and other incentives to promote industry upgrades (Chongqing & Chengdu)
	2. Tax relief plans or preferential treatment strategies for high-tech industry (Taiwan) Tax relief plans or preferential treatment strategies for high-tech industry (Mainland China)
	a. Tax relief plans or preferential treatment strategies for high-tech industry (Yangtze River Delta)
	b. Tax relief plans or preferential treatment strategies for high-tech industry (Pearl River Delta)

(Continued)

(*Continued*)

Main Index	Detailed Indices (Sources)	Trend for Recent Five Years (World Ranking)					Description of Indices
		1999	2000	2001	2002	2003	
	c. Tax relief plans or preferential treatment strategies for high-tech industry (Chongqing & Chengdu)						
	3. Rental tax relief for start-ups Rental tax relief for start-ups (Mainland China)						
	a. Rental tax relief for start-ups (Yangtze River Delta)						
	b. Rental tax relief for start-ups (Pearl River Delta)						
	c. Rental tax relief for start-ups (Chongqing & Chengdu)						
	4. Foreign enterprise tax burden (current taxes, deductions, and post-deduction taxes) (Taiwan)						

Foreign enterprise tax burden (Mainland China)

a. Foreign enterprise tax burden (Yangtze River Delta)
b. Foreign enterprise tax burden (Pearl River Delta)
c. Foreign enterprise tax burden (Chongqing & Chengdu)

5. Controls on international capital flow and time required for incoming and outgoing transfers (Taiwan)

Controls on international capital flow and time required for incoming and outgoing transfers (Mainland China)

a. Controls on international capital flow and time required for incoming and outgoing transfers (Yangtze River Delta)

(Continued)

(Continued)

Main Index	Detailed Indices (Sources)	Trend for Recent Five Years (World Ranking)					Description of Indices
		1999	2000	2001	2002	2003	
	b. Controls on international capital flow and time required for incoming and outgoing transfers (Pearl River Delta)						
	c. Controls on international capital flow and time required for incoming and outgoing transfers (Chongqing & Chengdu)						
	6. Bank credit rating (S&P) (Taiwan, S&P)	AA+ stable	AA+ negative	AA– negative	AA– negative	AA– stable	
	Bank credit rating (Mainland China)						
	a. Bank credit rating (Yangtze River Delta)						
	b. Bank credit rating (Pearl River Delta)						
	c. Bank credit rating (Chongqing & Chengdu)						

7. Exchange rate fluctuations (descriptive) (Taiwan)	CV = 1.55%	CV = 2.53	CV = 2.78	CV = 1.55	CV = 0.97
Exchange rate fluctuations (descriptive Mainland China)					
a. Exchange rate fluctuations (descriptive) (Yangtze River Delta)					
b. Exchange rate fluctuations (descriptive) (Pearl River Delta)					
c. Exchange rate fluctuations (descriptive) (Chongqing & Chengdu)					
8. Transparency of bank management (systematization and standardization) (S&P) (Taiwan)					
Transparency of bank management (Mainland China)					
a. Transparency of bank management (Yangtze River Delta)					
b. Transparency of bank management (Pearl River Delta)					

(Continued)

(*Continued*)

Main Index	Detailed Indices (Sources)	Trend for Recent Five Years (World Ranking)					Description of Indices
		1999	2000	2001	2002	2003	
	c. Transparency of bank management (Chongqing & Chengdu)						
	9. Transparency of corporate board management (systematization and standardization) (descriptive) (Taiwan) Transparency of corporate board management (Mainland China)						
	a. Transparency of corporate board management (Yangtze River Delta)						
	b. Transparency of corporate board management (Pearl River Delta)						
	c. Transparency of corporate board management (Chongqing & Chengdu)						

10. Ease with which foreign firms can obtain large sums of capital (listing, insofar as possible, numbers or using descriptions) (Taiwan)					
Ease with which foreign firms can obtain large sums of capital (Mainland China)					
a. Ease with which foreign firms can obtain large sums of capital (Yangtze River Delta)					
b. Ease with which foreign firms can obtain large sums of capital (Pearl River Delta)					
c. Ease with which foreign firms can obtain large sums of capital (Chongqing & Chengdu)					
11. Foreign investment as a proportion of all investment (Taiwan)	12.85	20.95	19.46	13.09	14.27

(*Continued*)

(Continued)

Main Index	Detailed Indices (Sources)	Trend for Recent Five Years (World Ranking)					Description of Indices
		1999	2000	2001	2002	2003	
	Foreign investment as a proportion of all investment (Taiwan China)						
	a. Foreign investment as a proportion of all investment (Yangtze River Delta)		31.6	43.92	50.3	58.5	
	b. Foreign investment as a proportion of all investment (Pearl River Delta)		110.86	157.55	165.5	189.41	
	c. Foreign investment as a proportion of all investment (Chongqing & Chengdu)		3.45	4.24	4.5	5.67	
	12. Tariffs (Taiwan)	12.85%	20.95%	19.46%	13.09%	14.27%	
	Tariffs (Mainland China)	35.6%	33%	16.2%	11.1%	9.9%	
	a. Tariffs (Yangtze River Delta)	35.6%	33%	16.2%	11.1%	9.9%	
	b. Tariffs (Pearl River Delta)	35.6%	33%	16.2%	11.1%	9.9%	
	c. Tariffs (Chongqing & Chengdu)	35.6%	33%	16.2%	11.1%	9.9%	

13	Individual income taxes (Taiwan)					
	Individual income taxes (Mainland China)	5–45%	5–45%	5–45%	5–45%	5–45%
	a. Individual income taxes (Yangtze River Delta)	5–45%	5–45%	5–45%	5–45%	5–45%
	b. Individual income taxes (Pearl River Delta)	5–45%	5–45%	5–45%	5–45%	5–45%
	c. Individual income taxes (Chongqing & Chengdu)	5–45%	5–45%	5–45%	5–45%	5–45%
14.	Corporate business credit and ethics in investment target area (Taiwan)					
	Corporate business credit and ethics in investment target area (Mainland China)					
	a. Corporate business credit and ethics in investment target area (Yangtze River Delta)					
	b. Corporate business credit and ethics in investment target area (Pearl River Delta)					

(*Continued*)

(Continued)

Main Index	Detailed Indices (Sources)	Trend for Recent Five Years (World Ranking)					Description of Indices
		1999	2000	2001	2002	2003	
	c. Corporate business credit and ethics in investment target area (Chongqing & Chengdu)						
	15. Foreign banks ability to engage in local currency business (Taiwan)	Yes	Yes	Yes	Yes	Yes	
	Foreign banks ability to engage in local currency business (Mainland China)	No	No	Yes	Yes	Yes	
	a. Foreign banks ability to engage in local currency business (Yangtze River Delta)	No	No	Yes	Yes	Yes	
	b. Foreign banks ability to engage in local currency business (Pearl River Delta)	No	No	Yes	Yes	Yes	
	c. Foreign banks ability to engage in local currency business (Chongqing & Chengdu)	No	No	No	No	Yes	

2. Internationalization and Liberalization of Society	1. Number and proportion of people going abroad (Ministry of Transportation and Communications) (Taiwan)	8.7%	10.4%	10.4%	10.7%	8.27%
	Number and proportion of people going abroad (Mainland China)			0.0096	0.013	
	a. Number and proportion of people going abroad (Yangtze River Delta)					
	b. Number and proportion of people going abroad (Pearl River Delta)					
	c. Number and proportion of people going abroad (Chongqing & Chengdu)					
	2. Number and proportion of people entering the area (Ministry of Transportation and Communications) (Taiwan)	10.9	11.8	11.7	13.2	9.9
	Number and proportion of people entering the area (Mainland China)		0.0081	0.0088	0.0106	0.0088

(*Continued*)

(Continued)

Main Index	Detailed Indices (Sources)	Trend for Recent Five Years (World Ranking)					Description of Indices
		1999	2000	2001	2002	2003	
	a. Number and proportion of people entering the area (Yangtze River Delta)	0.1254	0.1367	0.1537	0.1579	0.1578	
	b. Number and proportion of people entering the area (Pearl River Delta)	0.7679	0.8645	0.9323	1.0263	0.9611	
	c. Number and proportion of people entering the area (Chongqing & Chengdu)	0.0062	0.0089	0.0104	0.0139	0.0082	
	3. Number and proportion of legal foreign residents (Ministry of the Interior) (Taiwan)	1.55%	1.76%	1.71%	1.8%	1.79%	
	Number and proportion of legal foreign residents (Mainland China)						
	a. Number and proportion of legal foreign residents (Yangtze River Delta)						

b. Number and proportion of legal foreign residents (Pearl River Delta)					
c. Number and proportion of legal foreign residents (Chongqing & Chengdu)					
4. Number of international conferences and exhibitions held and number of attendees (Taiwan)	31,194	34,879	34,356	33,656	31,542
Number of international conferences and exhibitions held and number of attendees (Mainland China)					
a. Number of international conferences and exhibitions held and number of attendees (Yangtze River Delta)					
b. Number of international conferences and exhibitions held and number of attendees (Pearl River Delta)					

(*Continued*)

(Continued)

Main Index	Detailed Indices (Sources)	Trend for Recent Five Years (World Ranking)					Description of Indices
		1999	2000	2001	2002	2003	
	c. Number of international conferences and exhibitions held and number of attendees (Chongqing & Chengdu)						
	5. Number of foreign financial institutions (Taiwan)	41	39	38	36	36	
	Number of foreign financial institutions (Mainland China)	247					
	a. Number of foreign financial institutions (Yangtze River Delta)	57	66	69	69	89	
	b. Number of foreign financial institutions (Pearl River Delta)	18					
	c. Number of foreign financial institutions (Chongqing & Chengdu)	11		10			

6. Visas granted for students studying abroad — number of students and proportion of all university students (Ministry of Education) (Taiwan)	5.93%	5.66%	4.49%	4.15%	2.94%
Visas granted for students studying abroad — number of students and proportion of all university students (Mainland China)	0.58%	0.7%	1.17%	1.39%	
a. Visas granted for students studying abroad — number of students and proportion of all university students (Yangtze River Delta)					
b. Visas granted for students studying abroad — number of students and proportion of all university students (Pearl River Delta)					

(Continued)

(Continued)

Main Index	Detailed Indices (Sources)	Trend for Recent Five Years (World Ranking)					Description of Indices
		1999	2000	2001	2002	2003	
	c. Visas granted for students studying abroad — number of students and proportion of all university students (Chongqing & Chengdu)						
	7. Number and proportion of foreign students studying in the area (Ministry of Education) (Taiwan)	0.1%	0.1%	0.1%	0.1%	0.1%	
	Number and proportion of foreign students studying in the area (Ministry of Education) (Mainland China)						
	a. Number and proportion of foreign students studying in the area (Yangtze River Delta)						
	b. Number and proportion of foreign students studying in the area (Pearl River Delta)						

c. Number and proportion of foreign students studying in the area (Chongqing & Chengdu)						
8. Proportion of foreign laborers (IMD) (Taiwan)	3.05%	3.34%	3.1%	3.04%	2.98%	
Proportion of foreign laborers (IMD) (Mainland China)						
a. Proportion of foreign laborers (IMD) (Yangtze River Delta)						
b. Proportion of foreign laborers (IMD) (Pearl River Delta)						
c. Proportion of foreign laborers (IMD) (Chongqing & Chengdu)						
9. Entry and exit policies (Taiwan)						
Entry and exit policies (Mainland China)						
a. Entry and exit policies (Yangtze River Delta)						
b. Entry and exit policies (Pearl River Delta)						
c. Entry and exit policies (Chongqing & Chengdu)						

(*Continued*)

(Continued)

Main Index	Detailed Indices (Sources)	Trend for Recent Five Years (World Ranking)					Description of Indices
		1999	2000	2001	2002	2003	
	10. Transportation cost for entering or leaving the area (Taiwan) Transportation cost for entering or leaving the area (Mainland China) a. Transportation cost for entering or leaving the area (Yangtze River Delta) b. Transportation cost for entering or leaving the area (Pearl River Delta) c. Transportation cost for entering or leaving the area (Chongqing & Chengdu)						
	11. Extent of foreign language ability (English) (General English Proficiency Test, GEPT) (for China: proportion of Level 4 passes among university students) (Taiwan)			4.31%	7.64%	9.1%	

Extent of foreign language ability (English) (General English Proficiency Test, GEPT) (for China: proportion of Level 4 passes among university students) (Mainland China)	
a. Extent of foreign language ability (English) (General English Proficiency Test, GEPT) (for China: proportion of Level 4 passes among university students) (Yangtze River Delta)	
b. Extent of foreign language ability (English) (General English Proficiency Test, GEPT) (for China: proportion of Level 4 passes among university students) (Pearl River Delta)	

(*Continued*)

(*Continued*)

Main Index	Detailed Indices (Sources)	Trend for Recent Five Years (World Ranking)					Description of Indices
		1999	2000	2001	2002	2003	
	c. Extent of foreign language ability (English) (General English Proficiency Test, GEPT) (for China: proportion of Level 4 passes among university students) (Chongqing & Chengdu)						
	12. Degree of acceptance of foreigners by locals (European Chamber of Commerce) (Taiwan) Degree of acceptance of foreigners by locals (Mainland China)						
	a. Degree of acceptance of foreigners by locals (Yangtze River Delta)						
	b. Degree of acceptance of foreigners by locals (Pearl River Delta)						

c. Degree of acceptance of foreigners by locals (Chongqing & Chengdu)
13. Satisfaction with international schools (European Chamber of Commerce) (Taiwan) Satisfaction with international schools (Mainland China)
a. Satisfaction with international schools (Yangtze River Delta)
b. Satisfaction with international schools (Pearl River Delta)
c. Satisfaction with international schools (Chongqing & Chengdu)
14. Tuition for international schools (European Chamber of Commerce) (Taiwan) Tuition for international schools (Mainland China)

(*Continued*)

(*Continued*)

Main Index	Detailed Indices (Sources)	Trend for Recent Five Years (World Ranking)					Description of Indices
		1999	2000	2001	2002	2003	
	a. Tuition for international schools (Yangtze River Delta)						
	b. Tuition for international schools (Pearl River Delta)						
	c. Tuition for international schools (Chongqing & Chengdu)						
3. Living Environment, Medical Care and Social Order	1. Doctors of Western medicine/doctors per 10,000 (for China: please note if no solid figures for doctors of Western medicine) (Taiwan)	14.4	15.0	15.4	15.8	16.2	
	Doctors of Western medicine/doctors per 10,000 (Mainland China)	16.7	16.8	16.9	14.7	14.8	
	a. Doctors of Western medicine/doctors per 10,000 (Yangtze River Delta)	38.5	37.7	36.5	32.8	32.9	

b. Doctors of Western medicine/doctors per 10,000 (Pearl River Delta)	14.9	14.8	15.0	13.4	14.1
c. Doctors of Western medicine/doctors per 10,000 (Chongqing & Chengdu)	14.5	14.5	14.5	12.1	1.6
2. Hospital beds per 10,000 population (Taiwan)	55.6	56.8	57.0	59.2	60.3
Hospital beds per 10,000 population (Mainland China)	23.9	23.8	23.9	23.2	23.4
a. Hospital beds per 10,000 population (Yangtze River Delta)	53.7	55.3	57.5	57.0	55.7
b. Hospital beds per 10,000 population (Pearl River Delta)	20.7	21.0	21.4	21.4	22.3
c. Hospital beds per 10,000 population (Chongqing & Chengdu)	19.1	19.0	18.7	18.7	19.1
3. Prevalence of ordinary immunizations (Taiwan)	95%				
Prevalence of ordinary immunizations (Mainland China)	100	100	100	100	100

(*Continued*)

(Continued)

Main Index	Detailed Indices (Sources)	Trend for Recent Five Years (World Ranking)					Description of Indices
		1999	2000	2001	2002	2003	
	a. Prevalence of ordinary immunizations (Yangtze River Delta)	100	100	100	100	100	
	b. Prevalence of ordinary immunizations (Pearl River Delta)	100	100	100	100	100	
	c. Prevalence of ordinary immunizations (Chongqing & Chengdu)	100	100	100	100	100	
	4. Medical expenditures as proportion of GDP (Taiwan)	5.1%	5.3%	5.7%	5.8%	6%	
	Medical expenditures as proportion of GDP (Mainland China)	5.19	5.33	5.46	5.42		
	a. Medical expenditures as proportion of GDP (Yangtze River Delta)						
	b. Medical expenditures as proportion of GDP (Pearl River Delta)						
	c. Medical expenditures as proportion of GDP (Chongqing & Chengdu)						

5. Criminal offenses per 100,000 population (Taiwan)	1,748.3	1,968.5	2,190.2	2,235.2	2,188.7
Criminal offenses per 100,000 population (Mainland China)	1,788.2	2,869.8	3,509.9	3,388.1	3,376.9
a. Criminal offenses per 100,000 population (Yangtze River Delta)					
b. Criminal offenses per 100,000 population (Pearl River Delta)					
c. Criminal offenses per 100,000 population (Chongqing & Chengdu)					
6. Thefts per 100,000 population (Taiwan)	1,237.4	1,378.4	1513.2	1,516	1,237.4
Thefts per 100,000 population (Mainland China)	1,150.7	1,872.8	2,291.2	2,253.3	
a. Thefts per 100,000 population (Yangtze River Delta)					
b. Thefts per 100,000 population (Pearl River Delta)					
c. Thefts per 100,000 population (Chongqing & Chengdu)					

(*Continued*)

(Continued)

Main Index	Detailed Indices (Sources)	Trend for Recent Five Years (World Ranking)					Description of Indices
		1999	2000	2001	2002	2003	
	7. Rate of increase or decrease in fire accidents per year (Taiwan)	25.4%	–14.8%	–11.6%	–3.7%	–34.7%	
	Rate of increase or decrease in fire accidents per year (Mainland China)	27.4	5.13	14.58	19.2		
	a. Rate of increase or decrease in fire accidents per year (Yangtze River Delta)						
	b. Rate of increase or decrease in fire accidents per year (Pearl River Delta)						
	c. Rate of increase or decrease in fire accidents per year (Chongqing & Chengdu)						
	8. Vehicular accidents per 10,000 vehicles (Taiwan)	1.54%	31.76%	37.27%	48.77%		
	Vehicular accidents per 10,000 vehicles (Mainland China)	412,860	616,971	75,491	773,137		

	a. Vehicular accidents per 10,000 vehicles (Yangtze River Delta)					
	b. Vehicular accidents per 10,000 vehicles (Pearl River Delta)					
	c. Vehicular accidents per 10,000 vehicles (Chongqing & Chengdu)					
4. Government Implementation of Transparency and Administrative Efficiency	1. Degree of implementation of transparency (IMD) (Taiwan)	5.627	3.57(44)	4.92(26)	4.51(40)	5.42(25)
	Degree of implementation of transparency (IMD) (Mainland China)			4.69(28)	4.84(35)	4.15(43)
	a. Degree of implementation of transparency (IMD) (Yangtze River Delta)					
	b. Degree of implementation of transparency (IMD) (Pearl River Delta)				4.68(38)	5.67(19)
	c. Degree of implementation of transparency (IMD) (Chongqing & Chengdu)					
	2. Governmental administrative efficiency (IMD) (Taiwan)	51.864(18)	45.591(20)	60.897(21)		

(*Continued*)

(Continued)

Main Index	Detailed Indices (Sources)	Trend for Recent Five Years (World Ranking)					Description of Indices
		1999	2000	2001	2002	2003	
	Governmental administrative efficiency (IMD) (Mainland China)						
	a. Governmental administrative efficiency (IMD) (Yangtze River Delta)						
	b. Governmental administrative efficiency (IMD) (Pearl River Delta)						
	c. Governmental administrative efficiency (IMD) (Chongqing & Chengdu)						
	3. Complexity of administrative procedures (WEF) (Taiwan)						
	Complexity of administrative procedures (WEF) (Mainland China)						

(Continued)

a. Complexity of administrative procedures (WEF) (Yangtze River Delta)
b. Complexity of administrative procedures (WEF) (Pearl River Delta)
c. Complexity of administrative procedures (WEF) (Chongqing & Chengdu)
4. Transparency of approval process for foreign investment applications (descriptive) (Taiwan) Transparency of approval process for foreign investment applications (Mainland China)
a. Transparency of approval process for foreign investment applications (Yangtze River Delta)

(*Continued*)

(Continued)

Main Index	Detailed Indices (Sources)	Trend for Recent Five Years (World Ranking)					Description of Indices
		1999	2000	2001	2002	2003	
	b. Transparency of approval process for foreign investment applications (Pearl River Delta)						
	c. Transparency of approval process for foreign investment applications (Chongqing & Chengdu)						
	5. Communication mechanisms between government and foreign firms (descriptive) (Taiwan)						
	Communication mechanisms between government and foreign firms (descriptive) (Mainland China)						
	a. Communication mechanisms between government and foreign firms (descriptive) (Yangtze River Delta)						

	b. Communication mechanisms between government and foreign firms (descriptive) (Pearl River Delta)					
	c. Communication mechanisms between government and foreign firms (descriptive) (Chongqing & Chengdu)					
5. Completeness and Degree of Application of IP and Related Technology Laws	1. Degree of protection of patents and intellectual property rights (IMD) (Taiwan)	6.867	7.038(23)	6.47(25)	6.31(27)	6.43(27)
	Degree of protection of patents and intellectual property rights (IMD) (Mainland China)			4.95(38)	5.02(43)	5.31(41)
	a. Degree of protection of patents and intellectual property rights (IMD) (Yangtze River Delta)					
	b. Degree of protection of patents and intellectual property rights (IMD) (Pearl River Delta)				4.76(48)	5.26(42)

(Continued)

(Continued)

Main Index	Detailed Indices (Sources)	Trend for Recent Five Years (World Ranking)					Description of Indices
		1999	2000	2001	2002	2003	
	c. Degree of protection of patents and intellectual property rights (IMD) (Chongqing & Chengdu)						
	2. Selling and buying of intellectual property rights (hundred million) (Taiwan)	523.4	650.8	782.7	884.2		
	Selling and buying of intellectual property rights (Mainland China)	36.6	73.9	106.2	103.2		
	a. Selling and buying of intellectual property rights (Yangtze River Delta)	34.5	48.2	54.0	72.2		
	b. Selling and buying of intellectual property rights (Pearl River Delta)	32.5	29.6	28.9	33.2		

	c. Selling and buying of intellectual property rights (Chongqing & Chengdu)	523.4	650.8	782.7	884.2	
6. Degree of Trust and Mutual Respect in Society	1. Fairness (IMD) (Taiwan)	4.967	5.06(29)	5.31(26)	5.27(32)	5.93(36)
	Fairness (Mainland China)			5.18(25)	5.12(35)	5.26(42)
	a. Fairness (Yangtze River Delta)					
	b. Fairness (Pearl River Delta)				4.64(37)	7.26(20)
	c. Fairness (Chongqing & Chengdu)					
	2. Cohesion (IMD) (Taiwan)	5.833	5.392(33)	5.71(31)	4.51(50)	5.03(49)
	Cohesion (Mainland China)			7.06(16)	6.62(25)	7.15(19)
	a. Cohesion (Yangtze River Delta)					
	b. Cohesion (Pearl River Delta)				6.12(33)	7.26(17)
	c. Cohesion (Chongqing & Chengdu)					
	3. Protection of privacy (IMD) (Taiwan)			6.74(24)	6.806(30)	7.15(24)
	Protection of privacy (Mainland China)			4.75(44)	4.5(54)	5.67(48)

(*Continued*)

(Continued)

Main Index	Detailed Indices (Sources)	Trend for Recent Five Years (World Ranking)					Description of Indices
		1999	2000	2001	2002	2003	
	a. Protection of privacy (Yangtze River Delta)						
	b. Protection of privacy (Pearl River Delta)				3.8(58)	6.92(32)	
	c. Protection of privacy (Chongqing & Chengdu)						
	4. Cultural diversity and tolerance (descriptive) (Taiwan) Cultural diversity and tolerance (Mainland China)						
	a. Cultural diversity and tolerance (Yangtze River Delta)						
	b. Cultural diversity and tolerance (Pearl River Delta)						
	c. Cultural diversity and tolerance (Chongqing & Chengdu)						
	5. Harmony of labor-management relations (descriptive) (Taiwan)						

		Harmony of labor-management relations (Mainland China) a. Harmony of labor-management relations (Yangtze River Delta) b. Harmony of labor-management relations (Pearl River Delta) c. Harmony of labor-management relations (Chongqing & Chengdu)
7. Proactivity in Innovative Education and Educational Innovation	1.	Number and proportion of students participating in international science competitions (Taiwan) Number and proportion of students participating in international science competitions (Mainland China) a. Number and proportion of students participating in international science competitions (Yangtze River Delta)

(Continued)

(Continued)

Main Index	Detailed Indices (Sources)	Trend for Recent Five Years (World Ranking)					Description of Indices
		1999	2000	2001	2002	2003	
	b. Number and proportion of students participating in international science competitions (Pearl River Delta)						
	c. Number and proportion of students participating in international science competitions (Chongqing & Chengdu)						
	2. Budget for college and above education and proportion of total government education expenditures (Ministry of Education) (If no concrete figures available, describe related policies) (Taiwan) Budget for college and above education and proportion of total government education expenditures (Mainland China)						

		a. Budget for college and above education and proportion of total government education expenditures (Yangtze River Delta)					
		b. Budget for college and above education and proportion of total government education expenditures (Pearl River Delta)					
		c. Budget for college and above education and proportion of total government education expenditures (Chongqing & Chengdu)					
8.	Convenience and Quality of Transportation and Shipping Infrastructure	1. Number of international airlines (Taiwan)	—	—	291	273	292
		Number of international airlines (Mainland China)	128	133	134		192
		a. Number of international airlines (Yangtze River Delta)					

(*Continued*)

(Continued)

Main Index	Detailed Indices (Sources)	Trend for Recent Five Years (World Ranking)					Description of Indices
		1999	2000	2001	2002	2003	
	b. Number of international airlines (Pearl River Delta)						
	c. Number of international airlines (Chongqing & Chengdu)						
	2. Number of international flights weekly (Taiwan)	5,858	6,544	6,636	6,743	6,419	
	Number of international flights weekly (Mainland China)						
	a. Number of international flights weekly (Yangtze River Delta)						
	b. Number of international flights weekly (Pearl River Delta)						
	c. Number of international flights weekly (Chongqing & Chengdu)						
	3. On-time performance of international flights (MOTC, %) (Taiwan)	96.46	96.02	96.75	97.39	97.67	

On-time performance of international flights (Mainland China)				77.7	82
a. On-time performance of international flights (Yangtze River Delta)					
b. On-time performance of international flights (Pearl River Delta)					
c. On-time performance of international flights (Chongqing & Chengdu)					
4. Air cargo throughput (international and domestic) (Taiwan)	616,322	586,560	561,910	548,555	489,171
Air cargo throughput (Mainland China)	1,652,705	1,757,117	1,940,722		
a. Air cargo throughput (Yangtze River Delta)					
b. Air cargo throughput (Pearl River Delta)					
c. Air cargo throughput (Chongqing & Chengdu)					
5. Aviation accidents (MOTC) (Taiwan)	2.03	0	0	2.05	1.94
Aviation accidents (Mainland China)					
a. Aviation accidents (Yangtze River Delta)					

(*Continued*)

(Continued)

Main Index	Detailed Indices (Sources)	Trend for Recent Five Years (World Ranking)					Description of Indices
		1999	2000	2001	2002	2003	
	b. Aviation accidents (Pearl River Delta)						
	c. Aviation accidents (Chongqing & Chengdu)						
	6. Increase in highway mileage (DGBAS) (Taiwan)	2.5	0.44	2.13	1.64	—	
	Increase in highway mileage (Mainland China)	5.73	3.77	21.05	3.95	2.53	
	a. Increase in highway mileage (Yangtze River Delta)						
	b. Increase in highway mileage (Pearl River Delta)						
	c. Increase in highway mileage (Chongqing & Chengdu)						
	7. Port cargo loaded and unloaded (DGBAS) (Taiwan)	534,086	566,950	552,873	606,710	635,577	
	Port cargo loaded and unloaded (Mainland China)	10.5	12.6	24	28	33	

a. Port cargo loaded and unloaded (Yangtze River Delta)	1.86	2.04	2.21	2.64	3.16
b. Port cargo loaded and unloaded (Pearl River Delta)	1.64	1.87	2.13	2.63	4.24
c. Port cargo loaded and unloaded (Chongqing & Chengdu)	0.26	0.24	0.28	0.30	0.32
8. Mass rapid transit (subways and light rail systems) (Taiwan)	1,031,342	2,042,303	2,223,487	2,469,133	2,440,695
Mass rapid transit (subways and light rail systems) (Mainland China)					
a. Mass rapid transit (subways and light rail systems) (Yangtze River Delta)					
b. Mass rapid transit (subways and light rail systems) (Pearl River Delta)					
c. Mass rapid transit (subways and light rail systems) (Chongqing & Chengdu)	0	0	0	0	0

(*Continued*)

(Continued)

Main Index	Detailed Indices (Sources)	Trend for Recent Five Years (World Ranking)					Description of Indices
		1999	2000	2001	2002	2003	
9. Convenience and Quality	1. Number of telephone exchanges (DGBAS) (Taiwan)	12,044	12,642	12,858	13,099	13,355	
of Communi-cations	Number of telephone exchanges (Mainland China)	1.6	1.79	2.2	2.85	3.5	
Infrastructure	a. Number of telephone exchanges (Yangtze River Delta)	0.065	0.069	0.074	0.081	0.087	
	b. Number of telephone exchanges (Pearl River Delta)						
	c. Number of telephone exchanges (Chongqing & Chengdu)						
	2. Mobile phones per 1000 population (DGBAS) (Taiwan)	522.4	802.4	965.5	1,061.5	1,109.9	
	Mobile phones per 1000 population (Mainland China)	34.31	67.13	113.13	161.72	208.29	
	a. Mobile phones per 1000 population (Yangtze River Delta)	153.7	272.5	471.7	690.9	818.7	
	b. Mobile phones per 1000 population (Pearl River Delta)	83.9	161.1	311.1	413.49	521.0	

c. Mobile phones per 1000 population (Chongqing & Chengdu)	26.3	53.1	87.7	129.6	187.6
3. Fixed-line telephones per 1000 population (DGBAS) (Taiwan)	545.2	567.5	573.9	581.7	590.8
Fixed-line telephones per 1000 population (Mainland China)	87.3	114.27	141.48	167.19	204.11
a. Fixed-line telephones per 1000 population (Yangtze River Delta)		414.8	466.7	503.7	547.0
b. Fixed-line telephones per 1000 population (Pearl River Delta)		146.15	182.47	219.2	217.28
c. Fixed-line telephones per 1000 population (Chongqing & Chengdu)		89.3	112.4	125.4	161.2
4. Cellular telephone tariff per minute (Directorate General of Telecommunications) (Taiwan)	6	—	3.6	—	—
Cellular telephone tariff per minute (Mainland China)					
a. Cellular telephone tariff per minute (Yangtze River Delta)					

(*Continued*)

(Continued)

Main Index	Detailed Indices (Sources)	Trend for Recent Five Years (World Ranking)					Description of Indices
		1999	2000	2001	2002	2003	
	b. Cellular telephone tariff per minute (Pearl River Delta)						
	c. Cellular telephone tariff per minute (Chongqing & Chengdu)						
	5. International telephone tariff per minute (Directorate General of Telecommunications) (Taiwan)	23.99	20.8	11.63	7.92	—	
	International telephone tariff per minute (Mainland China)						
	a. International telephone tariff per minute (Yangtze River Delta)						
	b. International telephone tariff per minute (Pearl River Delta)						
	c. International telephone tariff per minute (Chongqing & Chengdu)						

6. Monthly broadband Internet fee for 512k/64k connection (Directorate General of Telecommunications) (Taiwan)	—	2,199	1,199	1,094	499
Monthly broadband Internet fee for 512k/64k connection (Mainland China)					92.3 RMB
a. Monthly broadband Internet fee for 512k/64k connection (Yangtze River Delta)					
b. Monthly broadband Internet fee for 512k/64k connection (Pearl River Delta)					
c. Monthly broadband Internet fee for 512k/64k connection (Chongqing & Chengdu)					
7. Internet openness and stability (descriptive) (Taiwan)					
Internet openness and stability (Mainland China)					

(*Continued*)

(Continued)

Main Index	Detailed Indices (Sources)	Trend for Recent Five Years (World Ranking)					Description of Indices
		1999	2000	2001	2002	2003	
	a. Internet openness and stability (Yangtze River Delta)						
	b. Internet openness and stability (Pearl River Delta)						
	c. Internet openness and stability (Chongqing & Chengdu)						
10. Public Environment and Infrastructure; Environmental Protection	1. Electrification (DGBAS) (Taiwan)	124.0	130.9	128.8	130.0	136.1	
	Electrification (Mainland China)						
	a. Electrification (Yangtze River Delta)						
	b. Electrification (Pearl River Delta)						
	c. Electrification (Chongqing & Chengdu)	2.11	2.12	2.09	2.07	2.07	
	2. Average price of electricity (Taipower, NT $/Kw) (Taiwan)						
	Average price of electricity (Mainland China)						

a. Average price of electricity (Yangtze River Delta)					
b. Average price of electricity (Pearl River Delta)					
c. Average price of electricity (Chongqing & Chengdu)					
3. Peak load and reserve capacity (Taipower) (Taiwan) Peak load and reserve capacity (Mainland China)	2,420.6/ 2,848	2,585.4/ 2,963.4	2,629/ 3,013.6	2,711.7/ 3,191.5	2,859.4/ 3,329
a. Peak load and reserve capacity (Yangtze River Delta)					
b. Peak load and reserve capacity (Pearl River Delta)					
c. Peak load and reserve capacity (Chongqing & Chengdu)					

(Continued)

(Continued)

Main Index	Detailed Indices (Sources)	Trend for Recent Five Years (World Ranking)					Description of Indices
		1999	2000	2001	2002	2003	
	4. Efficiency of operation for electrical equipment (including capacity utilization ratio, steam turbine net thermal efficiency, transmission line loss, number of automated shutdown events for nuclear reactors) (Taipower) (Taiwan)	91.1 35.37 5.92 4	94.1 35.30 5.58 2	92.4 35.26 5.45 4	90.3 35.13 5.68 4	93.5 35.22 5.03 2	
	Efficiency of operation for electrical equipment (including capacity utilization ratio, steam turbine net thermal efficiency, transmission line loss, number of automated shutdown events for nuclear reactors) (Mainland China)	83.8 33.2 8.1	83.8 33.2 7.81	84.6 33.1 7.5	78.1 33.3 7.45		

a. Efficiency of operation for electrical equipment (including capacity utilization ratio, steam turbine net thermal efficiency, transmission line loss, number of automated shutdown events for nuclear reactors) (Yangtze River Delta)
b. Efficiency of operation for electrical equipment (including capacity utilization ratio, steam turbine net thermal efficiency, transmission line loss, number of automated shutdown events for nuclear reactors) (Pearl River Delta)

(*Continued*)

(Continued)

Main Index	Detailed Indices (Sources)	Trend for Recent Five Years (World Ranking)					Description of Indices
		1999	2000	2001	2002	2003	
	c. Efficiency of operation for electrical equipment (including capacity utilization ratio, steam turbine net thermal efficiency, transmission line loss, number of automated shutdown events for nuclear reactors) (Chongqing & Chengdu)						
	5. Quality of electric supply (including distribution lines per household, time to restore power after outage, workers available to restore power after outage) (Taipower) (Taiwan)	93.1 8.2 84.9	80.4 7.2 73.2	77.3 6.1 71.2	63.8 5.7 58.1	39.7 4.8 34.9	

Quality of electric supply (including distribution lines per household, time to restore power after outage, workers available to restore power after outage) (Mainland China)

a. Quality of electric supply (including distribution lines per household, time to restore power after outage, workers available to restore power after outage) (Yangtze River Delta)
b. Quality of electric supply (including distribution lines per household, time to restore power after outage, workers available to restore power after outage) (Pearl River Delta)

(*Continued*)

(*Continued*)

Main Index	Detailed Indices (Sources)	Trend for Recent Five Years (World Ranking)					Description of Indices
		1999	2000	2001	2002	2003	
	c. Quality of electric supply (including distribution lines per household, time to restore power after outage, workers available to restore power after outage) (Chongqing & Chengdu)						
	Black-outs and suspended water supply per year (Taiwan)						
	Black-outs and suspended water supply per year (Mainland China)						
	Black-outs and suspended water supply per year (Yangtze River Delta)						
	Black-outs and suspended water supply per year (Pearl River Delta)						
	Black-outs and suspended water supply per year (Chongqing & Chengdu)						

6. Extent of connection to running water (DGBAS) (Taiwan)	90.3	90.5	90.5	90.8	90.3
Extent of connection to running water (Mainland China)	96.3	96.7	72.3	77.9	
a. Extent of connection to running water (Yangtze River Delta)	100	100	100	100	100
b. Extent of connection to running water (Pearl River Delta)	97.5	98.2	84.5	91.8	
c. Extent of connection to running water (Chongqing & Chengdu)	96.5	96.9	45	62.6	
7. Average unit cost for water (household and industrial use, respectively)	73.8	75.0	76.9	—	—
Average unit cost for water (household and industrial use, respectively) (Mainland China)	217.5	220	216	213	
a. Average unit cost for water (household and industrial use, respectively) (Yangtze River Delta)	268.4	340	272	266	

(Continued)

(Continued)

Main Index	Detailed Indices (Sources)	Trend for Recent Five Years (World Ranking)					Description of Indices
		1999	2000	2001	2002	2003	
	b. Average unit cost for water (household and industrial use, respectively) (Pearl River Delta)	344.7	379	335	333		
	c. Average unit cost for water (household and industrial use, respectively) (Chongqing & Chengdu)	167.2	172	178	159		
	8. Annual city water supply (Taiwan)	7.9	6.1	8.2	3.7	—	
	Annual city water supply (Mainland China)	467.5	469	466.1	466.5		
	a. Annual city water supply (Yangtze River Delta)	28.9	29.1	26.5	29.5		
	b. Annual city water supply (Pearl River Delta)	44.0	52.6	51.2	54.8		
	c. Annual city water supply (Chongqing & Chengdu)	6.0	6.1	6.3	6.1		
	9. Sanitary sewer connection rate (DGBAS) (Taiwan)	45.63	51.02	56.33	62.44	68.13	

Wastewater treatment capacity (Mainland China)	2,654.5	4,741	6,215	6,153	
a. Wastewater treatment capacity (Yangtze River Delta)	524.3	481.1	903	490	
b. Wastewater treatment capacity (Pearl River Delta)	176.6	327	327	360	
c. Wastewater treatment capacity (Chongqing & Chengdu)	55.0	60.6	57	58	
10. Storm sewer construction (DGBAS) (Taiwan)	49.3	50.7	53.0	55.3	—
Storm sewer construction (Mainland China)	134,486	141,758	158,128	173,042	
a. Storm sewer construction (Yangtze River Delta)	4,577	3,736	3,954	4,001	
b. Storm sewer construction (Pearl River Delta)	14,903	16,781	19,050	18,245	
c. Storm sewer construction (Chongqing & Chengdu)	1,954	2,223	2,436	2,671	
11. Cable TV penetration (DGBAS) (Taiwan)	67.9	72.0	72.3	74.8	—

(*Continued*)

(Continued)

Main Index	Detailed Indices (Sources)	Trend for Recent Five Years (World Ranking)					Description of Indices
		1999	2000	2001	2002	2003	
	Cable TV penetration (Mainland China)	7,700	7,920	8,803	9,638	10,508	
	a. Cable TV penetration (Yangtze River Delta)		300	320	327	367	
	b. Cable TV penetration (Pearl River Delta)						
	c. Cable TV penetration (Chongqing & Chengdu)						
	12. Extent of resource recycling (DGBAS) (Taiwan)	1.7	5.7	7.5	11.6	14.3	
	Extent of resource recycling (Mainland China)	45.6	45.9	52.1	52.0	55.8	
	a. Extent of resource recycling (Yangtze River Delta)						
	b. Extent of resource recycling (Pearl River Delta)						

c. Extent of resource recycling (Chongqing & Chengdu)					
13. Extent of appropriate handling of trash (DGBAS) (Taiwan) Neutralization of trash (Mainland China)	86.7	90.2	93.4	96.2	98.2
Neutralization of trash (Mainland China)				54.3	50.8
a. Neutralization of trash (Yangtze River Delta)					
b. Neutralization of trash (Pearl River Delta)					
c. Neutralization of trash (Chongqing & Chengdu)					
14. Annual natural disasters and damages caused (including typhoons, earthquakes, flooding) (Ministry of the Interior) (Taiwan)	4	11	9	4	8

(*Continued*)

(Continued)

Main Index	Detailed Indices (Sources)	Trend for Recent Five Years (World Ranking)					Description of Indices
		1999	2000	2001	2002	2003	
	Annual natural disasters and damages caused (including typhoons, earthquakes, flooding) (Mainland China) a. Annual natural disasters and damages caused (including typhoons, earthquakes, flooding) (Yangtze River Delta) b. Annual natural disasters and damages caused (including typhoons, earthquakes, flooding) (Pearl River Delta) c. Annual natural disasters and damages caused (including typhoons, earthquakes, flooding) (Chongqing & Chengdu)						

15. Quality of city environment (Taiwan)
    Quality of city environment (Mainland China)
    a. Quality of city environment (Yangtze River Delta)
    b. Quality of city environment (Pearl River Delta)
    c. Quality of city environment (Chongqing & Chengdu)

5. Cost of Business Analysis

Item	Cost
Total:	

References

Aghion, Philippe and Peter Howitt (1992), "A Model of Growth through Creative Destruction", *Econometrica*, 60: 323–351.

Aghion, Philippe and J. Tirole (1994), "The Management of Innovation", *Quarterly Journal of Economics*, 109: 1185–1209.

Becker, Gary S. (1993), *Human Capital: A Theoretical and Empirical Analysis, with Special Reference to Education*, 3rd edition. Chicago and London: University of Chicago Press.

Bell, Daniel (1999), "Communications? For Better or for Worse", in *The Information Resources Policy Handbook: Research for the Information Age*, pp. 303–320. Cambridge and London: MIT Press.

Buckley, Peter J. and M.C. Casson (1976), *The Future of the Multinational Enterprise*. London: Macmillan.

Buckley, Peter J. and M.C. Casson (2003), "The Future of the Multinational Enterprise in Retrospect and in Prospect", *Journal of International Business Studies*, 34(2): 219–222.

Chen, Chunlai (1997), "The Location Determinants of Foreign Direct Investment in Developing Countries", The University of Adelaide Working Paper No. 97/12.

Chen, T. and Y. Ku (2000), "The Effect of Foreign Direct Investment on Firm Growth: The Case of Taiwan's Manufacture", *Japan and the World Economy*, 12: 153–172.

Coase, Ronald H. (1997), "The Problem of Social Cost", in *Law and Economics*, Theoretical and Methodological Foundations, Vol. 1, pp. 433–476, Elgar Reference Collection; International Library of Critical Writings in Economics, Vol. 81. Cheltenham, UK and Lyme, NH: Edward Elgar.

Cull, Robert and Colin Lixin Xu (2003), "Who Gets Credit? The Behavior of Bureaucrats and State Banks in Allocating Credit to Chinese State-Owned Enterprises", *Journal of Development Economics*, 71(2): 533–559.

Djankov, Simeon and Peter Murrell (2002), "Enterprise Restructuring in Transition: A Quantitative Survey", *Journal of Economic Literature*, 40(3): 739–792.

Emery, James J., *et al.* (2000), "Administrative Barriers to Foreign Investment: Reducing Red Tape in Africa", Foreign Investment Advisory Service Occasional Paper Vol. 14. Washington, DC: World Bank.

Friedman, Andrew L. (2000), "Microregulation and Post-Fordism: Critique and Development of Regulation Theory", *New Political Economy*, 5(1): 59–76.

Gilly, Salmon (2000), "Developing Managers through Networked Technologies", in *The Current State of Business Disciplines*, Vol. 5, pp. 2475–2484. Rohtak, India: Spellbound Publications.

Gordon, Roger H. and Wei Li (2002), "Taxation and Economic Growth in China", Development Economics Working Paper No. 187. East Asian Bureau of Economic Research.

Hymer, Stephen (1960), "The International Operations of National Firms: A Study of Direct Foreign Investment", Ph.D. dissertation, MIT; published in 1976, Lexington, MA: Lexington Books.

IMD, *The World Competitiveness Yearbook 2004.*

IMF (1977), *Balance of Payment Manual*, Washington, DC.

Kang, Sung Jin and Yasuyuki Sawada (2000), "Financial Repression and External Openness in an Endogenous Growth Model", *Journal of International Trade and Economic Development*, 9(4): 427–443.

Kindleberger, Charles P. (1969), *American Business Abroad.* New Haven: Yale University Press.

Kojima, Kiyoshi (1973), "A Macroeconomic Approach to Foreign Direct Investment", *Hitotsubashi Journal of Economics*, 14: 1–12.

Kojima, Kiyoshi (1978), *Direct Foreign Investment: A Japanese Model of Multinational Business Operations.* London: Croom Helm.

Kojima, Kiyoshi (1982), "Macroeconomic versus International Business Approach to Direct Foreign Investment", *Hitotsubashi Journal of Economics*, 23(2): 1–19.

Lucas, Jr., Robert E. (1988), "On the Mechanics of Economic Development", *Journal of Monetary Economics*, 22(July): 3–42.

Magee, S.P. (1977), "Multinational Corporations, Industry Technology Cycle and Development", *Journal of World Trade Law*, 11: 297–321.

McMillan, John and Christopher Woodruff (2002), "The Central Role of Entrepreneurs in Transition Economies", *Journal of Economic Perspectives*, 16(3): 153–170.

Romer, Paul M. (1986), "Increasing Returns and Long-Run Growth", *Journal of Political Economy*, 94(October): 1002–1037.

Romer, Paul M. (1990), "Endogenous Technological Change", *Journal of Political Economy*, 98(October, Part 2): S71–S102.

Sachs, Jeffrey D. and Andrew M. Warner (1995), "Natural Resource Abundance and Economic Growth", NBER Working Paper No. 5398.

Schumpeter, Joseph A. (1942), *Capitalism, Socialism and Democracy.* New York: Harper-Row.

Shleifer, Andrei and Robert W. Vishny (1999), "Corruption", in *The Economics of Corruption and Illegal Markets*, The Economics of Corruption, Vol. 1, pp. 94–112. Edward Elgar.

Solow, Robert M. (1956), "A Contribution to the Theory of Economic Growth", *Quarterly Journal of Economics*, 70(February): 65–94.

Wei, Shang-Jin (2000), "How Taxing is Corruption on International Investors?" *Review of Economics and Statistics*, 82(1): 1–11.

Wei, Shang-Jin (2001), "Does Corruption Relieve Foreign Investors of the Burden of Taxes and Capital Controls?" in *International Taxation and Multinational Activity*, NBER Conference Report Series, pp. 73–87. Chicago and London: University of Chicago Press.

Wu, Se-Hwa (2001), *Knowledge-Based Economy Society Aggregate Index.* Taiwan: CEPD.

Index

A

ABS 99, 103, 105, 106
acoustic facilities 125
acrylic 227
adequate reward theory 5
administrative licensing 234
agglomeration effect 178
agriculture 18, 39, 42–44, 53, 210, 214, 232
anti-dumping cases 230
ASEAN (Association of Southeast Asian Nations) 25, 185, 186
Asian financial crisis 7, 14, 19
ASP 64
automotive safety system 103
automotive semiconductor service 98, 107

B

basic metal and products manufacturing industry 37
BGA (Ball Grid Array) 148, 149, 165, 167, 168, 171
bilateral 202, 240
bluetooth 172, 173
Bohai Rim Economic Area 31, 32, 87
border effect 219
Brown University 69, 322, 369

C

cable TV 62, 110, 315, 327, 427, 428
CAGR (Compound Annual Growth Rate) 61, 99, 106, 110, 165, 167, 168, 171
capacitors 163
capital movement 1
capital stock 7
CDMA20001X 172
cell phone, cellular telephone 108–110, 112–116, 119, 164, 172, 173, 175, 326, 415, 416
central-west area 32
CEPD 320, 352
CGI (Current Growth Index) 70, 71
chemical fertilizers 227, 228
CII (Current Impact Index) 257, 319, 331, 332
civil liberalization 62
clusters 28, 75, 78, 83, 143, 178, 275, 321, 360, 361
Coase theorem 4
color tubes, color TV sets 80, 81, 85, 86, 109, 125, 155
Commercial Affairs Department 178
commercial market 158, 171
commodity 172, 219, 232, 243

communication module 163
communication network 216
comparative advantage 7
components 25, 86, 87, 94, 97, 98, 103, 105–107, 131, 132–138, 150, 154, 163, 174, 175, 203, 220, 227, 251
compound internet system 158
computerized technology 158
connector 163
copper 77, 106, 156
copyright 113, 204, 205, 240
cotton 227
CPE 173
CPU 74, 75, 167
creative destruction 55
credit loans 188
criminal law 204
CSP 148, 150, 151, 165, 167, 168, 171
currency business 283, 382
currency exchange 195
currency value 191

D

debt-paying capability 194
debt-to-asset ratio 24
decision-making power 31, 226
degree of trust 325, 405
DELL 83, 130
desktop computer 159, 163
digital cameras 80, 109–111, 113, 125, 130, 144, 152, 175
DIP 149, 167, 168, 171
distance learning 123
distant medical treatment 123
downstream 73, 80, 118, 146, 147, 154–156, 160, 162, 178, 253
dragon countries 186
DRAM 148, 151–153, 164, 165, 167
DSL 173
DVD 81, 99, 100, 111, 144, 174
DVD recorders 174
DVD-R/+R 174

E

E-commerce 62, 64, 68, 123, 277, 322, 365
E-learning 60, 279, 322, 370
E-map 96, 99, 100
economic development zone 16
economic effects 6, 147, 226
economic growth 33, 53, 54, 57, 58, 71, 73, 87, 92–94, 127, 181, 190–192, 207, 216, 221, 248
economic rent 8
EI 63, 247, 248, 322, 367
EIU (Economist Information Unit) 247, 248, 322, 367
electric components 163
electro-optical 145, 163, 173–175
employment 33, 65, 220, 221, 232, 241, 248, 265
endogenous growth theory, new growth theory 54
equity 19, 22, 298
Ethernet 173
ETP 165, 167, 168
export growth rate 127
export-import value 127, 182, 183, 237, 240
export-processing zone 33
external economies 182
external threats 34, 177, 250

F

FDI 1, 3, 4, 6–9, 11–23, 25, 36, 49, 60, 181, 182, 185–187, 192–196, 198, 202, 207, 213, 214, 236, 248, 250, 296
fixed-line telephones 326, 415
flash memory card 112, 113, 115
flip chip 148, 149, 150
foreign capital 11–17, 19, 21–23, 25, 26, 28, 32–34, 40, 49, 50, 53, 73, 80, 87, 177, 192, 194, 248, 268, 282, 283, 317
foreign debt 193–195
foreign investment policy 177, 236, 240, 241, 243
four dragon 185, 196
FSK 105
FTTH 173, 175
FTTX 173

G

GaN devices 107
GCI (Growth Competitiveness Index) 70, 71
GDP 56, 88, 91, 92, 111, 127, 165, 181–184, 191, 192, 201, 208, 249, 263, 269–271, 292, 319, 320, 324, 339, 340, 349, 350, 353, 354, 396
GEPT 289, 324, 390, 391, 392
globalization 11, 65, 127, 130, 181, 215–217, 251, 252
goodwill 9
government budget 58, 357, 358
governmental efficiency 65, 127, 216, 224, 235, 248
GPRS 172

H

H-O principle 186
Harvard University 66
high-end color TVs 80, 86
high-skill workers 260
HR analysis 185
human capital 55, 57–59, 188
human literacy 57
human resource 31, 54, 55, 5759, 185, 258, 319, 333
hybrid fuel automobiles 107
hydro power 210

I

IA products 156
IC 74–80, 86, 88–93, 107, 108, 116–123, 143, 145–153, 160, 162–168, 182, 183, 205, 243, 251
IC industry 76, 88–93, 146, 147, 152, 153, 160, 162, 168, 182, 183
IC package 78, 146, 147, 149, 150, 160, 162, 165
IC test 146, 151
ICT 66, 67, 69, 71
ICT index 71
IDC (International Data Corporation) 61, 82, 109, 111, 128, 129, 322, 366, 367
IDM 147, 152, 162, 168, 169
IMF 7, 11
imperfect competitive market 1
inductors 163
industrial concentration 87
industrial economy 60

industrial electronic level 60
industrial organizational theory 1
Industrial Revolution 53
industrial scale 6, 86, 88, 89
industrial strategies 28
industrial technology cycle 6, 59
industrial transfer 87, 186
industry-academic cooperation 271, 321, 354, 355
Infineon 105, 108, 182
inflation rate 191
information industry 73, 84–87, 94, 110, 125, 127, 141, 156, 157
information technology 26, 53, 59, 61, 66, 73, 91, 94, 123, 124, 126, 143, 158, 177, 247, 249, 277, 322, 365
innovation 54–56, 59, 71, 183, 221, 249, 251, 252, 254, 275, 321, 325, 360–363, 407
innovation mechanism, innovative mechanism 55
innovative network 252
Intel 78, 108
intellectual property rights 29, 56, 118, 204, 205, 230, 239, 240, 242, 297, 298, 325, 403–405
internalization theory 4
international debt 194
international marketing 1, 2, 9, 33, 80, 92, 216, 217, 251, 253
international production 5
International Reserve 194, 195
international trade industry 37, 38
interrelatedness of patent and science 319, 332
intra-firm trade 5
Invention and Design Patent 56, 255, 319, 329
inward FDI 7–9
IP (intellectual property) 118
ISI (Information Society Index) 61

K

knowledge capital 53, 55, 56, 59, 255, 319, 329
knowledge circulation 57, 271
knowledge infrastructure 254
knowledge transfer 272, 321, 356
knowledge-based economy 54, 55, 57, 59, 60, 61, 249, 252, 253, 254
knowledge-based services 253
knowledge-intensive 25

L

labor cost 6, 40, 178, 185, 186, 205
labor force 33, 87, 178, 181, 185–188, 215, 218, 220, 221, 233, 248
labor market 6, 65, 221, 247
LAN 130, 172, 173
laser technologies 123
LCD 80, 86, 96, 110, 117, 130, 149, 150, 158, 159, 163, 164, 174, 182
LED 175
legal system 61, 201, 202, 204, 226, 234, 297
local theory 2, 179
Logic IC 149, 152, 160
low technology skill 23

M

market share 3, 6, 80, 81, 83, 107, 108, 111, 113, 114, 130, 132, 137, 156, 164, 180, 217, 219
marketing 1, 9, 69, 84–86, 88, 89, 112, 120, 131, 133, 140, 141, 155, 178, 203, 217, 228, 229, 241, 250–253
mass production 77, 107, 125, 158, 174
MCI 71
MCU 103, 106
mechanical manufacturing industry 37, 41
MEMS 117
MIC 159, 172, 182
micro electronics 117, 123
Micron 79, 90
microprocessor 78
midstream 156, 160
mobile office 123
mobile phone 62, 63, 80–82, 85, 114, 115, 172, 175, 213, 414, 415
monetary policy 190
monitor 85, 86, 98, 106, 150, 157, 163, 174
monopoly 3, 231, 242
MP3 109–112, 115, 116, 130
multinational companies, multinational enterprises 1–6, 23–32, 51, 81, 82, 87, 188, 197–199

N

nano-technology 253
neoclassical growth theory 53
network facilities 173
non-market power 2
northeast areas 32
notebook computers 80, 81, 84, 86
NRI (Networked Readiness Index) 66, 322, 367–369
nuclear power 34
NV memory 152, 153

O

OECD (Organization for Economic Cooperation and Development) 54, 275, 321, 359
OEMs 75, 77, 78, 83, 99, 117, 118, 121–123, 133, 134, 144, 146, 156, 157, 159, 163–165, 168, 170, 171, 173, 251
outsourcing 84, 89
outward FDI 8, 9, 185
overseas headquarters 31
overseas investment 32–34, 36, 38–39, 41, 51

P

P-chips 227
packaging type 148–150, 167, 168
passive components 105, 163
patent fee 81, 113
patent law 204, 205, 240
patent rights 3, 9, 56, 59, 258
PDA 111, 114, 115, 163, 164, 172, 202
PDP 81, 86
PGA 167, 168, 171
photo 115, 160, 161, 163

photoresist 160
pitch 150, 151
plasma 110, 112
PLCC 149
polyester 227
portfolio 4
power transformers DC-DC transformer 107
pre-tax profit 126
productivity 54, 56, 65, 71, 77, 78, 80–82, 123, 183, 221, 248
property rights 29, 56, 118, 181, 201, 202, 204, 205, 230, 239, 240, 242, 297, 298, 325, 403–405
protectionism 219, 231, 242

Q

QFP (quad flat package) 148–151, 167, 168, 171
quality 3, 55–59, 61, 67–69, 71, 118, 151, 155, 156, 158, 162, 172, 175, 176, 185–188, 196, 216, 218, 219, 230, 233, 248, 249, 251, 254, 255, 257–259, 262–264, 266–269, 300, 304, 310–312, 317, 319, 320, 325–327, 329, 333, 341, 345, 409, 414, 422–424, 431

R

R&D (research and development) 23, 25, 26, 28–31, 54–59, 75, 76, 84, 87, 95–97, 109, 118, 119, 153, 155, 160, 162, 183, 188, 218, 219, 240, 250–252, 259, 260, 264, 266–270, 273, 275, 319, 320, 321, 335, 336, 341, 345, 346, 349, 350, 356
radar system 103
rectifier diode 163
remote access 129
rental tax 323, 374
resistors 163
RF transmitter 103
RISC 75
ROM 153
rubber 38, 41, 45, 46, 228

S

scanner 163
SCI 58, 256, 257, 319, 330, 331
self-supply rate 77
semiconductor industry 27, 30, 32, 76, 144, 146, 160, 161, 163, 165, 168, 171
service industry 38, 40, 47, 48, 222, 223, 232, 239, 242, 243, 250–252
SiC 107
skills 9, 26, 50, 56, 68, 215, 218, 251
SLI (Science Linkage Index) 257, 258
SMEs 8, 9, 56, 129–131, 189, 221, 244, 251, 253, 254
Smiling Curve 250
SO 149, 151, 167, 168, 171
social capital 68
social economies 177
social infrastructure structure 53, 60, 254, 280, 323, 372
social security 60, 290
social trust 61, 298

software 26, 31, 62, 67, 86, 94, 123–130, 137, 156, 158, 205
SOHO router 173
sole proprietorship 27, 29, 30
Solow, Robert 53, 54
SOP 122, 149
special economic zones 15, 16, 206
spillover effect 55
SRAM 153, 164
substitute effect 175
System on Chip (SoC) 75, 76, 152
systemic 11

T

Taiwan 7–9, 11, 17, 20, 32–34, 36–41, 49–51, 56, 60, 74, 75, 80, 83, 121, 122, 126, 132–134, 139, 141, 143, 144, 146–157, 160, 161, 163, 164, 165, 171, 172, 173, 174, 175, 185, 186, 192, 207, 209, 213, 247–274, 276–280, 282–317, 319, 329–335, 337–339, 341–343, 345–390, 392–420, 422, 424–429, 431
technological property 9
technology 2, 3, 5, 9, 11, 17, 23, 25, 26, 28–30, 50, 53–56, 59–61, 63, 66, 68, 71, 73–78, 81, 83, 90, 91, 94, 95, 98, 100–102, 104–108, 111, 113, 116–119, 121–124, 126–129, 143–148, 150, 151, 155, 156, 158, 160, 162, 174, 177, 182, 187, 188, 205, 217, 218, 240, 244, 247–254, 260, 265, 266, 271–274, 277, 319–322, 325, 334, 356–358, 364, 365, 403
telecommunication industries 33, 37, 41, 228
textile industry 38, 41
TFT-LCD 80, 150, 158, 163, 174, 182
thermo power 210
tourism industry 241
TPMS 103, 105, 106
trademark law 204, 205, 240
trademarks 1204, 205, 240
TRIPS (Trade-Related Aspects of Intellectual Property) 204, 205

U

UNCTAD (United Nations Conference on Trade and Development) 15
upstream 97, 146, 155, 156, 160, 162, 163, 178
USPTO (U.S. Patent and Trademark Office) 319, 329

V

value-added 87, 88, 101, 130, 132, 136, 140
VCD 125
VIA 161
VoIP 129, 173

W

wafer 77–79, 118, 148, 151, 160
WCDMA 172
WEF (World Economic Forum) 66, 70, 71, 248, 249, 319, 321, 325, 332, 360–362, 364, 400, 401

wire and cable industry 154, 156, 175
WLAN 172, 173
World Bank 182, 188, 207, 208, 216, 219, 224
WTO 139, 176, 178, 181, 198, 203–205, 226, 227, 229–232, 234, 236–239, 241, 242

X

XDSL 173

Y

Yanzhijiang Delta Region 31, 32, 78, 82, 87, 90, 255–270, 273–275, 278–280, 282, 284–308, 311–315, 317

Z

Zhujiang Delta Economic Area 31, 32, 81, 83, 87, 90, 255–280, 282–308, 311–315, 317